# Nanotechnologies:
# The Physics of Nanomaterials
## Volume I

*The Physics of Surfaces and*
*Nanofabrication Techniques*

# Nanotechnologies:
# The Physics of Nanomaterials
## Volume I

*The Physics of Surfaces and Nanofabrication Techniques*

**David S. Schmool**

APPLE ACADEMIC PRESS

First edition published [2021]

**Apple Academic Press Inc.**
1265 Goldenrod Circle, NE,
Palm Bay, FL 32905 USA

4164 Lakeshore Road, Burlington,
ON, L7L 1A4 Canada

**CRC Press**
6000 Broken Sound Parkway NW,
Suite 300, Boca Raton, FL 33487-2742 USA

2 Park Square, Milton Park,
Abingdon, Oxon, OX14 4RN UK

© 2021 Apple Academic Press, Inc.

*Apple Academic Press exclusively co-publishes with CRC Press, an imprint of Taylor & Francis Group, LLC*

**Library and Archives Canada Cataloguing in Publication**

Title: Nanotechnologies : the physics of nanomaterials, volume I : the physics of surfaces and nanofabrication techniques. / David S. Schmool.

Names: Schmool, D. S. (David S.), author.

Description: Includes bibliographical references and indexes. | Contents: Volume I. The physics of surfaces and nanofabrication techniques.

Identifiers: Canadiana (print) 20200297066 | Canadiana (ebook) 20200297120 | ISBN 9781771889476 (set) | ISBN 9781771889483 (v. I ; hardcover) | ISBN 9781003100218 (v. I ; ebook)

Subjects: LCSH: Nanotechnology. | LCSH: Nanostructured materials. | LCSH: Nanoscience. | LCSH: Physics.

Classification: LCC QC176.8.N35 S36 2021 | DDC 620/.5—dc23

**Library of Congress Cataloging-in-Publication Data**

........................................................................................................................................................

CIP data on file with US Library of Congress

........................................................................................................................................................

Nanotechnologies: The Physics of Nanomaterials, 2 Volume (Set)
ISBN: 978-1-77188-947-6 (hbk)
ISBN: 978-1-00310-019-5 (ebk)

Nanotechnologies: The Physics of Nanomaterials, Volume I
ISBN: 978-1-77188-948-3 (hbk)
ISBN: 978-1-00310-021-8 (ebk)

# Dedication

*For Ulysse*

# About the Author

**David Schmool**, PhD, is a former Director of the Groupe d'Etude de la Matière Condensée GEMaC at CNRS (UMR 8635) and is currently Professor in Physics at the Université de Versailles/Saint-Quentin-en-Yvelines and Université Paris-Saclay, France. Prior to this he has held teaching and research positions at the University of Perpignan/Laboratoire PROMES – CNRS, Perpignan, France; University of Porto, Portugal, University of the Basque Country, Bilbao, Spain; Istituto IMEM, Parma, Italy; University of Exeter, UK and University of Liverpool, UK. David Schmool has also been a visiting fellow to several institutions, including Simon Fraser University (Canada), the University of Versailles (France), the University of Duisburg-Essen (Germany), and the University of Glasgow (UK). He obtained his DPhil in Physics from the University of York in 1994. In addition to his research experience, he has lectured on physics since 2000 on a broad range of subjects. He has over 25 years of research and teaching experience in areas related to nanosciences and nanotechnologies. He has also developed master's and PhD level courses in nanotechnologies and related subjects, which he has also taught. He has published widely, including over 80 journal papers, 12 book chapters, a textbook and has given many conference presentations, including 20 invited talks and over 20 invited seminars.

# Contents

# Preface

This two-volume book stems from a master's level course in *Nanotechnologies* that I originally gave at the University of Porto in Portugal from 2008 to 2013. It was morphed in various forms to cover subjects in parts of courses I gave in *Nanomedicine* (2011–2013) (at the Faculty of Engineering of the University of Porto), *Characterization of Nanomaterials* and *Innovative Materials* (2015) at the University of Perpignan, France. More recently, I have used some of these chapters in the preparation of a master's course on *Nanomaterials and Characterization* at the University of Versailles/the University of Paris-Saclay. I have also adapted the course to a doctoral level taught course, which I also gave at the University of Porto. This was a shorter more focused module that aimed at an introduction for first-year graduates. Since leaving the University of Porto, my colleague Dr. André Pereira has taken up the role of teaching the course that I initially designed. He has extensively used my lecture notes. For the most part, he has maintained the structure of the course and has incorporated some innovations and adaptations of his own. I am very grateful to Dr. Pereira for his useful suggestions in the compilation of this two-volume book. In addition to this, I have greatly benefited from feedback from students who have taken my courses, with some helpful suggestions, particularly in its early development.

In recent years, there has been a rapid expansion of the literature in all areas *nano*. This ranges from the fabrication techniques to specialized texts in nanomagnetism and nanoelectronics, for example. There are also many textbooks on general introductions to nanotechnology. In this two-volume textbook, my main objective has been to provide a reasonably in-depth account of the subject, aimed principally at university students at the postgraduate level. It is intended for students of physics and engineering-based courses, though it may also serve as a general introduction to the expert newcomer.

As will be noted from the Contents, I have tried to start at the beginning. Or at least where I think the beginning should be. Chapter 1 is meant to provide a gentle introduction to the subject. I have opted to provide a general overview of why nanotechnologies are of such importance in modern technologies and today the groundwork for the context in modern science and technologies. As with the whole book, the emphasis being on Physics. I feel a little uncomfortable discussing biological issues as it is not an area I specialize in. That said, I am well aware that biological issues are extremely important in modern developments in nanoscience. However, this is intended to be a book on the physics of nanotechnologies.

I have chosen to divide the book into three parts and in two volumes. The first part of Volume 1 aims to introduce some of the basic physics required for understanding where nanotechnologies came from. After discussing vacuum physics in Chapter 2, much of the rest of the following chapters in Part I are dedicated to surface physics. Here I introduce the main concepts regarding the nature of surfaces and in particular the modification of the surface crystalline structure due to symmetry changes, which leads to the formation of surface reconstructions. The chapter on thin films is a basic introduction to the main concepts of film growth and methods of preparation. I make no pretense at being exhaustive. My main aim is to provide enough basic physics that the reader can use these concepts to understand other techniques when she/he is confronted with them. Chapter 5 aims at providing an introduction to a broad spectrum of methods used in the analysis of the characterization of surfaces. Many of these methods are widely used in the study of nanomaterials. I have partitioned this into sections of the study of structures via diffraction methods, electron spectroscopies for chemical analysis and microscopies.

The fabrication of nanostructured materials is a large and important topic in nanotechnologies. I have used Part II of Volume 1 to introduce some of the main methods employed to prepare nanostructured materials. Chapter 6 looks at one of the main classes of methods; lithographies. There are a number of techniques available and these are discussed in some detail. In more recent times other techniques have come to the fore, such as the replication methods, this is the subject of Chapter 7. Nanoparticles and nanowires are of huge importance and particularly in terms of their applications, notably in biomedical sciences. In Chapter 8, I have provided an overview of the main methods of fabrication. Chapter 9 brings together other specialized subjects that didn't fit into the previous chapters of Part II. Here we discuss the important class of carbon allotropes, such as the fullerenes and carbon nanotubes (we also discuss graphene, but more in terms of its physical properties). Self-assembly is a fascinating method for the preparation of nanostructures and comprises many approaches. This chapter also provides a brief introduction to some of these methods.

The last part of this two-volume book, i.e., Volume 2, is dedicated to an in-depth study of the physical properties of nanomaterials. I have selected to treat these as the main physical properties of materials: mechanical properties, electronic properties, optical properties, and magnetic properties. The approach I have used in general is to provide a basic introduction to the subject at a level which anyone with a general grasp of solid-state physics should have no problem understanding and is meant as a revision of the main issues. This provides the context for understanding the specific properties of materials with reduced physical dimensions. I have tried to provide ample examples and applications throughout these chapters. It is inevitable that I have missed some topics.

As I mention in the opening paragraph, this book emerged from my preparations for a master's level course in physics on the subject of nanotechnologies. In the preparation of my notes, I used a number of texts on all areas covered in the book. Since turning those notes into the book form, I have expanded and extended many of the chapters. I have also added some subjects that were missing. The book maybe a little too long for one single course on nanotechnologies. My intention is not to provide an exhaustive compendium on the subject, nor is it meant to be a definitive course. I see it more as a guide to the basic principles of the subject. As such I think the book can be used in a number of ways. Principally as a guide to the main subjects. The course tutor/lecturer can choose those topics she/he finds the most pertinent to their approach. They may also wish to provide more in-depth coverage of certain topics. The student can use the book in a similar way, by picking and choosing the subjects of interest and those required in their course. The general reader may wish to delve into specific topics related to their interests. The book may also provide the basis for specialized courses. The chapters on Volume 2, for example, could be used as the basis for specialized topics. In which case, the course provider may use the chapter as a guide upon which further material can be added and treated in more detail. In all of these chapters, I have attempted to provide a general introduction to the principal physical properties of solids. I then attempt to show how the physical properties are modified by the size of the objects. Often we will consider the characteristic length scales which determine at what size of object these modifications will occur. Where possible I have added examples taken from recent research literature which should help demonstrate the main physics involved as well as to provide some examples of how these properties can be harnessed in the plethora of applications and devices that researchers the world over have dedicated their research. Part II of Volume 1 can also be seen as an overview of fabrication techniques, which while not being exhaustive, should provide a solid basis upon which a course on this subject can be prepared.

The use of mathematics is essential for our understanding of the physical properties of materials. I have tried to keep this to the essential. Anyone with undergraduate level mathematics will have no problem following the mathematics in this book. In each chapter, I have tried to end with a brief summary of the subject as well as giving some problems at the end for the student to test their knowledge and understanding. I also list the main references and some further reading at the end of each chapter.

I must admit that when I started the writing of this book, I had some romantic notions of what I would like the book to look like, etc. Reality hits fast! One of the hardest tasks for anyone writing on subject matters which are as vast as those of nanotechnologies is what to include and importantly, where to stop! I can't claim that I have made a good job at this, but at some point, you have to stop. Nanotechnologies are a broad-ranging subject and indeed each chapter could be turned into a specialized book. The literature is extensive and seemingly infinite. A choice must be made. I have opted to try to cover the main topics and provide some pointers for further reading. Despite these difficulties, I have to say that I really enjoyed writing this book. First of all, it is a subject I have come to love and am always discovering some new aspect or recent application. It is indeed an amazing area of science. I hope you will enjoy it as much as I do. I am frequently struck by the ingenuity of scientists and the clever tricks they can do.

I would like to thank the editors of this book for their patience, particularly with respect to the extensions to the extensions of deadlines... I have had to balance the writing of this book with a number of tasks over recent years. As well as the ever urgency of teaching duties, I have also been occupied by the demands of the administrative tasks of being the Director of a research laboratory. Inevitably there will be some errors and mistakes. I am hugely indebted to the dedication and assistance of my colleague Dr. Daniel Markó, who has generously dedicated a large amount of time to read the entire draft of the book, making many corrections and providing countless useful comments and suggestions. I am sincerely grateful for his careful and extensive corrections. Thanks, Daniel! I am grateful to my family for their love and support, especially Mike and Debbie. Thanks are also due to Louise, for being Louise. I cannot begin to estimate the debt I owe to my partner Virginie, she has suffered my extended absences with good humor and my life would infinitely more difficult without her love and support. I would also like to mention Ulysse, our son, who has provided us with so much joy over the last two years and to whom I dedicate this two-volume book.

*—David Schmool*
*Versailles, France, 2019*

# Chapter 1

# Introduction to Nanotechnologies

## 1.1   What is Nanoscience and Why Are We So Interested in It?

Nanotechnology has become a buzzword in modern science and has grabbed the attention of both the media and the general public. Despite this, the general publics' notion of what Nanotechnologies really means is still rather limited, as frequent non-scientific comments attest. While we should really concern ourselves directly with such issues, we should be aware that poor forms of publicity can be detrimental to the publics' notion of what is meant by Nanotechnology and this goes for science in general.

In this chapter, we aim to provide a broad overview of the extent of the science involved in this multi- and a cross-disciplinary branch of modern science. Current research in nanoscience is strongly rooted in all of the three of the main branches of science: Physics, Chemistry, and Biology. Indeed, we are at a particularly unique and interesting stage of scientific evolution, which starting from the period of modern scientific development in the 16th and 17th centuries saw the diversification of the various areas of science into the main branches of Physics, Chemistry, and Biology. Each has witnessed a strong branching of domains into more specialized areas. What makes the area of nanotechnology unique is the potential for cross-linking between the three principal topics of science? Inter-disciplinary research has become the norm in recent years, with each subject bringing a different perspective to the broad range of challenges presented by nanoscience. Of the emerging fields, the study of nanobiomedicine is one that particularly stands out. Scientists from all branches of science are engaged in research on a vast range of issues that concern biomedical application which has the potential to touch the lives of virtually all humanity.

In this book, we will deal principally with the Physics branch of Nanoscience, where we are concerned with objects ranging from small clusters of atoms to objects on the scale of 100s of nanometers and even reaching the micron scale. One of the main issues that we need to address is the "why?" Why is the nanoscale so important in physics? Or more precisely the physics of materials. We will address this issue on various levels and in particular in the final part of this book, where we will discuss the physical properties of materials at the nanoscale. The answer has a number of issues that need to be addressed as there is no single answer to this question, but a multitude which arises from the nature of materials and in particular the dependence of the different properties on their characteristic length scales. This will be discussed in Section 1.2.

Nanotechnology can be generally thought of as the technology concerning the nanoscale and applications for real-world measurable quantities. The subject matter concerns both the production of materials as well as the study of their physics properties. Included in this are the many and varied applications of nanoscale materials and devices. There is no doubt that Nanotechnology will have increasing importance on society, with a profound impact on the global economy in the 21st century and well beyond. Its impact is likely to be more profound than that of the semiconductor revolution of the mid-twentieth century, which has influenced society in a pervading manner, from information technology, computing, and mobile communications to name a few. Nanotechnology will impact further fields such as nanoelectronics, including spintronics, medicine, healthcare, biotechnologies, security technologies, energy applications, etc.

The influence of nanoscience, as we have noted, has been very broad in scientific terms. In Physics, the effects of size reduction have caused nothing less than a revolution, both on the research level and also in its influence in industrial applications. Arguably this impact is felt most acutely in the area of electronic applications. In this area, current devices are approaching the scale of 10 nm or less. This has a knock-on effect on the functionality of these devices as well as the computational capacities associated with them and is directly linked with the effort to keep up with Moore's Law prediction of approximately doubling the number of transistors on a chip each year. We can expect a crunch in this situation, which has been envisaged for over a decade, where the devices reach the dimensions of clusters of atoms. This provides much food for thought.

The discovery of the transistor in 1947 by John Bardeen, William Shockley et Walter Brattain, was the first step along the road to the electronics revolution that saw the widespread use of semiconductors in everyday applications and leads to the 1956 Nobel prize being awarded to this American trio. Their discovery and work can be seen as an initial step in the microelectronics revolution that could possibly be one of the most important technological advances of the 20th century, with an inestimable impact on today's society. Processing and fabrication methods underwent great progress in the 1960s, allowing the development of integrated circuits, which emerged in the late 1950s, with the work of Jack Kilby and Robert Noyce and others. The evolution of IC technology is emblematic of the developments in nanoscience in general and illustrates the enormous efforts the have been undertaken in the miniaturization and optimization of device technologies. In Figure 1.1 we illustrate this evolution, with the original IC developed by Kilby (a) as well as the complexity of modern IC designs (b–d) and their realization. The designation of the level of integration denotes the number of transistors per chip has illustrated this development, from SSI (small-scale integration) from 1964 to ULSI (ultra-large-scale integration) in 1984, with the former indicating from 1 to 10 transistors and the latter over one-million. In 2018 Graphcore boasts a count of over $23 \times 10^9$ transistors in the GC2 IPU, while Intel's Stratix 10 has a count of over 30 billion transistors. In terms of miniaturization, the smallest process size for semiconductor manufacturing is the Qualcomm Snapdragon 855 (octa-core ARM64 "mobile SoC"), with a size of 7 nm. This figure is expected to reduce to 5 nm in the next couple of years (at the time of writing in 2018). Samsung announced that they plan to use Gate-All-Around technology to produce 3 nm FETs in 2021 (McGrath, 2018). Moore's law for the number of transistors on a chip is shown in Figure 1.2 for the 40 year period 1971–2011. The evolution is quite staggering. With the reduction of the physical dimensions of a device, the processing time will also reduce. This can be thought of as being related to the time for an electron to traverse the device, hence the smaller its size the quicker the electron can travel across it. This means processing speeds reduce and operational frequencies will increase.

Developments in the microelectronics industry are probably the most sophisticated in terms of device dimensions and performance. However, other technologies have also shown stunning progress. The optoelectronics area closely related to that of the microelectronics industry has taken advantage of much of the same technologies. The earliest solid-state lasers date from

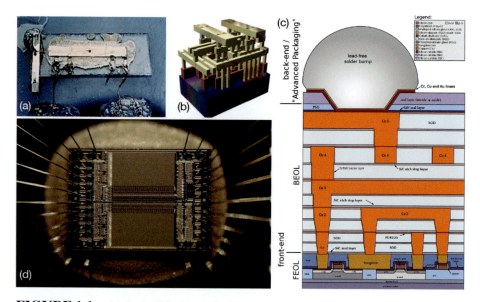

**FIGURE 1.1**    (a) Jack Kilby's original integrated circuit. (b) Rendering of a small standard cell with three metal layers (dielectric has been removed). The sand-colored structures are metal interconnect, with the vertical pillars being contacts, typically plugs of tungsten. The reddish structures are polysilicon gates, and the solid at the bottom is the crystalline silicon bulk. (c) Schematic structure of a CMOS chip, as built in the early 2000s. The graphic shows LDD-MISFET's on an SOI substrate with five metallization layers and solder bump for flip-chip bonding. It also shows the section for FEOL (front-end-of-line), BEOL (back-end-of-line) and first parts of the back-end process. (d) Integrated circuit from an EPROM memory microchip showing the memory blocks, the supporting circuitry and the fine silver wires which connect the integrated circuit die to the legs of the packaging.

1962 using GaAs. In magnetic materials, one area of particular technological importance is that of data storage, where the measure used is the bit size or the number of bits per square inch. This parameter is also observed to follow Moore's law, and while doubling at a faster rate than that for transistors, the so-called Kryder rate is somewhat optimistic. Kryder (2009) projected that if hard drives were to continue to progress at their then-current pace of about 40% per year, then in 2020 a two-platter, 2.5-inch disk drive would store approximately 40 terabytes (TB) and cost about $40. In 2014, a single 2.5-inch platter stored around 0.3 terabytes in 2009 and this reached 0.6 terabytes.

The trends in nanotechnologies were categorized in 2006 by Renn and Roca (2006), who gave a roadmap for the development of industrial prototyping for the commercialization of nanotechnologies. This prediction, shown

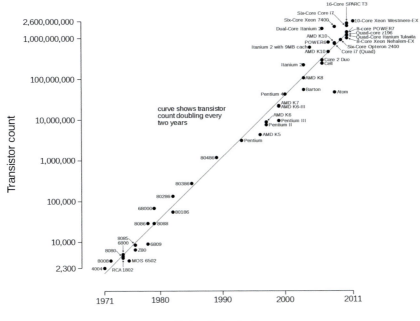

**FIGURE 1.2** Transistor count as a function of date. This variation indicates a doubling every two years.

in Figure 1.3, envisages the progression from passive to active nanostructures, leading to the assembly of nanosystems and the eventual development of molecular nanosystems. Looking at the evolution, indeed we are a long way along this path, though the third stage may not be as well developed as predicted. We have certainly come a long way, especially considering the state of technology in the late 1950s, when Richard Feynman made his now-famous talk at the annual meeting of the American Physical Society. His presentation, entitled *"There's Plenty of Room at the Bottom,"* concerned the potential for miniaturization and proved quite visionary considering that the first steps towards the integrated circuit had only been developed in the previous year or so. The essay is indeed fascinating and reading it today one is struck by the force of Feynman's predictions. It is a great place to start your adventure into this fast-moving subject, and I would invite all students to type this title into their search engine, where they will easily find the essay. I am certain that Feynman would be amazed by the progress that has been made over the previous decades. In the following section, we will outline some of the principal reasons why nanotechnologies have such an important

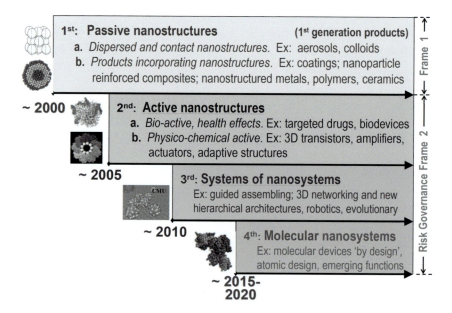

**FIGURE 1.3**   Timeline for the beginning of industrial prototyping and nanotechnology commercialization: Four overlapping generations of products and processes. Reprinted by permission from O. Renn, & M. C. Roca, (2006). ©Springer, Nature, *J. Nanoparticle Research, 8*, 153.

impact on science and technology, starting with the fundamental concepts of how material properties can be modified and controlled by the size of a solid object.

## 1.2   Modification of Physical Properties

One of the principal reasons that nanomaterials are of interest is related to the ability to modify the physical properties of solids by controlling their size (and shape). In the first instance, the reduction of the size of a body will mean that the number of atoms located at and in the region of the surface will become a significant proportion of the atoms in the body. In Chapter 3, we will discuss the physical surface, where we will note that the reduced coordination of atoms located at surfaces and in their vicinity, will effectively alter their physical properties. Indeed, as any textbook on Solid State Physics will attest, the physical properties of solids depend on the type of atoms present in the solid and their physical arrangement or crystalline structure. This leads to the intrinsic physical properties of solids. In the surface region of a solid, the atoms can alter their coordination with neighboring atoms as well as undergo a change in their periodic structure. This is referred to as the surface e reconstruction of atoms and will be discussed in more detail in Section 3.3. Both the change of atomic coordination and the modification of the crystalline order can be expected to have a significant influence on the physical properties of the solid. This is also true to a certain extent to the alteration of the interatomic spacing, which is a common occurrence in surface reconstructions.

It is instructive to calculate the variation of the ratio of the number of surface atoms to those in the bulk as a function of the size of a solid structure. To do this we will use a simple model allowing a direct calculation. We note that the crystalline structure will also be important for specific cases. In our model, we will consider a hypothetical material with a simple cubic structure and a lattice parameter of $a = 3.0$ Å. The body will be assumed to have a cubic shape. This model is illustrated in Figure 1.4 Changing any one of these details will alter our calculation. The chosen model allows us to perform simple calculations and obtain a general form for the variation of the surface to volume atoms as a function of the nanostructure size. A simple consideration of the number of atoms in surface and volume (bulk-like) positions leads us to the following expressions:

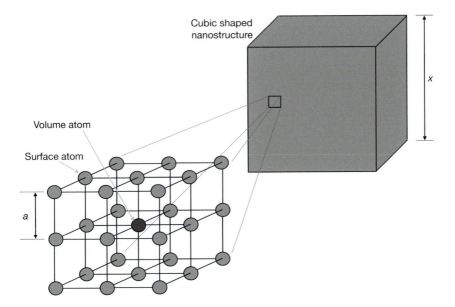

**FIGURE 1.4** Simple model to illustrate the variation of the ratio of surface to volume atoms in a cubic shaped nanostructure with a simple cubic crystalline structure.

$$n_v = (n-2)^3 \tag{1.1}$$

for the number of volume atoms, which have full atomic coordination, where $n$ denotes the number of atoms along the edge of the structure, and

$$n_s = n^3 - (n-2)^3 = n_t - n_v \tag{1.2}$$

for the number of surface atoms. Here $n_t = n^3$ is the total number of atoms in the nanostructure. The edge length of the latter is given by $l = (n-1)a$. The surface to volume ratio of atoms should provide a reasonable approximation of where we may expect surface dominated properties to dominate. This can be written as:

$$\frac{n_s}{n_v} = \left(\frac{n}{n-2}\right)^3 - 1 \tag{1.3}$$

While this is an oversimplified model, it shows the kind of considerations we can make in establishing where the physical properties of the body might be influenced by its size. The variation for this model of the surface to volume ratio of atoms is illustrated in Figure 1.5. While at a linear size of around 2.5

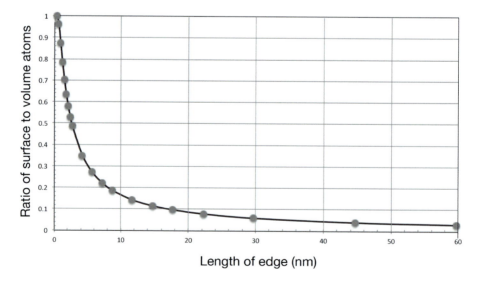

**FIGURE 1.5**     Variation of the ratio of surface to volume atoms in a cubic shaped nanostructure with a simple cubic crystalline structure as a function of its linear dimension, $l$.

nm, around about 50% of the atoms in the nanoparticle will be located at its surface, we can expect that the properties of the particle will be noticeably different for a lower proportion of surface atoms. In part, this will depend on the particular property in question, but a good rule of thumb would be around about the 10% region. This would correspond to a size of about 20 nm. It should be noted that the surface reconstruction can extend a few atomic layers into the solid, which would mean that the physical properties will be modified for even larger nanoparticles and thus be measurable experimentally.

As an example of the variation of the physical properties as a function of the particle size we will consider melting temperature of a solid. The melting temperature of a solid is proportional to its cohesive energy. This basically means that the melting temperature is a direct measure of the total bonding energy in a solid. For a nanosolid this energy can be expressed as (Qi et al., 2005):

$$E_{Cn} = \frac{1}{2}(n_v\beta_V + n_s\beta_S)\varepsilon = \frac{1}{2}[n_t\beta_V - (\beta_V - \beta_S)n_s]\varepsilon \qquad (1.4)$$

Here $\beta_{V,S}$ denote the coordination numbers for atoms in the volume (V) and at the surface (S), while $n_{s,v,t}$ are indicated above and $\varepsilon$ is the energy associated with each interatomic bond. For a bulk material we can neglect surface effects and the cohesive energy will be given by:

$$E_0 = \frac{1}{2}n_t\beta_V\varepsilon \qquad (1.5)$$

As we have mentioned above, the melting temperature for the solid is proportional to the cohesive energy. This means that we can associate the ratio of the melting temperatures for the nanostructure, $T_{mN}$ and the bulk, $T_{mV}$, with the ratio of their cohesive energies. This allows us to establish the following relation (Safaei et al., 2007):

$$\frac{T_{mN}}{T_{mV}} = \frac{E_{Cn}}{E_0} = 1 - (1-q)\frac{n_s}{n_t} \qquad (1.6)$$

where $q = \beta_S/\beta_V$ is the surface-to-volume coordination number ratio. This equation should be valid for all nanostructures independently from their shape. Clearly the nanostructure shape will affect the value of $n_s/n_t$, so the shape of the particle is also a factor which needs to be taken into account for calculations as will the crystalline structure of the material in question. By introducing a surface and bulk packing fraction, denoted as $P_S$ and $P_V$, respectively, it is possible to account for different shaped particles and structures.

The explicit relation in either case will depend on the crystalline structure of the material. In this manner it is possible to express a general form for the melting temperature ratio, which can be expressed as:

$$\frac{T_{mN}}{T_{mV}} = 1 - 2(1-q)\left(\frac{3-\lambda}{3}\right)\frac{aX}{b+\left(\frac{3-\lambda}{3}\right)aX} \qquad (1.7)$$

where $a = 2P_S d$, $b = P_L$ and $X = 1/(\text{size})$, which can be the diameter for a particles or the thickness for a film. We further note that here $d$ denotes the atomic diameter and $\lambda$ is a shape parameter for a particular shaped structure ($\lambda = 0$ for a spherical particle, $\lambda = 1$ for nanowires and $\lambda = 2$ for films). In Figure 1.6, we illustrate the variation of the normalized melting temperature for nanoparticles of In and Au. Also illustrated in these figures are the model fits for $\lambda = 0$ with various values of $q$ for Eq. (1.7). This is a nice demonstration of the impact of object size on the physical behavior of a nanosystem. It is interesting to note the broad agreement between the variations indicated in Figures 1.5 and 1.6, despite the simplifying assumptions used in the former.

Confinement effects can also have a profound effect on the physical response of a nanostructure. The confinement in such cases frequently refers to the limited space in which free electrons can move in the solid, and arise from the boundary affects the spatial limits of the nanoparticle. This has

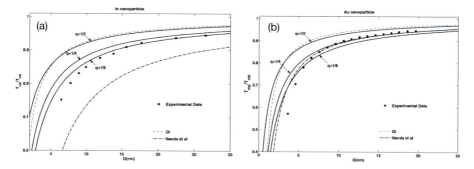

**FIGURE 1.6**  Variation of melting temperature on the diameter of (a) In nanoparticles and (b) Au nanoparticles. The lattice type of In is body-centred tetragonal (bct) ($P_L = 0.68, P_S = 0.78$), $r = 0.1843$ nm and $T_{mV} = 429.8$ K (Nanda et al., 2002). Experimental data for the In nanoparticles is taken from (Skripov et al., 1981). (b) The lattice type of Au is fcc ($P_L = 0.74, PS = 0.91$), $r = 0.1594$ nm and $T_{mV} = 1337.6$ K (Nanda et al., 2002). Experimental data for Au particles is taken from (Schmidt, 2001). Also illustrated are the models developed by Nanda et al., (2002) and Qi (2005) (Safaei et al., 2007). ©IOP Publishing. Reproduced with permission. All rights reserved.

a strong influence, for example, on the electronic and optical properties of solids. This subject will be treated in detail in Chapters 2 and 3 of Volume 2. The confinement of particles is a well-known problem in quantum mechanics and can be observed in the macroscopic physical measurements of confined systems. Due to the scaling laws, this often means that quantum effects are important on the nanoscale. This frequently leads to the discretization of electronic energy levels leading to characteristic optical properties arising from the allowed transitions between these states. We can also note that in metallic systems, the confinement of the electron gas by the ionic potential of the atoms leads to localized plasmon resonances in metallic nanoparticles (see Chapter 3, Section 3.4), a phenomenon which has been shown to give rise to specific colors in glasses, some of which have been known from antiquity. The famous example being the Lycurgus Cup, which is a 4th-century Roman glass cage cup made of dichroic glass. The cup exhibits a different color depending on whether or not light is passing through it; red when lit from behind and green when lit from in front. These properties are related to the inclusion of Au nanoparticles in the glass which arises from the plasmonic response of the nanoparticles, see Figure 1.7. Such effects also account for the colors observed in many ancient glasses used in the stained glass windows of churches and cathedrals across Europe for example.

Confinement effects in magnetism are also well known. The boundary created at a surface leads to magnetic anisotropies which differ from those in the volume of a magnetic material. The boundary conditions due to these anisotropies can allow systems with specific dimensions to support standing spin-wave modes. Typically for the transition metal ferromagnets, the critical dimensions lie in the region of tens of nanometers. Such effects have been observed in ferromagnetic thin films and are discussed in detail in Chapter 4 (Section 4.7.4).

Interactions between nanoparticles and nanostructures can also have an important influence on the physical properties of assemblies of nanoparticles and nanostructures. Examples of such effects are given by the dipolar interaction in both magnetic nanoparticle systems and in the plasmonic response of assemblies of metallic nanoparticles. The physical interactions between the nanoparticles can be controlled by their density in the assembly. This results from the variation of the average interparticle separation in the assembly as a function of the particle density. While these effects are secondary, they also provide a means for manipulating the physical properties of nanosystems. Other forms of interaction will depend on the nature of the materials and the

**FIGURE 1.7**    The Lycurgus cup photographed in transmission (left) and in reflection (right), where it exhibits characteristic red and green colors, respectively.

media in which they are embedded. The mediation of interactions can occur via a number of mechanisms, such as the free electrons in metallic systems, an example being the RKKY interaction between ferromagnetic layers separated by a non-magnetic metallic spacer layer. This was one of the effects that lead to the discovery of the giant magnetoresistance effect first observed at the end of the 1980s and ultimately giving rise to a number of applications, notably in the data storage industry and whose importance was recognized in the awarding of the 2007 Nobel prize to Albert Fert and Peter Grúnberg.

## 1.3    Laws of Scaling in Physics

In Physics, it is customary to consider the behavior of systems by taking the extreme limits. For materials, these limits would naturally be in terms of the size scale. The large extreme would correspond to bulk material, while at the other end of the scale we have the atom. We are interested in defining where the limit of the physical properties change as a function of size. Clearly, this problem suffers from the difficulty of finding the crossover from bulk to the nanoscale since the limits are not abrupt from one to the other. Furthermore, the limit will also depend on the physical effect under consideration and specific material. As we noted above we can make some progress by defining structures and shapes of nanostructures to see where there is a crossover in terms of the surface-to-volume ratio. Another approach is to consider the scaling laws, which are relevant for the various domains of physics; mechanical properties, fluid mechanics, electromagnetism, optics, electronic properties, magnetism, etc.

The goal of this approach is to show how the different physical effects scale with size in terms of the linear dimensions of an object. This can be developed in principle for all know physical properties (Wautelet et al., 2006). We will further outline some of the consequences of this approach at the beginning of Section 1.1 of Volume 2.

## 1.4 Methodologies: Bottom-Up/Top-Down

The formation of nanostructures is one of the fundamental requirements in nanotechnologies. We require nano-objects so that we can study them and exploit their physical properties in a broad range of applications. This can be considered as our starting point. The fabrication of nanostructures and nano-objects, in general, can be viewed as a *bottom-up* or a *top-down* approach to the problem. Bottom-up assembly refers to the process of fabricating nanostructures starting from fundamental processes, which in their nature are random events. Most biological structures are formed in this way. They rely on a set of random events which are controlled by environmental factors to perform a *self-assembly* process which ends with a full-formed structure. While many lessons can be learned from such processes, man-made objects are generally less complex.

Bottom-up processes include a broad range of techniques and methodologies. Generally speaking, they depend on the agglomeration of a large number of components, or building blocks, which must come together and interact. This interaction is required to form a cohesive structure which is stable. On the most elemental level, we can consider the formation of crystals, such as those which form nanoparticles or thin films in epitaxial growth, as being bottom-up processes. Here, the atoms are brought together either in solution as precursors that decompose to allow certain elements to coalesce in the formation of a nanoparticle, or they can be evaporated from a solid target in a vacuum to condense onto a substrate. In either case, the individual atoms must form bonds with other atoms to form the solid. In doing so entropy will play an important role in determining the equilibrium to form the solid phase with a particular crystalline structure. Non-equilibrium conditions can also be used in processes where the system does not have time to thermalize, however, the structure formed is unlikely to be the same as the equilibrium structure. The key factor here is the interaction, which forms the bond between the building blocks. Self-assembled structures, as will be discussed in Chapter 9, are also classed as bottom-up processes since they depend on the organized assembly of the principal components.

Top-down processes consider the problem from the other direction. Here we start will a large block of material and etch away at it to produce the structure we desire. Sculpting a large statue from a marble block would be a good example of a top-down process. For the purposes of nanotechnologies, top-down processes encompass all forms of lithography, such as photo-lithography, electron lithography, etching processes, focused ion-beam milling methods, etc. We will provide in-depth outlines of these and other related technologies for the fabrication of nanostructures, see Chapters 6 and 7.

## 1.5   The Study of Nanomaterials

Once we have found the method for preparing our nanosized objects we must consider the nature and composition of the samples themselves. To study the physical properties of materials, we rely on a number of techniques, depending on the physical properties of interest. Since the physical properties of materials are intimately related to the structure and composition of the sample, this is usually a starting point for many characterization studies. Often we aim to correlate the physical behavior with the structure, composition and occasionally the morphology of the sample. Structures can be studied using diffraction methods, such as x-ray or electron diffraction, while compositions can be studied using spectroscopic means, for example, Auger spectroscopy is a standard technique thatch be used. The study of morphologies can be performed using a number of microscopies, such as scanning electron and atomic force microscopies. In Chapter 5, we will provide detailed introductions to these and other related techniques used in the study of thin films and nanomaterials. Many of these methods have been specialized over the years to study specific types of samples and are highly adapted for low dimensional systems.

The physical properties, such as optical-electronic and magnetic, are specialized subjects in nanomaterials. These properties are specific to the materials used as well as their dimensions. Each has its own specific characteristic length scales, and many techniques have been developed and adapted to the study of nanostructures. We will provide in-depth discussions into these subjects in the final chapters of Volume 2.

As we have noted above, the number of applications in Nanoscience is enormous. It would be impossible to provide even a cursory overview of all of these. However, we will provide many details of some of the more important applications of nanomaterials and structures. These are given in the specialized chapters on the physical properties of nanosystems.

## 1.6 Problems

(1) Demonstrate that the ratio of surface-to-bulk atoms in a cubic structure with a simple cubic lattice can be expressed as:

$$\frac{n_s}{n_v} = \frac{6n^2 - 12n + 8}{n^3 - 6n^2 + 12n - 8}$$

where $n$ denotes the number of atoms along the cubic edge of the structure.

(2) Explain how you would derive the expression for the ratio of surface-to-bulk atoms in nanoparticles with fcc structure with cubic and spherical shapes.

(3) Show that the surface area to volume ratio for cubic and spherical structures have similar size dependencies.

(4) Prove that the melting temperature of a nanoparticle can be expressed as;

$$T_{mN} = T_{mV} \left[ 1 - (1 - q)\frac{n_s}{n_t} \right]$$

where $T_{mV}$ denotes the bulk melting temperature. Further show that the corresponding expression for a cubic structure with a simple cubic lattice can be expressed as:

$$T_{mN} = T_{mV} \left\{ 1 - \frac{1}{3}\left[ 1 - \left(\frac{n-2}{n}\right)^3 \right] \right\}$$

(5) The semiconductor CdSe has a bulk melting temperature of 1678 K. However if this material is formed into nanoparticles of 3 nm diameter, this melting temperature significantly reduces to 700 K. Explain this empirical observation. Estimate the melting temperature for 2 nm diameter particles. What assumptions have you made in this estimation?

(6) Estimate the time that a thermal electron would need to traverse the active region of a FET of 10 nm. What assumptions have you made in your calculation?

## References and Further Reading

Agrawal, D. C. (2013). *Introduction to Nanoscience and Nanomaterials*, World Scientific, Singapore.

Bhushan, B. (Ed.), (2010). *Springer Handbook of Nanomaterials*, Springer Verlag, Berlin Heidelberg.

Dresselhaus, M. S. (2016). *Nat. Rev. Mat., 1*, 1.

Dupas, C. Houdy, P., & Lahmani, M. (Eds.), (2007). *Nanoscience: Nanotechnologies and Nanophysics*, Springer, Berlin, Heidelberg.

Feynman, R. P. J. (1992). *Microelectromech. Systems, 1*, 60; $http://calteches.library.caltech.edu/1976/1/1960Bottom.pdf$ (accessed on 30 March 2020).

Kryder, M. H., & Chang Soo, K. (2009). *IEEE Trans. Magn., 45*, 3406.

Lindsay, S. M. (2010). *Introduction to Nanoscience*, Oxford University Press, Oxford.

McGrath, D. (2018). EETimes, $https://www.eetimes.com/document.asp?doc\_id = 1333318$ (accessed on 30 March 2020).

Nalwa, H. S. (Ed.), (2000). *Encyclopedia of Nanoscience and Nanotechnology*, Elsevier, Amsterdam.

Nanda, K. K., Sahu, S. N., & Behera, S. N. (2002). *Phys. Rev. A, 66*, 13208.

Poole, C. P. Jr., & Owens, F. J. (2003). *Introduction to Nanotechnology*, Wiley, New York.

Qi, W. H. (2005). *Physica B, 368*, 46.

Qi, W. H., Wang, M. P., Zhou, M., & Hu, W. Y. (2005). *J. Phys. D: Appl. Phys., 38*, 1429.

Renn, O., & Roco, M. C. (2006). *J. Nanoparticle Research, 8*, 153.

Safaei, A., Attarian Shandiz, M., Sanjabi, S. & Barber, Z. H. (2007). *J. Phys.: Condens. Matter., 19*, 216216.

Schmidt, G. (2001). *Nanoscale Materials in Chemistry*, K. J. Klabunde (Ed.), pp. 23–24, Wiley, New York.

Schmool, D. S. (2016). *Solid State Physics*, Essentials of Physics Series, Mercury Learning and Information, Stylus Publishing LLC, USA.

Sengupta, A., & Kumar Sarkar, C. (Eds.), (2015). *Introduction to Nano*, Springer, Berlin.

Skripov, V. P., Koverda, V. P., & Skokov, V. N. (1981). *Phys. Status Solid,* *66*, 109.

Vajtai, R. (2013). *Springer Handbook of Nanomaterials*, R. Vajtai (Ed.), pp. 1–36, Springer Verlag, Berlin Heidelberg.

Wautelet, M. et al., (2006). *Les Nanotechnologies* (2e), Dunod, Paris.

**PART I**

**THE BASICS: SURFACE SCIENCE, THIN FILMS, AND SURFACE ANALYSIS**

# Chapter 2

# Vacuum Science and Technology

## 2.1 Introduction and Orders of Magnitude

The physical vacuum is an essential condition for many experimental studies in Physics. This is particularly true for the study of surfaces and many nano-objects. Indeed, many of the analytical tools used to study surfaces and thin films (as will be discussed in Chapter 5) require high and ultra-high vacuum conditions for them to be able to probe the outermost atomic layers of an object. In fact, the very systems themselves require a vacuum for their functioning, such as electron guns and analyzers. The reason a vacuum is required in these types of the experiment is related to the composition of the atmosphere and the interaction of these gases with the surfaces of films and nanostructures. Objects in the air are continually bombarded by atoms and molecules in our everyday environment effectively coat all matter. To allow the efficient study of surfaces at the atomic level and for the preparation of thin films and nanostructures, it is necessary that we control the environment of the outer limits of the material.

In terms of the physical environment, a vacuum is defined as a region of space in which the number of atoms or molecules is reduced with respect to that of normal atmospheric conditions. As such, we use a rather loose definition of the vacuum, which allows us to take into account the various levels of a vacuum. A vacuum is typically characterized in terms of the gas pressure in the vacuum chamber. Since we measure the physical pressure as the force caused by the impact of atoms and molecules in a gas per unit area of the surface containing the vacuum, the unit of pressure $p$, the Pascal (Pa), is equivalent to $1 \text{ N m}^{-2}$. We can relate the number density of atoms or molecules, $n$, in the vacuum with the pressure using the *ideal gas equation*, which allows us to establish:

$$p = nk_B T \qquad (2.1)$$

where $k_B = 1.380\,648 \times 10^{-23}$ J K$^{-1}$ is the Boltzmann constant and $T$ is the temperature. Typically we take the latter as room temperature (295 K). It is often useful to consider the standard conditions of temperature and pressure (with $T = 0$ K $= 273°$C and $p = 101.3$ kPa), from which we calculate, with the aid of Eq. (2.1), that the number of molecules in a cubic meter of gas is $2.5 \times 10^{25}$.

Our broad definition of what a vacuum is can now be put a little in perspective since it in no way means "nothing at all." In a vacuum produced in a laboratory, with a pressure of 15 orders of magnitude lower than atmospheric pressure, there will still be $2.5 \times 10^{10}$ molecules per m$^3$. Deep space is thought to have an average of less than one atom of hydrogen per cubic meter.

In practice, we use different units of pressure measurement. For example, atmospheric pressure is typically quoted in units of *millibar* (mbar). One bar being defined as 100,000 Pa exactly. Thus we have 1 mbar = 100 Pa and therefore standard atmospheric pressure will be 1013.25 mbar. The mbar is the main standard pressure unit for vacuum practice in Europe, though the unit of torr (from Torricelli) is also common, particularly in North America. 1 torr (= 101,325/760) = 133 Pa, the 760 refers to the mm of Hg measured from Torricelli's experiment, being the height of a column of mercury supported by atmospheric pressure, this is derived from the pressure defined by $h\rho g$, $h$ being the height, $\rho$ the density of mercury and $g$ the acceleration due to gravity. The three main units of pressure can now be related as: 1 mbar = 100 Pa = 0.75 torr.

The practice of vacuum technology spans a wide range of pressures, about 15 orders of magnitude below atmospheric pressure, corresponding to pressures in the range of around 1000 mbar down to $10^{-12}$ mbar. We subdivide this into smaller ranges as defined in Table 2.1.

As was previously stated, the level of vacuum required in practice will depend on the physical conditions necessary for a particular measurement or the level of acceptable contamination in the preparation of samples, thin film or otherwise.

**TABLE 2.1**   Classification of Vacuum

| Vacuum | Pressure range |
|---|---|
| Low (rough) vacuum | Atmospheric pressure to 1 mbar |
| Medium vacuum | 1 to $10^{-3}$ mbar |
| High vacuum (HV) | $10^{-3}$ to $10^{-8}$ mbar |
| Ultrahigh vacuum (UHV) | $10^{-8}$ to $10^{-12}$ mbar |
| Extreme high vacuum (XHV) | Below $10^{-12}$ mbar |

## 2.2   Physical Principles of the Vacuum

Our understanding and description of the physical vacuum is derived principally from the *kinetic theory of gases*. In this section we will revise some of the main points of the theory to give an outline for the guiding principles of vacuum physics.

The *ideal gas law* defines as the equation of state the relation between pressure, volume, and temperature and is expressed as:

$$pV = NRT \tag{2.2}$$

Here, the volume of the gas is given as $V$, $N$ is the number of moles of atoms or molecules in the gas (1 mole being the Avogadro constant, $N_A = 6.022 \times 10^{23}$ mol$^{-1}$) and $R$ is the ideal, or universal, gas constant, equal to the product of the Boltzmann constant and the Avogadro constant; $R = k_B N_A = 8.314\,4598(48)$ J K$^{-1}$ mol$^{-1}$. Equation (2.2) is a good approximation for gases at low pressure. Corrections to this equation are necessary for gases in which there are appreciable interactions between the constituent atoms or molecules, see for example the van der Waals equation.

### 2.2.1   The Kinetic Theory of Gases

The kinetic theory of gases was primarily developed by Clausius, Maxwell, and Boltzmann, though it is also derived from many other contributions, which mark the evolution of thermodynamics. The underlying principle of the kinetic theory is that gases are made up of atoms and molecules in constant motion, whose average velocity is related to the temperature of the gas. The random motion of the gas molecules means that they regularly collide with the walls of the container and thus transfer momentum to the walls. The averaging of all collisions between the molecules and the walls give rise to the notion of pressure in the gas. For a single molecule, of mass $m$ and an instantaneous velocity $v$ in an elastic collision (i.e., without losing energy) with the wall, the momentum transfer of this event can be expressed as:

$$\Delta(mv) = mv - m(-v) = 2mv \tag{2.3}$$

For a container of linear dimension $L$, and assuming a normal motion to the walls, the molecule will take a time of $L/v$ to reach the far wall, where it will again impart a momentum of $2mv$. Assuming that the walls have an area $A$, then the force associated with these collisions will be:

$$F = \frac{\Delta p}{\Delta t} = \frac{2mv}{L/v} = \frac{2mv^2}{L} \tag{2.4}$$

from which we obtain the pressure on the walls as:

$$p = \frac{F}{2A} = \frac{mv^2}{LA} = \frac{mv^2}{V} \tag{2.5}$$

Extending this argument to $\mathcal{N}$ particles in a container, the pressure will now be:

$$p = \mathcal{N} \frac{m \langle v_\perp \rangle^2}{V} \tag{2.6}$$

where we have specified that the velocity is averaged for its perpendicular motion with respect to the walls. Since we have specified that the motion is random, we cannot distinguish directions, and therefore $\langle v \rangle^2 = \langle v_x \rangle^2 + \langle v_y \rangle^2 + \langle v_z \rangle^2 = 3 \langle v_\perp \rangle^2$. Using the number density of molecules $n = \mathcal{N}/V$, we obtain:

$$p = n \frac{m \langle v \rangle^2}{3} \tag{2.7}$$

Using Eq. (2.1), we can now establish the relation:

$$3k_B T = m \langle v \rangle^2 \tag{2.8}$$

Since the average kinetic energy, $\langle E_K \rangle$, of a gas molecule is given by $m \langle v \rangle^2 / 2$, we find that the average kinetic energy can be expressed in terms of the gas temperature:

$$\langle E_K \rangle = \frac{3}{2} k_B T \tag{2.9}$$

This result can also be obtained from the *equipartition of energy*, which states that each independent degree of freedom has an equal amount of energy of $k_B T / 2$. The total energy of the ideal gas is then expressed as:

$$E_{internal} = n_{deg}\frac{1}{2}k_BT \tag{2.10}$$

where $n_{deg}$ is the number of degrees of freedom. The thermal equation of state in 3D for a monoatomic gas is then:

$$E_{internal} = n_{deg}\frac{3}{2}k_BT \tag{2.11}$$

from which we can obtain Eq. (2.9).

It is worth noting that the ideal gas law comprises the laws of Boyle and Charles, which are expressed as: (i) $pV = (2/3)(E_K)_{TOT} = $ const. at any temperature (Boyle's law); (ii) $pV = $ const. $\times T$ (Charles' law). A consequence of Boyle's law is that, at a constant temperature, as volume increases, the pressure of the gas decreases in proportion. Similarly, as volume decreases, the pressure of the gas increases. This can be expressed for a container with movable (but impermeable) walls as:

$$p_1V_1 = p_2V_2 \tag{2.12}$$

Initial and final volumes and pressures of a fixed quantity of gas must, therefore, be maintained at the same temperature (heating or cooling will, in general, be required) to comply with Boyle's law.

### 2.2.2 *The Maxwell-Boltzmann Velocity Distribution*

We consider an ideal gas in which $\mathcal{N}$ molecules are contained within a volume $V$, i.e., with a number density $n = \mathcal{N}/V$. The kinetic theory of gases assumes that the number of molecules is sufficiently large that in the small volume $dV$, there are still enough molecules ($dN = n\,dV$) that it is still typical of the gas as a whole. Molecules are presumed to act as hard spheres which interact elastically, with no other forces involved. Collisions with the container walls are also presumed to be elastic. From the exchange of energies between the molecules, the velocities of the particles achieve a smooth distribution which is expressed as the quantity $f(v)dv$, defined as the probability that a molecule will have a velocity between $v$ and $v + dv$. Therefore of the total $\mathcal{N}$ molecules, considering all directions, a fraction $d\mathcal{N}/\mathcal{N}$ will have velocities between $v$ and $v + dv$, such that $d\mathcal{N} = \mathcal{N}f(v)dv$. The classical model, developed by Maxwell and Boltzmann, derives the probability distribution function of molecular velocities as:

$$f(v) = \frac{4}{\sqrt{\pi}} \left( \frac{m}{2k_BT} \right)^{\frac{3}{2}} v^2 e^{-mv^2/2k_BT} \tag{2.13}$$

The form of this velocity distribution is illustrated in Figure 2.1. The maximum of the curve corresponds to the most probable velocity, $v_p$. This can be simply obtained from the condition $\partial f(v)/\partial v = 0$, which gives $v_p = \sqrt{2k_BT/m}$. The average velocity, $\langle v \rangle$, is evaluated from the integration:

$$\langle v \rangle = \frac{\int_0^\infty v f(v) dv}{\int_0^\infty f(v) dv} = \sqrt{\frac{8k_BT}{\pi m}} \tag{2.14}$$

The root mean square of the velocity, $\sqrt{\langle v^2 \rangle}$, is given by:

$$v_{rms} = \sqrt{\frac{\int_0^\infty v^2 f(v) dv}{\int_0^\infty f(v) dv}} = \sqrt{\frac{3k_BT}{m}} \tag{2.15}$$

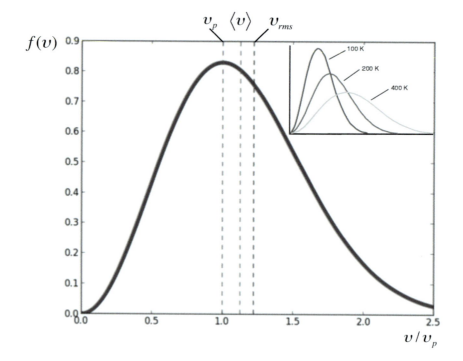

**FIGURE 2.1** Maxwell-Boltzmann distribution of molecular velocities for an ideal gas. Illustrated are the most probable, average and root mean square velocities, $v_p$, $\langle v \rangle$ and $v_{rms}$, respectively.

It is a simple matter to obtain the relative values of the velocities: $\langle v \rangle = 1.128v_p$ and $v_{rms} = 1.225v_p$, as are illustrated in Figure 2.1. The nature of this distribution means that a large proportion of molecules will have velocities close to that of the most probable value, with less than 5% of molecules with a velocity greater than $4v_p$. It is important to remember that the distribution represents an average for a large number of molecules, with molecules changing their particular kinetic energies, but the rates at which they are scattered into and out of a particular range of velocities due to collisions are always equal. The distribution itself is temperature-dependent, with the most probable velocity increasing with temperature, reducing in absolute peak height and broadening with an increase of temperature, see inset of Figure 2.1.

### 2.2.3 Rate of Incidence

A useful quantity in vacuum science is the rate of incidence of the components of a gas on a given surface. This can be ultimately used to calculate the pressure of the gas and will also be related to the mean free path of the molecules between collisions. We start by considering the geometry of our enclosing chamber and we define the solid angle (measured in steradians) subtended by an area $A$ at some distance $r$, this can be expressed as:

$$\Omega = \frac{A}{r^2} \tag{2.16}$$

for small angles/areas, we write:

$$d\Omega = \frac{dA}{r^2} \tag{2.17}$$

and for an area taken at some angle $\theta$ from its normal, we have:

$$d\Omega = \frac{dA \cos \theta}{r^2} \tag{2.18}$$

Now we consider the impact of the various molecules, with their corresponding velocity distribution, arriving from the volume ($n$ per unit volume) at the surface of the container. Since we consider a random distribution, all angles of arrival will be equally probable. As such, a small solid angle, centered on any given direction, will have $n \, dq(d\Omega/4\pi)$ molecules per unit volume arriving from directions within the solid angle $d\Omega$. The number of molecules arriving with velocities between $v$ and $v + dv$ will be:

$$dn_v = nf(v)\left(\frac{d\Omega}{4\pi}\right) dv \tag{2.19}$$

We can further consider the number of molecules arriving at a surface of area $dA$ from within a volume defined by $dA\,v\,dt\,\cos\theta$, which takes into account that only those molecules with a velocity between $v$ and $v+dv$ striking the surface $dA$ in time interval $dt$. This number is: $dA\,v\,dt\,\cos\theta\,dn_v\,(d\Omega/4\pi)$. To account for all possible directions of arrival at the surface, we need to integrate over the whole solid angle between $\theta$ and $\theta+d\theta$ and around the azimuthal angle $\phi$, such that the total number of molecules with velocities between $v$ and $v+dv$ arriving at area $dA$ in time $dt$ will be

$$\frac{1}{2}v\,dn_v\sin\theta\cos\theta\,d\theta\,dA\,dt \qquad (2.20)$$

We now define the impact rate, per unit area, of molecules with velocity $v$, with angle $\theta$ as:

$$J_{v,\theta}=\frac{1}{2}v\,dn_v\sin\theta\cos\theta\,d\theta \qquad (2.21)$$

Integrating with respect to $\theta$ between $\theta=0$ to $\theta=\pi/2$, to cover the whole $2\pi$ solid angle above the surface, yields:

$$J_v=\frac{1}{4}v\,dn_v=\frac{1}{4}\int vn\,f(v)\,dv=\frac{1}{4}n\langle v\rangle \qquad (2.22)$$

This rate of impact is, as stated previously, very important for the calculation of various quantities, being employed in the evaluation of condensation and evaporation rates, adsorption times and the flow properties of gases. If we substitute $n$ from $p=nk_BT$ and the average velocity $\langle v\rangle$ from Eq. (2.14), we can write:

$$J=\frac{p}{\sqrt{2\pi mk_BT}} \qquad (2.23)$$

which now expresses the rate of impact or collisions of the molecules with the gas pressure.

## 2.2.4　*Mean Free Path*

Since the particles in the gas will suffer collisions with each other, we can define the distance traveled between such events; this is defined as the *mean free path*, and will be a function of the particle density and their mean velocity. To evaluate the mean free path we must consider the effective size of the molecules in the gas. We can do this by imagining them as solid or hard spheres, with a radius $d$ and thus an effective area or cross-section of $\sigma=\pi d^2$. In its motion, the molecule will sweep out a cylinder of volume $\sigma l$, where $l$ is the length of the cylinder, which can be expressed as $l = v \, dt$. If the center of the sphere moves within a distance $d$ of another molecule, they will collide. We can thus define the free volume swept out, on average, by a molecule as: $V_{free} = \pi d^2 v \, dt = \sigma v \, dt$. Per unit time, we can consider the number of collisions in the volume to be $n\sigma\langle v \rangle$. The mean free path is the distance traveled in one second divided by the number of collisions, which yields: $\lambda = 1/n\sigma$. This estimation neglects the motion of other particles. Allowing for the Maxwell-Boltzmann distribution of velocities gives rise to a factor of $1/\sqrt{2}$, such that:

$$\lambda = \frac{1}{\sqrt{2}n\sigma} = \frac{1}{\sqrt{2}n\pi d^2} \tag{2.24}$$

Using Eq. (2.1), we can express the mean free path in terms of the pressure and temperature of the gas:

$$\lambda = \frac{k_B T}{\sqrt{2}p\pi d^2} \tag{2.25}$$

The product $\lambda p$ will be a constant for a given gas at a specific temperature. If we know the effective molecular diameters of gas components, it is possible to estimate the values of $\lambda$. For example, at room temperature and pressure, and taking the diameter of a nitrogen molecule to be 0.376 nm, we obtain $\lambda \simeq 65$ nm. For the common gases, at a given pressure, the mean free paths are of similar magnitude. For our purposes, the results of this model are adequate for vacuum practice.

The common mental image of a gas is that of numerous molecules in motion, frequently colliding with one another and changing their direction. We generally ignore the nature of these collisions and simply consider them as hard-sphere collisions like snooker balls. The density and speed of the particles being then related to the pressure of the gas and its temperature. At standard conditions of temperature and pressure, the gas is thought of as a fluid, offering resistance to motion. As we saw, the mean free path is quite short under these conditions, and considerably shorter than the length scale of the container ($D$) (which can be thought of as being on the cm scale; $\lambda \ll D$) and hence the gas behavior will be dominated by molecule-molecule interactions. As the pressure is reduced, the mean free path length will increase and collisions between molecules will decrease in frequency and at low pressures, the dominant interactions will be those between the molecules and the container walls, at which point we have $\lambda \geq D$. Such a condition is referred to as the *molecular state* of the gas. Indeed, for a gas pressure of $10^{-3}$ mbar, assuming $N_2$ molecules at 295 K, $\lambda \simeq 6.5$ cm.

A useful criterion for distinguishing the behavior of gas as fluid or molecular is provided by the *Knudsen number*, $K_n$, which is defined by:

$$K_n = \frac{\lambda}{D} \tag{2.26}$$

This is clearly a dimensionless quantity. In practice, while there is no abrupt transition from the fluid or viscous state to the molecular flow regime, the following provides a useful indication of the gas behavior:

$$K_n < 0.01 \text{ Viscous flow}$$

$$K_n > 1 \text{ Molecular flow}$$

$$0.01 < K_n < 1 \text{ Transition}$$

Clearly, the Knudsen number engenders the mean free path as well as the container dimensions. This is a useful concept since the gas pressure does not always provide a reliable measure of the gas state. For example, for gas at $10^{-3}$ mbar with $\lambda \simeq 6.5$ cm, the gas would be in a molecular flow state in a tube of 1 cm diameter, but in a large vacuum chamber of 1 m, it would be more accurately described in the viscous state.

## 2.3   Pumping and Gas Flow

To obtain a stable vacuum, it is not sufficient to pump down a chamber to a specific pressure. While we must indeed remove the air from the vessel, there are (virtual) sources of gas within the chamber itself, which release molecules into the system, such as outgassing, leaks, and sources of vapor within the vacuum system. We must therefore continually pump the system to remove as many of the contaminants as possible. In this section, we will review the pumping process as well as discuss other factors of importance for the maintenance of a vacuum.

An important consideration of any gas flow system is the movement of the gas molecules. Since gases are compressible, a gas will expand in flowing from a higher to a lower pressure through a pipe or conduit. The size of this expansion will depend on the ambient conditions within the system. Let us consider the flow of a gas through a pipe connected to two reservoirs. We will take the left reservoir to be at a higher pressure, $P_U$, than the volume on the right, $P_D$. Given this condition, $P_U > P_D$, gas will flow through the pipe at a steady rate and at a constant temperature. In the pipe, at cross-section point 1, the entrance, the pressure is $p_1$, while downstream, at the exit point, position 2, the pressure is $p_2$. This is depicted in Figure 2.2. The mass of gas flowing per second across section 1, at pressure $p_1$ will have a volume associated to it, given as $V_1$. Likewise at position 2, at the lower pressure $p_2$, we have an associated volume of the mass of gas flowing, $V_2 > V_1$. Under conditions of steady isothermal flow and assuming the gas acts ideally, we have $p_1 V_1 = p_2 V_2$. From this relation we can establish the relationship between the *volumetric flow rates*, a quantity given by the time derivative of the volume, $\dot{V}$, at the positions 1 and 2 as: $p_1 \dot{V}_1 = p_2 \dot{V}_2$.

Under static isothermal conditions, the product $pV$ gives a direct measure of the quantity of gas. In a similar manner, the product $p\dot{V}$ defines the throughput, $Q$, of the gas at that point, and is a measure of the rate at which the gas is flowing. Thus we define the throughput as:

$$Q = p\dot{V} \tag{2.27}$$

For steady flow, $Q$ will be continuous, i.e., it will have the same value at all positions along the pipe, which is another form of the principle of conservation of mass. We can therefore write: $Q_{in} = Q_{out}$. The SI units of $Q$ is Pa m$^3$ s$^{-1}$, however, a more practical unit would be mbar l s$^{-1}$. The quantity $Q$ is generally easy to establish and to manipulate in practice.

When mass flow rates are required in units of kg per second, conversion can easily be made. Designating $\dot{W}$ as the mass flow rate of a gas, measured

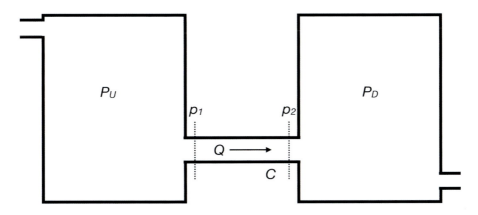

**FIGURE 2.2** Gas flow in a pipe between two reservoirs.

in kg s$^{-1}$, where $W = n_M M$ is the product of the number of moles of the gas and the molar mass. The number of moles is simply expressed by the ideal gas equation:

$$n_M = \frac{pV}{RT} \tag{2.28}$$

where $R$ is the universal gas constant. From the above, we can write the flow rate in moles as $\dot{n}_M$, in units of mol s$^{-1}$. At a particular plane in the pipe, where the pressure is given by $p$, we can express this as:

$$\dot{n}_M = \frac{p\dot{V}}{RT} = \frac{Q}{RT} \tag{2.29}$$

or

$$Q = \dot{n}_M RT \tag{2.30}$$

Now since $\dot{W} = \dot{n}_M M$, we can also write:

$$\dot{W} = \frac{M}{RT}Q \tag{2.31}$$

It can be useful to express the throughput with the number of molecules flowing per second, $(dN/dt)$, which is referred to as the particle flow rate. From Eq. (2.31), we can divide numerator and denominator by Avogadro's constant, $N_A$, which yields:

$$\dot{W} = \frac{m}{k_B T}Q \tag{2.32}$$

where $m$ denotes the mass of a molecule. Since we also have $\dot{W} = m(dN/dt)$, we obtain:

$$Q = k_B T \frac{dN}{dt} \tag{2.33}$$

The volumetric flow rate $\dot{V}$ is often expressed by the symbol $S$, which is also referred to as the *pumping speed*. This quantity refers to the intake of a pump or the entrance to a pipe that is connected to a pump. In practice, the units are expressed as liters per second or liters per minute. Since $S$ is a volumetric flow rate, from Eq. (2.27), we can write:

$$Q = Sp \tag{2.34}$$

This is a common form of the first of two basic defining equations that describe gas flow in vacuum practice. This expresses the quantity of gas flowing as a product of the pressure and the volumetric flow rate at that pressure.

The other fundamental equation of flow relates the throughput to the difference between the upstream and downstream pressures, $p_U$ and $p_D$, in the two volumes the pipe connects to and serves to define the *conductance*, $C$, of the pipe:

$$Q = C(p_U - p_D) \qquad (2.35)$$

It will be clear that the conductance will have the same units as $S$, i.e., volume per second, usually expressed as $1 \text{ s}^{-1}$. There is a useful analogy with dc electrical circuits, where the current is associated with the conductance and the potential difference with the difference in pressure. While Ohm's law connects the electric quantities via the resistance, in vacuum practice, this is made with the conductance, being the inverse of resistance. According to this analogy, the simple rules, that can be found for the combination of conducting elements in series and parallel, can also be transposed onto the vacuum quantities. For a combination of elements linked in parallel, we can express the equivalent conductance as:

$$C_{P,eq} = C_1 + C_2 + C_3 + \ldots = \sum_n C_n \qquad (2.36)$$

while for a series combination we have:

$$\frac{1}{C_{S,eq}} = \frac{1}{C_1} + \frac{1}{C_2} + \frac{1}{C_3} + \ldots = \sum_n \frac{1}{C_n} \qquad (2.37)$$

where the $C_n$ denote the conductances of the various components. A note of caution should be added when using these formulae for the molecular flow regime. For Eq. (2.37) to be accurate in describing the flow between two pipes in series, we require that they are separated by an intermediate buffer volume which is large enough to ensure the existence of near-equilibrium and the random entry conditions into the second pipe.

Equations (2.34) and (2.35) can, in principle, be used to solve all problems regarding the flow of the gas and correspond to the statement that currents are continuous and also to Ohm's law. We can use these expressions for example to determine the effect of the pipes between the pump and chamber on the pumping speed. Let us consider a vacuum chamber, at pressure $p_1$, which is connected via a pipe of conductance $C$, to a pump with a pumping speed, $S_2$, see Figure 2.3. The corresponding pumping speed at the vessel is $S_1$, while the pressure at the exit of the pipe and entrance to the pump is $p_2$. The throughput from the vessel, though the pipe into the pump is given by:

$$Q = C(p_1 - p_2) = S_1 p_1 = S_2 p_2 \qquad (2.38)$$

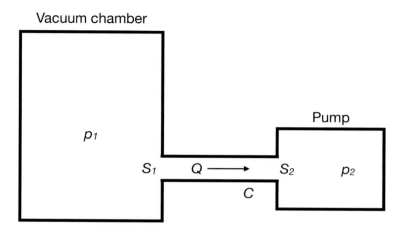

**FIGURE 2.3**    The effect of conductance on the pumping speed in a vacuum system.

This can be used to obtain the pumping speed in the vessel as:

$$S_1 = \frac{CS_2}{C+S_2} \tag{2.39}$$

Clearly, $S_1$ is less than $S_2$. The result of this then shows that only for values of the conductance, $C$, which are appreciably greater than the pumping speed at the pump, $S_2$, does the pumping speed at the vessel become comparable with those of the pump. This is illustrated in Figure 2.4, where we show the pumping speed at the outlet of the vessel versus the pipe conductance, both normalized to the pumping speed at the pump. When $C = S_2$, the pumping speed at the outlet of the vacuum chamber will be half that of the speed at the pump itself. This can be seen then as the difference between the pumping speeds at the vessel and at the pump. The gas expands as it moves downstream through the pipe into the pump. The volumetric flow rate that is fixed at the pump end of the pipe by the pump's speed must, therefore, be smaller at the vessel.

## 2.4   The Evacuation Process

Using the above principles, we can now evaluate the general evacuation procedure for the production of a vacuum. In our discussion we shall consider the vacuum chamber to be of volume, $V$, which is connected to a pipe of conductance, $C$, to a pump with pumping speed, $S_2$. The pumping speed at the vessel exit being, $S_1$, as expressed in Eq. (2.39). In addition to the gas

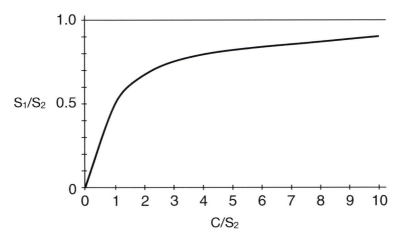

**FIGURE 2.4** Variation of the pumping speed at the outlet of the vacuum chamber as a function of the conductance of the pipe leading to the vacuum pump (Adapted from A. Chambers, 2000).

in the chamber, there are other sources of gas in the system that must also be taken into account. These can include the possible existence of leaks, represented as an effective throughput, $Q_L$, process gas due to operating procedure, such as film deposition, etc., expressed as a throughput, $Q_P$ as well as outgassing from the interior surfaces of the vessel, which we can express in terms of the throughput, $Q_G$. As such the total throughput will be of the form, $Q_T = Q_L + Q_P + Q_G + \ldots$.

The pumping equation assumes isothermal conditions, expressing the fact that the change in the quantity of the gas in the volume, $V$, is associated with the change $dp$ in the pressure, $p$, in the time interval, $dt$, and must express the difference in the quantities entering and leaving the volume. This leads to the relation, (Chambers, 2000):

$$V dp = Q_T dt - S_1 p dt \qquad (2.40)$$

In this case, what we require, is that the rate of exit of the gas must exceed the rate of entry in order to evacuate the system. This means, that the quantity $dp$ and therefore $dp/dt$ must be negative. We can, therefore, rewrite Eq. (2.40) in the following form:

$$-V \frac{dp}{dt} = S_1 p - Q_T \qquad (2.41)$$

This shows that the rate of change of the amount of gas in the systems is the difference between the rate of its removal, $Sp$, and the influx rate, $Q_T$. In practice, it is quite difficult to use this expression since some of the parameters are difficult to evaluate. For example, the pumping speed depends on the pressure and the influx from say outgassing is very difficult to measure, and varies significantly over time. Despite this, Eq. (2.41) can allow us to evaluate the ultimate pressure achievable and the pump-down time when the pumping speed can be considered to be a constant.

We note from Eq. (2.41), that when the pressure eventually reaches a steady-state, such that $dp/dt = 0$, the constant pressure in the chamber, the system's *ultimate* or *base pressure*. This can thus be expressed as:

$$p_u = \frac{Q_T}{S_1} \tag{2.42}$$

This follows logical thinking in that we would expect the best (lowest) pressures for small gas loads and large pumping speeds. This equation essentially arises from the balance achieved in the steady-state where the gas load and pumping capacity are in equilibrium. In the molecular flow regime, it may depend on the type of pump used, the pumping speed can depend on the nature of the gas. In this case, the expression in Eq. (2.42) would need to be expressed as the sum of the ultimate partial pressures of each gas species.

In Eq. (2.41), we note that for the early stages of pumping, where we consider the contribution from outgassing and leaks to be negligible (at pressures near atmosphere). Here, the term $S_1 p$ will be very much larger than $Q_T$, which we can ignore. This allows us to approximate the relation as:

$$\frac{dp}{p} = -\frac{S_1}{V} dt \tag{2.43}$$

Taking the pumping speed of the primary pump to be a constant, a reasonable approximation at high pressure, we can deduce the initial variation of the pressure as:

$$p(t) = p_0 e^{-(S_1/V)t} \tag{2.44}$$

which shows an exponential decrease of the pressure and where the time constant of the decay is given by $\tau = V/S_1$. Rewriting this equation can allow us to evaluate the time taken for the pressure to drop from its initial value, $p_0$ to $p$:

$$t = \frac{V}{S_1} \ln\left(\frac{p_0}{p}\right) \tag{2.45}$$

Alternatively, we can determine the pumping speed required to pump a system of specific volume, to a given pressure:

$$S_1 = \frac{V}{t} \ln\left(\frac{p_0}{p}\right) \tag{2.46}$$

These expressions are of practical use in the workings of vacuum systems. In Figure 2.5, we show the variation to the pressure as a function of pumping time for the case of a constant pumping speed. The variation of pressure is initially linear with the logarithmic pressure. In curve A, the variation is illustrated for the hypothetical case with no leaks. After the linear phase, the pressure decrease slows down as the system reaches its base pressure, where the outgassing from the interior surfaces begins to dominate the gas load. In the case illustrated, the ultimate base pressure is around $10^{-3}$ mbar, slightly above that of the pump. At these pressures, a relatively large gas load $Q_L$ and pressure remain within the range of constant pumping speed, and we can write, $Q_L S_1 = p_L$, as the base pressure of the system. In this case, Eq. (2.43) can be expressed as:

$$\frac{dp}{p - p_L} = -\frac{S_1}{V} dt \tag{2.47}$$

The solution of this yields:

$$p(t) = p_L + (p_0 - p_L)e^{-(S_1/V)t} \tag{2.48}$$

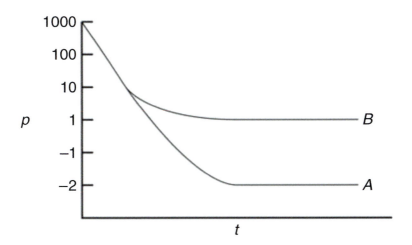

**FIGURE 2.5** Variation of pressure with pumping time for a system without (A) and with (B) a leak. (Adapted from A. Chambers, 2000).

For a system with a leak, curve B, the pressure initially falls with the same rate as for the system without a leak, however, the pressure falls more quickly, stabilizing at a higher ultimate pressure.

For vacuum systems designed to reach a high vacuum, a secondary pump is usually required to take over the pumping as the pressure reduces to the lower limit of the primary pump. This secondary pump will be required to pump the outgassing load to reduce the base pressure of the system. The choice of pumping system will largely depend on the nature of the vacuum system and the desired ultimate pressure for the specific environment required. In the following section, we will outline some of the more common pumps used for primary and secondary pumping.

## 2.5   Production of Vacua: Vacuum Pumps

As we stated in the Introduction, the purpose of the vacuum is to control the environment in which we perform certain experiments. To obtain a vacuum we require a pumping system, the choice of which will depend on the vacuum system and the degree of vacuum required for our experiment. In terms of working principles, we can define three types of pump: (i) the positive displacement pump, (ii) the momentum transfer pump, and (iii) the capture pump (or gas-binding). The categories (i) and (ii) can also be collectively referred to as gas-displacement pumps. While gas-displacement vacuum pumps can be used without limitation, gas-binding vacuum pumps have a limited gas absorption capacity and must be regenerated at certain process-dependent intervals. Vacuum pumps can be further categorized by their operating pressure range and as such are classified as: primary pumps, booster pumps or secondary pumps. Within each pressure range are several different pump types, each employing a different technology, and each with some unique advantages in regard to pressure capacity, flow rate, cost and maintenance requirements. In the following, we will provide a short outline of some of the most common pumps.

Gas-displacement pumps, which are also referred to as gas transfer pumps, are classified either as positive displacement pumps or kinetic vacuum pumps. Positive displacement pumps displace gas from sealed areas to the atmosphere or to a downstream pump stage. Kinetic pumps displace gas by accelerating it in the pumping direction, either via a mechanical drive system or through a directed vapor stream that is condensed at the end of the pumping section. Gas-binding vacuum pumps either bind the gas to an especially active substrate through gettering or condense the gas at a suitable temperature. Chemisorption is performed technically by a pump type known as getter pumps which constantly generate pure getter surfaces through vaporization and/or sublimation or sputtering. If the gas particles to be bound are ionized in an ion getter pump before interacting with a getter surface, they can at the same time clean the getter surface by sputtering and be buried by sputtered material. Non-evaporable getters (NEG) consist of highly reactive alloys, mainly of zirconium or titanium, and have a very large specific surface. Gases can penetrate into deeper layers of the getter material through micropores and be bound there into stable chemical compounds.

Regardless of their design, the basic principle of operation is the same. The vacuum pump functions by removing the molecules of air and other gases from the vacuum chamber (or from the outlet side of a higher vacuum pump if connected in series). While the pressure in the chamber is reduced, removing additional molecules becomes exponentially harder to remove, as discussed in the previous section. As a result, a vacuum system must be able to operate over a portion of an extraordinarily large pressure range, typically varying from 1 to $10^{-6}$ mbar of pressure. In research and scientific applications, this is extended to $10^{-9}$ mbar or lower. In order to accomplish this, several different styles of pumps are used in a typical system, each covering a portion of the pressure range, and operating in series at times. In Figure 2.6, we show a schematic illustration of a typical vacuum system consisting of both a primary and secondary pump.

Transfer pumps operate by transferring the gas molecules by either momentum exchange (kinetic action) or positive displacement. The same number of gas molecules are discharged from the pump as enter it and the gas is slightly above atmospheric pressure when expelled. The ratio of the exhaust pressure (outlet) to the lowest pressure obtained (inlet) is referred to as the *compression ratio*. Kinetic transfer pumps work on the principle of momentum transfer, directing gas towards the pump outlet to provide an increased probability of a molecule moving towards the outlet using high-speed blades or introduced vapor. Kinetic pumps do not typically have sealed volumes but can achieve high compression ratios at low pressures. Positive displacement transfer pumps work by mechanically trapping a volume of gas and moving it through the pump. They are often designed in multiple stages on a common drive shaft. The isolated volume is compressed to a smaller volume at a higher pressure, and finally, the compressed gas is expelled to the atmosphere (or to the next pump). It is common for two transfer pumps to be used in series to provide a higher vacuum and flow rate. Capture pumps operate by capturing (trapping) the gas molecules on surfaces within the vacuum system. Capture pumps operate at lower flow rates than transfer pumps, but can provide ultra-high vacuum (UHV), down to $10^{-12}$ mbar, and generate an oil-free vacuum. Capture pumps operate using cryogenic condensation, ionic reaction, or chemical reaction and have no moving parts. These different pumping mechanisms are illustrated along with the corresponding pumps in Figure 2.7.

## 2.5.1    Primary (Backing) Pumps

### 2.5.1.1    Oil Sealed Rotary Vane Pump

(Wet, Positive Displacement)

In the rotary vane pump, illustrated in Figure 2.8, the gas enters the inlet port and is trapped by an eccentrically mounted rotor. This compresses the gas and transfers it to the exhaust valve. The valve is spring-loaded and allows the gas to discharge when atmospheric pressure is exceeded. Oil is used to seal and cool the vanes. The pressure achievable with a rotary pump is determined by the number of stages used and their tolerances. A two-stage

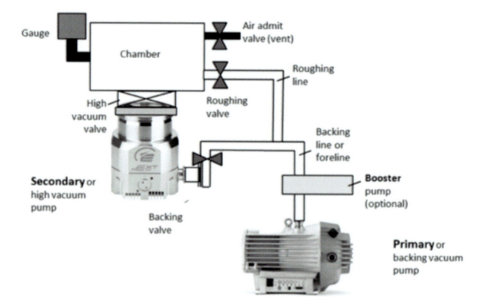

**FIGURE 2.6**    A typical vacuum system for attaining vacua in the HV, or better, regime.

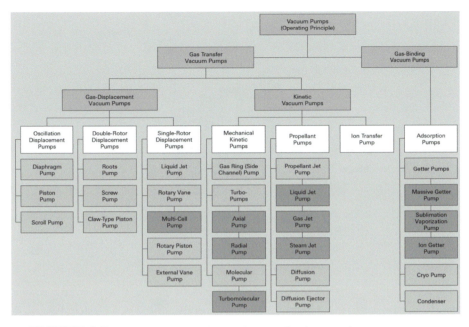

**FIGURE 2.7**    Classification of pumping mechanisms and vacuum pumps.

design can achieve a pressure of $1 \times 10^{-3}$ mbar. It has a pumping speed of 0.7 to 275 m$^3$ h$^{-1}$.

## 2.5.1.2   Liquid Ring Pump

(Wet, Positive Displacement)

The liquid ring pump, illustrated in Figure 2.9, compresses the gas by rotating a vaned impeller located eccentrically within the pump housing. The liquid is fed into the pump and, as a result of centrifugal acceleration, forms a moving cylindrical ring against the inside of the casing. This liquid ring creates a series of seals in the space between the impeller vanes, which form compression chambers. The eccentricity between the impeller's axis of rotation and the pump housing results in a cyclic variation of the volume enclosed by the vanes and the ring, which compresses the gas and discharges it through a port at the end of the housing. This pump has a simple, robust design as the shaft and impeller are the only moving parts. It is very tolerant of process upsets and features a large capacity range. It can provide a pressure of 30 mbar using 15°C water and lower pressures are possible with other liquids. It has a pumping speed range of 25 to 30,000 m$^3$ h$^{-1}$.

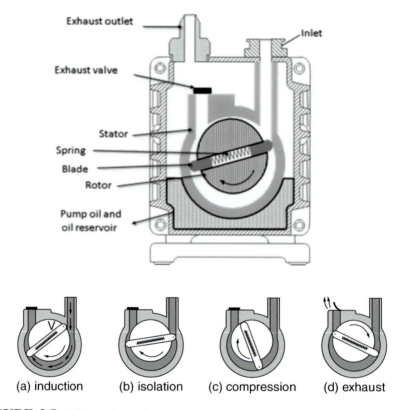

(a) induction    (b) isolation    (c) compression    (d) exhaust

**FIGURE 2.8**    Illustration of a rotary vane pump.  Also shown is the pumping cycle: inlet, isolation, compression, and outlet.

**FIGURE 2.9**    Illustration of a cross-section of a typical ring pump.

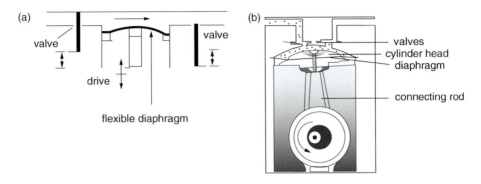

**FIGURE 2.10**   Cross-section of a typical diaphragm pump.

### 2.5.1.3   Diaphragm Pump

(Dry, Positive Displacement)

In this pump, a diaphram is rapidly flexed by a rod riding on a cam rotated by a motor. This produces a gas transfer from one valve (input) to another (output). This pump is compact and of low maintenance. The lifetime of the diaphragms and valves is typically over 10,000 operating hours. The diaphragm pump, see Figure 2.10, is used for backing small compound turbo-molecular pumps in clean, high vacuum applications. It is a small capacity pump widely used in R&D labs for sample preparation. A typical ultimate pressure of $5 \times 10^{-8}$ mbar can be achieved when using the diaphragm pump to back a compound turbo-molecular pump. It has a pumping speed range of 0.6 to 10 m$^3$ h$^{-1}$.

### 2.5.1.4   Scroll Pump

(Dry, Positive Displacement)

The scroll pump, as illustrated in Figure 2.11, uses two scrolls that do not rotate, but where the inner one orbits and traps a volume of gas and compresses it in an ever decreasing volume; compressing it until it reaches a minimum volume and maximum pressure at the spiral's center, where the outlet is located. A spiral polymer (PTFE) tip seal provides axial sealing between the two scrolls without the use of a lubricant in the swept gas stream. A typical ultimate pressure of $1 \times 10^{-2}$ mbar can be achieved. It has a pumping speed range of 5 to 46 m$^3$ h$^{-1}$.

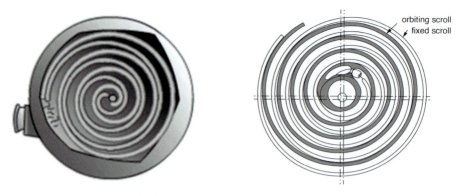

orbiting scroll
fixed scroll

**FIGURE 2.11**    Illustration of a scroll pump.

## 2.5.2    *Booster Pumps*

### 2.5.2.1    *Roots Pump*

(Dry, Positive Displacement)

The Roots pump, illustrated in Figure 2.12, is primarily used as a vacuum booster and is designed to remove large volumes of gas. Two lobes mesh without touching and counter-rotate to continuously transfer the gas in one direction through the pump. It boosts the performance of a primary/backing pump, increasing the pumping speed by approximately 7:1 and improves ultimate pressure by approximately 10:1. Roots pumps can have two or more lobes. A typical ultimate pressure of $< 10^{-3}$ Torr can be achieved (in combination with primary pumps). It can achieve pumping speeds in the order of 100,000 m$^3$ h$^{-1}$.

### 2.5.2.2    *Claw Pump*

(Dry, Positive Displacement)

The claw pump, as illustrated in Figure 2.13, features two counter-rotating claws and operates similarly to the Roots pump, except that the gas is transferred axially, rather than top-to-bottom. It is frequently used in combination with a Roots pump, that is, a Roots-claw primary pump combination in which there are a series of Roots and claw stages on a common shaft. It is designed for harsh industrial environments and provides a high flow rate. A typical ultimate pressure of $1 \times 10^{-3}$ mbar can be achieved. It has a pumping speed range of 100 to 800 m$^3$ h$^{-1}$.

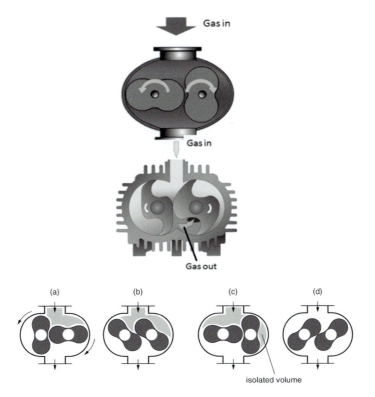

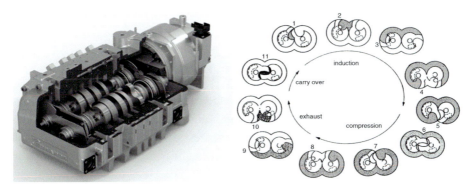

**FIGURE 2.12** Illustration of a cross-section of a typical roots pump. Also shown is the pumping cycle: inlet, isolation, compression, and outlet.

**FIGURE 2.13** Illustration of a cross-section of a claw pump. Also shown is the pumping cycle: inlet, compression, and exhaust.

### 2.5.2.3 Screw Pump

(Dry, Positive Displacement)

The screw pump, as illustrated in Figure 2.14, utilizes two rotating screws,

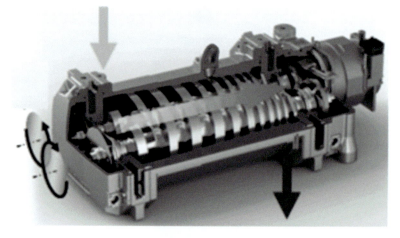

**FIGURE 2.14**   Illustration of a cross-section of a screw pump.

one left-handed and one right-handed, that mesh without touching. The rotation transfers the gas from one end to the other. The screws are designed such that the space between them becomes reduced as the gas passes along, and it becomes compressed, causing a reduced pressure at the entrance end. This pump features a high throughput capacity, good liquid handling, and tolerates dust and harsh environments. A typical ultimate pressure of approximately $1 \times 10^{-2}$ mbar can be achieved. It has a pumping speed range of up to 750 m$^3$ h$^{-1}$.

### 2.5.3   Secondary Pumps

#### 2.5.3.1   Turbomolecular Pump

(Dry, Kinetic Transfer)

The turbomolecular pump, as illustrated in Figure 2.15, works by transferring kinetic energy to gas molecules using high-speed rotating, angled blades that propel the gas at high speeds: the blade tip speed is typically 250–300 m/s (670 miles/hr.) By transferring momentum from the rotating blades to the gas, it can provide a greater probability of molecules moving towards the outlet. Turbo pumps provide low pressures and have low transfer rates. A typical ultimate pressure of less than $1 \times 10^{-10}$ mbar can be achieved. It has a pumping speed range of 50–5000 l/s. The bladed pumping stages are often combined with drag stages, that enable turbomolecular pumps to exhaust to higher pressures ($>$ 1 mbar).

**FIGURE 2.15**    Illustration of a cross-section of a turbomolecular pump.

### 2.5.3.2    *Vapor Diffusion Pump*

(Wet, Kinetic Transfer)

Vapor diffusion pumps, see Figure 2.16, transfer kinetic energy to gas molecules using a high velocity heated oil stream that "drags" the gas from the inlet to the outlet, providing a reduced pressure at the inlet. These pumps feature an older technology, largely superseded by dry turbomolecular pumps. They have no moving parts and provide high reliability at a low cost. A typical ultimate pressure of less than $1 \times 10^{-10}$ mbar can be achieved. It has a pumping speed range of 10–50,000 l/s.

### 2.5.3.3    *Cryopump*

(Dry, Entrapment)

The Cryopump, see Figure 2.17, operates by capturing and storing gases and vapors, rather than transferring them through the pump. They use cryogenic technology to freeze or trap the gas to a very cold surface (cryocondensation or cryosorption) at 10 to 20 K. These pumps are very effective but have limited gas storage capacity. Collected gases/vapors must periodically be removed from the pump by heating the surface and pumping it away through another vacuum pump (known as regeneration). Cryopumps require a refrigeration compressor to cool the surfaces. These pumps can achieve a pressure of $1 \times 10^{-10}$ mbar and have a pumping speed range of 1200 to 4200 l/s.

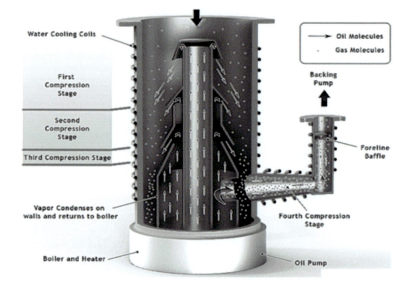

**FIGURE 2.16**   Illustration of a cross-section of a diffusion pump.

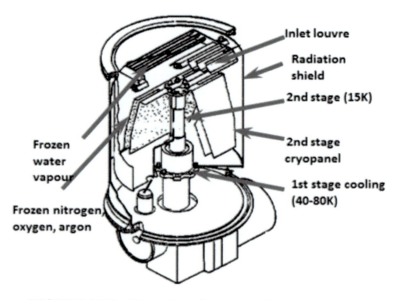

**FIGURE 2.17**   Illustration of a cross-section of a cryopump.

### 2.5.3.4   Sputter Ion Pump

(Dry, Entrapment)

The sputter ion pump, see Figure 2.18, traps gases using the principles of gettering (whereby chemically active materials combine with gases to remove them) and ionization (gas molecules are made electrically conductive and captured). A high magnetic field combined with a high voltage (4 to 7 kV), creates a cloud of electrons-positive ions (plasma), which are deposited onto a titanium cathode and sometimes a secondary additional cathode composed of tantalum. The cathode captures the gases, resulting in a better film. This phenomenon is referred to as sputtering. The cathode must be periodically replaced. These pumps have no moving parts, are low maintenance, and can achieve a pressure as low as $1 \times 10^{-11}$ mbar. They have a maximum pumping speed of 1000 l/s.

## 2.6   Summary

We have seen in this chapter that the treatment of a vacuum can be well described from the classical kinetic theory of gases. Here we envisage the molecules of the gas as hard spheres undergoing elastic collisions and changing their direction. The pressure is well approximated as a perfect gas and depends on the density of gas molecules and the temperature (kinetic energy) of the gas. We have described the different pressure ranges as well as the flow regimes in terms of the Knudsen number. Within the classical theory, we have

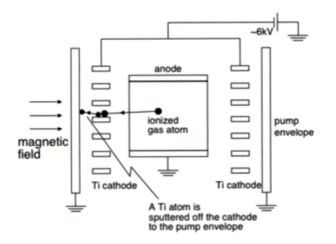

**FIGURE 2.18**   Illustration of a cross-section of a sputter ion pump.

**TABLE 2.2**    Number Density, Mean Free Path and Impact Rate As a Function of Gas Pressure [Calculations are based on nitrogen molecules at 295 K (Table adapted from A. Chambers, 2000.)]

| $p$ (mbar) | $n$ (m$^{-3}$) | $\lambda$ | $J$ (cm$^{-2}$ s$^{-1}$) |
|---|---|---|---|
| $10^3 = 1$ atm | $2.5 \times 10^{25}$ | 65 nm | $2.9 \times 10^{23}$ |
| 1 | $2.5 \times 10^{22}$ | 65 m | $2.9 \times 10^{20}$ |
| $10^{-3}$ | $2.5 \times 10^{19}$ | 65 mm | $2.9 \times 10^{17}$ |
| $10^{-6}$ (HV) | $2.5 \times 10^{16}$ | 65 m | $2.9 \times 10^{14}$ |
| $10^{-10}$ (UHV) | $2.5 \times 10^{12}$ | 650 km | $2.9 \times 10^{10}$ |

used the Maxwell-Boltzmann velocity distribution of the gas molecules to evaluate the most probable, average and root-mean-squared velocities. This along with the geometrical considerations of the molecular trajectories with respect to the vacuum chamber walls allows us to calculate the rate of impact of the gas molecules with the walls as well as the average distance (mean free path) traveled by the particles in the rarefied gas. By substituting the constants in our equations and assuming room temperature for nitrogen ($N_2$) gas, we can obtain practical expressions for the number density, Eq. (2.1), impact rate, Eq. (4.10) and the mean free path, Eq. (2.24), which we express as:

$$n = 2.5 \times 10^{22} p \tag{2.49}$$

with $p$ in mbar and $n$ in m$^{-3}$.

$$J = 2.9 \times 10^{22} p \tag{2.50}$$

with $p$ in mbar and $J$ in cm$^{-2}$ s$^{-1}$, and

$$\lambda = \frac{6.6 \times 10^{-3}}{p} \tag{2.51}$$

with $p$ in mbar and $\lambda$ in cm. It is instructive to illustrate the different levels of vacuum numerically, which based on the above equations are shown in Table 2.2.

The numbers are a spectacular demonstration of what a vacuum really is, particularly, when we reach the UHV regime, where the mean free path between molecular collisions is hundreds of km!

## 2.7 Problems

(1)(a) Specify the amount of air (mbar l) in an empty 1-liter bottle of whiskey at room temperature and at a pressure of 1000 mbar.

(b) How many such bottles are required to contain this gas at a pressure of $10^{-2}$ mbar?

(2) How many molecules are there in 1 mbar l of an ideal gas at $22°C$?

(3) To how many mbar l does 1 $cm^2$ of a monolayer of adsorbed gas ($10^{15}$ molecules $cm^{-2}$) correspond to?

(4) Show that the throughput in a pipe of diameter D can be written in the form:

$$Q = \frac{npD^2 \langle v \rangle k_B T}{4}$$

where $\langle v \rangle$ is the average drift velocity of the gas.

(5) A fan moves atmospheric air through a room at a rate of 0.9 $m^3$/minute. What is the throughput (in mbar l $s^{-1}$)?

(6) Derive the conductance formulae for pipes in series and parallel.

(7) Calculate the molecular flow conductances for $N_2$ at 295 K of the following:

(a) Circular aperture of diameter 5 cm.

(b) Pipe of diameter 5 cm and length 20 cm.

(c) Pipe of diameter 5 cm and length 1 m.

(d) Pipe of diameter 2.5 cm and length 1 m.

NB: $C_{pipe}$ acts like a series connection of an aperture, $C_0$, and a tube, $C_L$, with

$$C_L = \left( \frac{D^3}{6L} \right) \sqrt{\frac{2pRT}{M}}$$

$$C_0 = A \sqrt{\frac{RT}{2pM}}$$

(8) For an aperture of D = 5 cm and length 20 cm, calculate the conductance for hydrogen gas at 295 K.

(9) Calculate the time required for a $N_2$ molecule to travel its mean free path in a vacuum of $10^{-10}$ mbar at 300 K.

(10) Design a vacuum system, using any components you deem necessary, which requires a base pressure of $10^{-10}$ mbar. Explain the function of all components used and why they are necessary.

## References and Further Reading

Chambers, A., Fitch, R. K., & Halliday, B. S. (1998). *Basic Vacuum Technology*, IOP, Bristol, U.K.

Chambers, A. (2000). *Modern Vacuum Physics*, Chapman and Hall, CRC, Boca Raton, Florida.

John F. O'Hanlon, (2003). *A User's Guide to Vacuum Technology*, John Wiley and Sons, Hoboken, New Jersey.

Lafferty, J. Ed., (1998). *Foundations of Vacuum Science and Technology*, John Wiley and Sons, New York.

Venables, J. A. (2000). *Introduction to Surface and Thin Film Processes*, Cambridge University Press, Cambridge, U.K.

Vac Aero Website (2016). https://vacaero.com/information-resources/ vacuum-pump-technology-education- and-training/1039-an-introduction-to-vacuum-pumps.html (accessed on 30 March 2020).

# Chapter 3

# The Physical Surface

## 3.1 Introduction

Surface science is a vast area of study covering many topics, which are not generally thought of as being specific to what appears at first glance as a narrow and limited area of interest. Indeed, surface science has implications in many fields of research and applications. In terms of applications, we can note that oxidation studies and catalysis are very much in the domain of surface science.

In the present context, we are interested in the physical properties of the surfaces of condensed matter objects and in particular when those objects have at least one dimension on the nanoscale.

As we have previously commented, the physical properties at low physical dimensions tend to be modified with respect to those in bulk solid materials. The modification of physical behavior being generally correlated to the ratio of surface to bulk atoms, or more precisely, the number of atoms with reduced coordination with respect to those with full bulk coordination. This can be thought of as a reasonable rule of thumb and specifics will depend also on the type of atoms present as well as the shape of the structure. In the case of thin films, we can usually reduce our first approach analysis of the number of atomic planes in the film. For atomic monolayers, this will mean that all atoms have reduced coordination and the physical properties can be expected to be significantly modified with respect to those of the bulk material. Of course, this is just a simplified first approach and as we have mentioned specific cases need to be considered on their own merits and conditions. In this chapter, we will discuss some basics with regard to what we exactly mean by the term "surface" and what the physical implications are once we have defined our surface.

In terms of physically measuring the properties of a surface, we can only consider experimental techniques, which are sensitive to the surface, i.e., surface sensitivity becomes the criterion which we must apply when defining surface measurement techniques. Such considerations will be discussed in Chapter 5. In general, a surface-sensitive technique will mean that the experimental method is capable of detecting just a few atomic layers.

## 3.2  The Physical Surface: Perfect Surfaces and Real Surfaces

We can commence our discussion with the consideration of what we mean by an "ideal" surface. An ideal or perfect surface can be thought of as an atomically flat crystalline plane as cut from a perfect, defect-free crystal. This bulk terminated plane will have all atomic sites occupied by the same type of atom, or molecule. The crystalline order and crystalline plane will define the atomic positions and their respective periodicity.

This is all well and good, but real surfaces are rarely such models of perfection. Indeed, most real surfaces will have a host of imperfections, which we will discuss presently. Now, since we know that the physical properties of a surface will depend on the coordination of atomic bonds between neighboring atoms, even for perfect surfaces, there can be an expected modification of physical properties with respect to those of the bulk crystal. With the addition of crystalline imperfections, we can further expect changes to the physical properties, the degree of which will depend on the concentration and type of defect.

Let us first consider the surface as a perfect cleavage of a perfect crystal along some well-defined crystalline planes. In the simplest case, where there is no alteration of the crystalline structure of the surface atomic plane, the atoms continue to occupy the same relative equilibrium positions and the surface is said to be a *bulk exposed plane* (Figure 3.1(a)). In this case, there is minimal disturbance of the crystalline symmetry. Despite this, many properties related to the electronic and magnetic properties of the bulk depend on the three-dimensional symmetry of the crystal and its subsequent electronic periodic potential. The loss of symmetry at the surface can result in the alteration of the electronic states at the surface and in its proximity. A lack of near-neighbor atoms on one side of the surface atoms can also make available chemical bonds, which extend into the space above the crystal. Such free chemical states are often referred to as *dangling bonds*.

In the formation of a perfect surface, we can consider that "half" of the crystal, i.e., that above the surface plane, is missing. It is more likely, that the disturbance of the missing portion of the crystal due to the absence of bonding between nearest neighbors will result in the reestablishment of new equilibrium positions for the atoms in the vicinity of the surface. One simple model of the surface results in the relaxation of the surface atomic layers parallel to the outermost atomic plane, in the direction of its normal. Here, the interplanar separation increases and can vary over several atomic planes, decreasing in magnitude as we move further and further into the crystal. The entire region, over which there is a deviation from the bulk crystalline symmetry, including the variation of interatomic planar separation, is referred to as the *selvedge*. This type of surface relaxation retains the atomic symmetry within the atomic plane parallel to the surface, changing just the spacing normal to the surface. This is illustrated in Figure 3.1(b). A more complex perturbation of the crystal surface occurs when the surface atoms rearrange themselves into a new symmetry and nearest neighbor configuration which can be entirely different from that of the bulk solid. This is called a *surface reconstruction* and is represented in Figure 3.1(c). The reconstruction of the surface can occur over several atomic layers of the selvage and can significantly affect the structure sensitive properties of the crystal, including atomic vibrations as well as chemical, optical, magnetic, and electronic properties.

There are a large number of processes which are of great scientific interest that are related to surfaces. These include: thermionic emission, crystal growth, chemical reactions, catalysis, semiconductor interfaces, low dimensional magnetic structures, colloids. Some of these subjects will be met in later chapters of this book, where they are relevant to the scientific topics related to nanotechnologies and low dimensional structures.

In addition to the formation of the physical surface discussed above, surface defects can also play an important role in their properties, in much the same way as they do in bulk crystals. For example, surface roughness can be considered as a deviation from the perfectly flat surface, even if the atoms still occupy the correct atomic sites of a particular crystalline structure. Indeed, surface roughness is quite the norm in many crystal and film surfaces, especially those of metals. In Figure 3.2, we illustrate some of the more common forms of the surface defect. The terrace corresponds to a region of the flat crystal surface, which can have missing atoms, or vacancies as well as adatoms on its surface which can be of the same type as the surface itself or of another atomic species, defined as an impurity adatom. The terrace edge

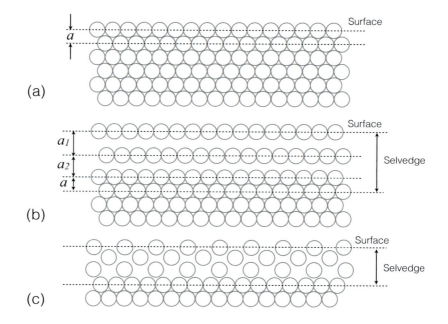

**FIGURE 3.1**    Representations of a solid surface with hexagonal close-packed atoms.  (a) Bulk exposed plane.  (b) Surface relaxation of surface planes outwards from the bulk. (c) Hypothetical surface reconstruction of the outer atomic planes.

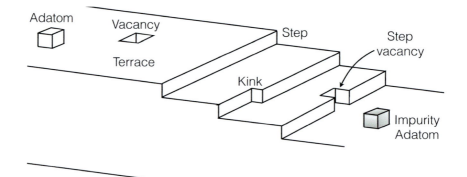

**FIGURE 3.2**    Representation of the types of defect that can be encountered at a surface: adatom, vacancy, step, kink, step vacancy and impurity atom.

is defined as a step, which can be continuous have kinks or vacancies. Other forms of defect can exist, such as edge dislocations terminating at a surface or an emerging screw dislocation.

We can further note that the "atmospheric" conditions will also affect the surface since atoms in the surrounding region above the surface can "land" on the surface and become adsorbed. This may be followed by desorption with the foreign atom returning to the atmosphere above the surface. This leads to a definition of the "sojourn time" discussed in Chapter 2. The number (or concentration) and type of foreign atoms on a surface will affect many physical properties of a surface, since the uppermost atoms of the crystal will have modified chemical or electronic states due to the bonding with these foreign or impurity atoms.

## 3.3   Surface Crystallography

Crystallography is the branch of physics that deals with atomic ordering in regular periodic three-dimensional arrays. Most textbooks on Solid State Physics will outline the principal elements of crystallography by defining the lattice type and basis. In three dimensional crystals, all regular arrays of atoms (objects) can be reduced to one of the 14 Bravais lattices. These take into account the restricted number of translation and rotational symmetry operations, and the finite number of point groups and space groups can be used to define further symmetry-related properties for all possible crystals. As with standard Solid State Physics, Surface Physics attempts to account for the intrinsic physical properties of surfaces via the symmetry of atoms at the surface of a crystal. In this section, we will outline the principal elements for the corresponding situation for a two-dimensional regular array of atoms.

The crystalline surface, as we have seen, can have exactly the same symmetry as the corresponding plane, defined by its Miller indices (*hkl*), in the three-dimensional crystal. If the symmetry is altered in some manner, we define this as surface reconstruction, as will be discussed shortly. This notion of a surface only considers a clean atomically clean surface. Often a surface structure may concern the surface structure of *adsorbates*, which refers to a region typically above the selvage containing a localized excess of foreign or impurity adatoms. Indeed the adsorbate may have a different structure to that of the selvage. The addition of these atoms can also alter the physical properties of the surface.

In Chapter 5, we will outline some of the principal experimental techniques used for studying the surface arrangement of atoms. Such techniques typically have a very limited range of penetration such that we can ignore bulk structures of the interior of the crystal. For the moment, we will limit

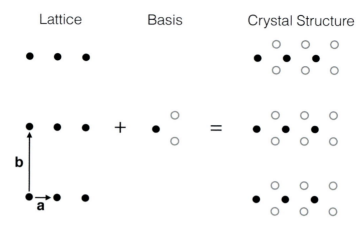

**FIGURE 3.3**   Illustration of crystal structure in terms of the lattice symmetry and the basis.

our discussion to the main points regarding the surface structure. A consideration of the symmetry properties of a two-dimensional net or lattice of points will lead to just five possible symmetrically different Bravais lattices. As with its 3D counterpart, the crystalline structure can be considered as the sum of the lattice plus its basis (see Figure 3.3). Following the analogy with the bulk situation, we can consider the translation vector as being described in terms of the lattice vectors **a** and **b** as defined in Figure 3.3. As such a translation vector of the form:

$$\mathbf{r}' = \mathbf{r} + n\mathbf{a} + m\mathbf{b} = \mathbf{r} + \mathbf{R} \tag{3.1}$$

where $n$ and $m$ are integers, will leave the crystal invariant. This means the crystal would look identical after such an operation. We define the vector **R** as a *lattice vector*. The parallelogram with sides **a** and **b** is referred to as the *unit cell*, while the unit cell of minimum area is called the *primitive cell*. There are many alternatives for choosing the unit cell, which usually one of convenience, such as is the case for square and rectangular cells as well as the so-called Wigner-Seitz cell. The five Bravais lattices for surface lattices are illustrated in Figure 3.4. The combination of the five Bravais lattices with the ten different point groups (as expressed from the International notation of the symmetries as: 1, 2, 1*m*, 2*mm*, 4, 4*mm*, 3, 6, 3*m*, and 6*mm*) gives rise to a possible 17 two-dimensional space groups. The Bravais lattice properties are summarized in Table 3.1.

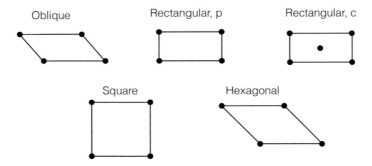

**FIGURE 3.4**    The five two-dimensional Bravais lattices.

**TABLE 3.1**    Summary of the Five Two-Dimensional Bravais Lattices

| Bravais Lattice | Mesh symbol | Mesh shape | Nature of axes and angles |
|---|---|---|---|
| Oblique | p | General parallelogram | $a \neq b, \gamma \neq 90°$ |
| Rectangle | p | Rectangle | $a \neq b, \gamma = 90°$ |
| Rectangle | c | Rectangle | $a \neq b, \gamma = 90°$ |
| Square | p | Square | $a = b, \gamma = 90°$ |
| Hexagonal | p | 60° rhombus | $a = b, \gamma = 120°$ |

Typically the low-index crystalline planes are those which are most familiar to us. For example, in cubic crystalline structures, the planes such as (100), (110) and (111) are the most common ones that are considered. For hexagonal crystals, we are most familiar with the hexagonal plane perpendicular to the c-axis, (0001). If a crystal is cut slightly off-axis to one of these low-index planes, the resulting surface will appear to be stepped. Such a surface is often referred to as a *vicinal plane*. For example, in Figure 3.5, we illustrate the fcc (755) vicinal plane. We can also express the vicinal plane as a combination of the terrace plane (t) and the step plane (s), using the notation: $[n(h_t k_t l_t) \times (h_s k_s l_s)]$, where $n$ gives the number of atomic rows in the terrace parallel to the step edge. From this, we can see that the fcc(755) vicinal plane can also be designated as fcc [6(111)x(100)].

Point and space groups are outlined in Table 3.2.

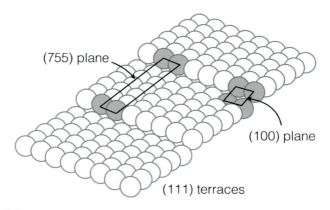

**FIGURE 3.5**    The vicinal fcc(755) plane which can also be designated as fcc [6(111)x(100)].

**TABLE 3.2**    Point and Space Groups for Two-Dimensional Crystals

| Crystal System | Bravais Lattice | Point Group | Space Group |
|---|---|---|---|
| Oblique | *p*-oblique | 1 | *p*1 |
| 2 | | 2 | *p*211 |
| Rectangle | *p*-rectangular | *m* | *p*1*m*1 |
| 2*mm* | | | |
| | | | *p*1*g*1 |
| | | 2*mm* | *p*2*mm* |
| | | | *p*2 *m g* |
| | | | *p*2 *g g* |
| | *c*-rectangular | *m* | *c*1*m*1 |
| | | 2*mm* | *c*2*mm* |
| Square | *p*-square | 4 | *p*4 |
| 4*mm* | | | |
| | | 4*mm* | *p*4*mm* |
| | | | *p*4 *g m* |
| Hexagonal | *p*-hexagonal | 3 | *p*3 |
| 6*mm* | | | |
| | | 3*m* | *p*3*m*1 |
| | | 6 | *p*31*m* |
| | | | *p*6 |
| | | 6*mm* | *p*6*mm* |

### 3.3.1 Surface Reconstruction

The outermost atomic layers form the physical surface of a crystal or solid. As we have noted previously, the arrangement of the atoms into periodic arrays often display symmetries which differ from that of the bulk of the solid. Such a situation is referred to as the *surface reconstruction* of a particular layer and is often expressed in terms of the symmetry of the bulk terminated plane that would occur for the unreconstructed surface. We will, in the following, describe the notations that are typically used to describe this reconstruction, and this can be done for the selvedge or the adsorbate at the outermost surface. The first approach was formulated by Park and Madden (1968) and consists in the vectorial construction of the primitive translation vectors of the selvedge or adsorbate, denoted by **a'** and **b'**, with respect to those of the substrate, **a** and **b**. These are expressed in the following relations:

$$\mathbf{a'} = G_{11}\mathbf{a} + G_{12}\mathbf{b} \tag{3.2}$$

$$\mathbf{b'} = G_{21}\mathbf{a} + G_{22}\mathbf{b} \tag{3.3}$$

where the four coefficients, $G_{ij}$, form the matrix:

$$G = \begin{pmatrix} G_{11} & G_{12} \\ G_{21} & G_{22} \end{pmatrix} \tag{3.4}$$

thus we can express the meshes of substrate and adsorbate/selvedge as:

$$\begin{pmatrix} \mathbf{a'} \\ \mathbf{b'} \end{pmatrix} = G \begin{pmatrix} \mathbf{a} \\ \mathbf{b} \end{pmatrix} \tag{3.5}$$

The values of the matrix elements $G_{ij}$ can be used to classify the properties of the surface reconstruction and determine whether it is *commensurate* or *incommensurate* with respect to the substrate lattice. *Commensurability* means that a rational relationship between the vectors **a**, **b** and $\mathbf{a'}$, $\mathbf{b'}$ can be established. We can distinguish three cases:

(1) When all matrix elements $G_{ij}$ are integers—the lattices of the surface and bulk substrate are simply related—the surface lattice is called a simple superlattice.
(2) When all matrix elements $G_{ij}$ are rational numbers—the two lattices are rationally related—the surface lattice is called commensurate surface superstructure.

(3) When at least one matrix element $G_{ij}$ is an irrational number, the two lattices are irrationally related and the superstructure is said to be in-commensurate. An incommensurate superstructure is registered in-plane incoherently with the underlying substrate lattice.

An alternative to the matrix notation is that of Wood (1964), which is frequently used to express a label of the surface reconstruction occurring in superstructures. Firstly, the characterization will note the substrate surface crystallographic orientation $(hkl)$ with chemical composition: $S(hkl)$. There is then a simple notation for the surface reconstruction based on the transla-tion vectors:

$$|\mathbf{a}'| = m|\mathbf{a}| \tag{3.6}$$

$$|\mathbf{b}'| = n|\mathbf{b}| \tag{3.7}$$

and a rotation angle $\phi$ if necessary. As such the surface superstructure can be characterized as:

$$S(hkl)K(m \times n) - R\phi \tag{3.8}$$

where K is either $p$ for primitive or $c$ for centered, according to how the sur-face Bravais lattice is defined. When $p$ is dropped, the primitive notation is understood implicitly. The quantity $(m \times n) = (|\mathbf{a}'|/|\mathbf{a}| \times |\mathbf{b}'|/|\mathbf{b}|)$ indicates the ratios of the (usually) primitive basis vectors of the selvedge/adsorbate, $(\mathbf{a}', \mathbf{b}')$ lattice and the bulk lattice beneath, $(\mathbf{a}, \mathbf{b})$. Usually, one speaks of the $(m \times n)$ reconstruction. Typical notations could be:

$$S(hkl)(m \times n); S(hkl)c(m \times n); S(hkl)K(m \times n) - R\phi$$

Most surface superstructures or reconstructions can be represented in the Wood notation. However, there are some specific cases where this is not possible and the matrix notation is required for an unambiguous description. Some examples are given in Figures 3.6 and 3.7 and illustrate the correspon-dence between the Wood and matrix notations.

In addition to the representation of surface reconstruction, the Wood nota-tion can also express the adsorbate; for example, CO adsorbed molecularly on the Ni(100) surface at a fractional surface coverage of 1/2 forms an overlayer, as illustrated in Figure 3.7(c), which can be expressed as Ni(100) c(2 × 2) −

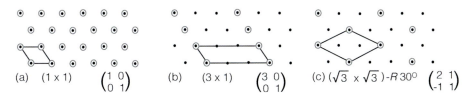

**FIGURE 3.6** Wood and matrix notations for some superlattices on a hexagonal 2D surface.

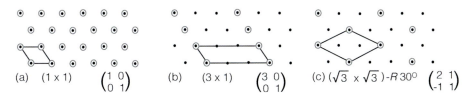

**FIGURE 3.7** Wood and matrix notations for some superlattices on a square 2D surface.

CO or Ni(100) $(\sqrt{2} \times \sqrt{2})R45^o$ – CO. We note here the plurality of the Wood notation in this example. In some cases, multiple domain reconstructions are possible and will depend on the surface symmetry. For example, on the bcc (100) surface, domains of $(1 \times 2)$ and $(2 \times 1)$ reconstructions should occur with equal frequency, all other factors being equal. While the Wood notation is less rigorous than the matrix representation for surface reconstructions, it is generally favored due to its conciseness and practicality.

### 3.3.2 The 2D Reciprocal Lattice

As in the case of 3D crystalline structures, the 2D lattice structures will have a reciprocal lattice which is useful when dealing with structural analysis and diffraction techniques. We will further discuss diffraction methods and surface analysis in Chapter 5. The construction of the reciprocal lattice in 2D is very similar to that in 3D and we can define the 2D reciprocal lattice as a set of points whose coordinates are given by the reciprocal lattice vectors:

$$\mathbf{G}_{hk} = h\mathbf{a}^* + k\mathbf{b}^* \tag{3.9}$$

where $h, k$ are integers $(0, \pm 1, \pm 2, ...)$ and the primitive translation vectors in reciprocal space, $\mathbf{a}^*$ and $\mathbf{b}^*$, are related to the primitive translation vectors of the real-space lattice, $\mathbf{a}$ and $\mathbf{b}$, by the relationships:

$$\mathbf{a}^* = 2\pi \frac{\mathbf{b} \times \mathbf{n}}{|\mathbf{a} \cdot \mathbf{b} \times \mathbf{n}|} \tag{3.10}$$

$$\mathbf{b}^* = 2\pi \frac{\mathbf{n} \times \mathbf{a}}{|\mathbf{a} \cdot \mathbf{b} \times \mathbf{n}|} \tag{3.11}$$

where $\mathbf{n}$ is the unit vector normal to the surface. From the above equations we can distinguish the following properties of vectors $\mathbf{a}^*$ and $\mathbf{b}^*$:

– Vectors $\mathbf{a}^*$ and $\mathbf{b}^*$ lie in the same surface plane as the real-space vectors $\mathbf{a}$ and $\mathbf{b}$.

– Vector $\mathbf{a}^*$ is perpendicular to $\mathbf{b}$; $\mathbf{b}^*$ is perpendicular to $\mathbf{a}$, see Figure 3.8.

The lengths of vectors $\mathbf{a}^*$ and $\mathbf{b}^*$ are:

$$|\mathbf{a}^*| = \frac{2\pi}{a \sin \theta} \tag{3.12}$$

$$|\mathbf{b}^*| = \frac{2\pi}{b \sin \theta} \tag{3.13}$$

where $\theta$ is the angle between vectors $\mathbf{a}$ and $\mathbf{b}$. We note that the lengths of these vectors are given in units of $m^{-1}$, i.e., inverse length, which defines our reciprocal space.

Figure 3.8 shows the reciprocal lattices for the five 2D Bravais lattices. From this, we see two general features:

– Each pair consisting of a real-space lattice and the corresponding reciprocal lattice belong to the same type of Bravais lattice. – The angle between the reciprocal lattice unit vectors $\mathbf{a}^*$ and $\mathbf{b}^*$, expressed as $\theta^*$, is related to that between the real-space unit vectors $\mathbf{a}$ and $\mathbf{b}$, $\theta$, as given by:

$$\theta^* = 180 - \theta \tag{3.14}$$

Thus, for rectangular and square lattices, these angles are the same, $90°$, but for the real-space and reciprocal hexagonal lattices, the angle is $120°$ and $60°$, respectively.

In the following section, we will see how the reciprocal lattice construction is useful in the interpretation of diffraction methods.

### 3.3.3 The Brillouin Zone for Crystalline Surfaces

The Wigner-Seitz cell of the reciprocal lattice is referred to as the first Brillouin zone. The concept of the Brillouin zone is of prime importance for the analysis of the electronic band structure of crystals and the propagation of phonons in solids. The concept of the Brillouin zone is discussed in all introductory texts on Solid State Physics, see Schmool (2016), for example.

In translationally invariant systems, the wave vector $\mathbf{k}$ defines a set of good quantum numbers for each type of elementary excitation. In the case of a surface of an ordered crystal, such a wave vector $\mathbf{k}$ is restricted to two dimensions, i.e., it is parallel to the surface. In the reduced zone scheme it is restricted to a 2D Brillouin zone (BZ). The entire 2D reciprocal space can be covered by the vectors $\mathbf{k} + \mathbf{g}$, where $\mathbf{g}$ is a surface reciprocal lattice vector of the form of Eq. (3.9). The surface BZ is defined as the smallest polygon in the

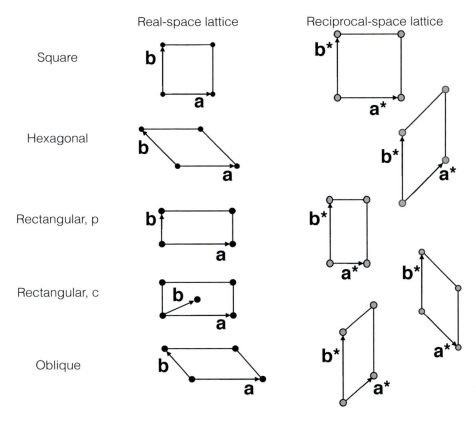

**FIGURE 3.8** Real-space (left) and corresponding reciprocal-space (right) lattices for the five Bravais lattice types.

2D reciprocal space situated symmetrically with respect to a given reciprocal space lattice point (used as coordinate zero) and bounded by points **k**, which satisfy the equation:

$$\mathbf{k} \cdot \mathbf{g} = \frac{1}{2}|\mathbf{g}|^2 \qquad (3.15)$$

The set of points defined by this equation gives a straight line at a distance $|\mathbf{g}|/2$ from the zero point which bisects the connection to the next lattice point **g** at right angles. This situation is illustrated in Figure 3.9.

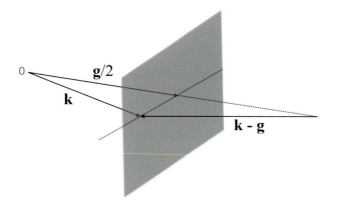

**FIGURE 3.9**    Real-space (left) and corresponding reciprocal-space (right) lattices for the five Bravais lattice types.

As there are five different Bravais lattices, there will be five different reciprocal surface lattices, and correspondingly five different 2D or surface Brillouin zones. These are illustrated in Figure 3.10.

In order to use Bloch-like eigenvalues of a bulk elementary excitation, the relationship between the eigenvalues of the bulk crystal and the wave vector must be altered. To represent all eigenstates, usually, the component $\mathbf{k}_\parallel$ of the 3D vector parallel to the surface can be fixed, while the perpendicular component $\mathbf{k}_\perp$ can be varied. An illustration of bulk directions and points of high symmetry in the 3D-BZ projected onto the 2D surface BZ are shown in Figure 3.11.

In this figure, the bulk BZ is projected onto the plane of the surface BZ. We denote the component of the wave vector **k** of the bulk BZ parallel to the surface as $\mathbf{k}_\parallel$. The boundary points of the projected bulk BZ are located on straight lines determined by:

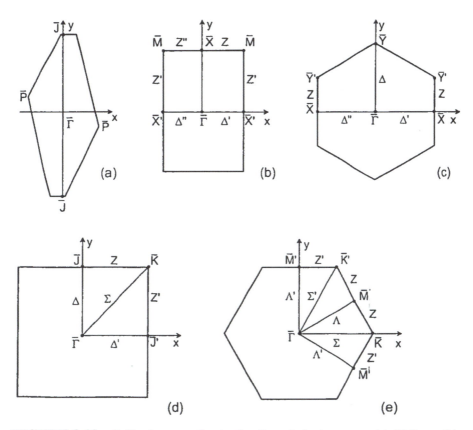

**FIGURE 3.10** Brillouin zones for the five Bravais lattice types: (a) Oblique, (b) *p*-rectangular, (c) *c*-rectangular, (d) Square, and (e) Hexagonal. Symmetry lines and points are also given.

$$\mathbf{k}_{\parallel} = k_{\parallel}^{(1)}\mathbf{a} + k_{\parallel}^{(2)}\mathbf{b} \tag{3.16}$$

$$\mathbf{k}_{\parallel} \cdot \mathbf{g} = \frac{1}{2}|\mathbf{g}|^2 \tag{3.17}$$

In general, the projected bulk BZ does not coincide with that of the surface BZ, usually being larger, see Figure 3.12, which shows the (100) surface of an fcc crystal, and one must fold back part of the projected bulk BZ not contained in the surface BZ. Since these parts of the projected bulk BZ agree with neighboring 2D BZ belonging to reciprocal lattice vectors $\mathbf{g}(\mathbf{k}_{\parallel})$, the folding is identical with a displacement by $g(k_{\parallel})$. Consequently, all wave

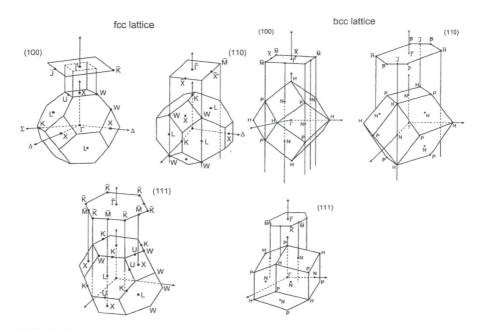

**FIGURE 3.11**    Relationship between the 2D Brillouin zones for low-index surfaces and the 3D Brillouin zone for fcc and bcc lattice structures.

vectors **k** in the surface BZ can be expressed as:

$$\mathbf{k} = \mathbf{k}_{\parallel} + \mathbf{g}(\mathbf{k}_{\parallel}) \qquad (3.18)$$

## 3.4  Summary

The physics of surfaces can be thought of in the same manner as that of bulk 3D solids. Here we are concerned with the physical properties related to an ordered assembly of atoms with the sole limitation that the atoms are ordered in only two dimensions. Indeed, we can consider this as an extension of regular solid-state physics. We have defined crystallography related to the surface array of atoms, and is based on the same rules that are applied to three-dimensional crystals. This allows us to define the allowed Bravais lattices that can occur. In the case of surfaces, there are five distinct lattices that can be defined. This then can be used as the basis to form the point and space groups for a complete parallel to the bulk situation.

In defining the crystalline formation for a surface, we have also considered the defects the can occur on surfaces. These can consist of points like

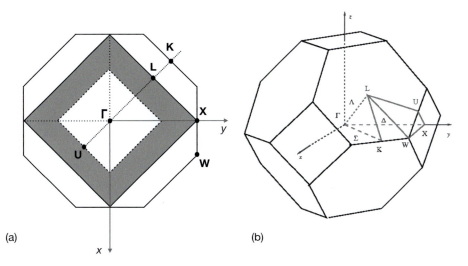

(a)

(b)

**FIGURE 3.12** (a) Brillouin zone of a (100) surface (shaded area)together with the projected bulk BZ for an fcc crystal. The projected critical points of the 3D BZ are indicated. (b) Bulk BZ for the fcc crystals for comparison.

defects, such as vacancies, adatoms, and impurity atoms. Line defects are also possible, such as steps or terraces. We can further note that kinks and vacancies can also occur along the step edge itself. Bulk defects such as edge and screw dislocations can end at a surface giving rise to a surface defect.

As we have seen, the regular bulk terminated lattice is a useful approximation as a first step in considering the nature of the physical surface. However, the lack of full atomic coordination that occurs in the bulk of the crystal often means that dangling bonds are formed. These unoccupied states can give rise to a reconfiguration of the surface structure to form a surface reconstruction. This often involves a reorganization of the atomic ordering and is frequently accompanied by a modification of the nearest neighbor separation between atoms. To define the surface reconstruction we consider the symmetry of the surface lattice with respect to the bulk terminated plane from which it is derived. We can express the surface reconstruction using a matrix formulation, though Wood's notation is more commonly used in practice.

Since we have defined the crystalline structure of the surface in a similar manner to that used for bulk crystals, we have also extended the definition of the reciprocal lattice for surface structures. This serves the same purpose as for bulk crystals, i.e., in the study of surface diffraction (see Chapter 5) and the electronic properties of surfaces. As an extension of this, we can also consider the formation of the Brillouin zone for surface structures.

## 3.5 Problems

(1) Identify the following superstructures indicated with the filled circles:

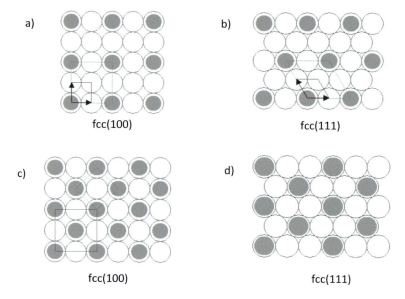

a) fcc(100)

b) fcc(111)

c) fcc(100)

d) fcc(111)

(2) Illustrate the form of the following surface reconstructions:
   (a) bcc (110)c(4×2)
   (b) fcc (110) (3×2)
   (c) zinc-blende (111)c(4×2)

(3) Use the matrix notation to express the surface reconstructions from the Wood's notation for the structures of Questions 1 and 2.

(4) Illustrate the form of the reciprocal lattice for the Si(111)-(7×7) surface.

(5) Sketch the following surface structure Ni(111)($\sqrt{3} \times \sqrt{3}$)R30-CO. The CO molecules are located in bridge positions between the surface Ni atoms.

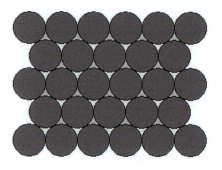

(6) Demonstrate the validity of Eq. (3.15) using Figure 3.9 as a reference.

## References and Further Reading

Park, R. L., & Madden Jr. H. H., (1968). *Surface Science*, *11*, 188.

Prutton, M. (1998). *Introduction to Surface Physics*, Oxford University Press.

Schmool, D. S. (2016). *Solid State Physics*, Essentials of Physics Series, Mercury Learning and Information, Stylus Publishing LLC.

Venables, J. A. (2000). *Introduction to Surface and Thin Film Processes*, Cambridge University Press, Cambridge, U.K.

Wood, E. A. (1964). *J. Appl. Phys.*, *35*, 1306.

Woodruff, D. P., & Delchar, T. A. (1994). *Modern Techniques of Surface Science*, Second Edition, Cambridge University Press.

Zangwill, A. (1988). *Physics at Surfaces*, Cambridge University Press.

# Chapter 4

# Thin Films

## 4.1 Introduction

Thin-film growth refers to the process of deposition of one layer of material upon another. When such growth is crystallographically oriented upon a single crystal substrate, the growth is said to be epitaxial; this can be subdivided into homoepitaxy, when the film and substrate are of the same material, or heteroepitaxy, when the film and substrate are different. The nature of the film growth is controlled by a subtle interplay of thermodynamic and kinetic processes. The general trends in film growth can be understood in terms of the relative surface and interface energies within the thermodynamical approach. On the other hand, film growth is a non-equilibrium kinetic process in which the rate-limiting steps affect the net growth mode. In this chapter, the surface phenomena involved in thin film growth and their effect on the growth mode, as well as on the structure and morphology of the grown films, will be discussed.

## 4.2 Deposition and Growth Modes

There are three principal growth modes that are distinguished in thin-film growth; these are illustrated in Figure 4.1 and are indicated in the following:

– **Layer-by-layer or Frank van der Merwe (FM)**: This refers to the case when the film atoms are more strongly bound to the substrate than to each other. Consequently, each layer is fully completed before the next layer starts to grow. This is the case of two-dimensional growth.

– **Island or Volmer-Weber (VW)**: This growth mode corresponds to the situation when film atoms are more strongly bound to each other than to the

73

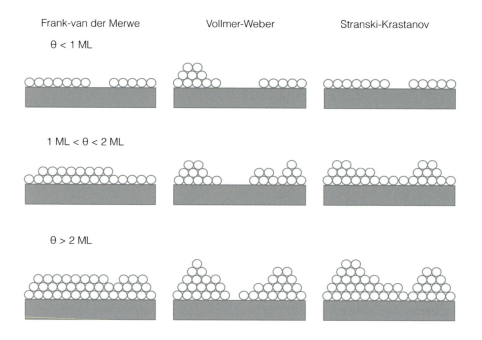

**FIGURE 4.1**　Schematic illustration of the three different epitaxial growth modes for different coverages, $\Theta$: less than one monolayer (ML), between one and two monolayers and greater than two monolayers. The Frank-van der Merwe growth mode is 2D, with each monolayer being complete before the next starts. For the Volmer-Weber mode, each monolayer can commence on a previous layer which does not need to be a complete monolayer. This corresponds to 3D growth. The Stranski-Krastanov mode is characterized by 2D growth of the first ml (wetting layer) followed by 3D island growth.

substrate. In this case, three-dimensional islands will nucleate and grow directly on the substrate surface.

**– Layer-Island or Stranski-Krastanov (SK)**: This case represents the intermediate situation between FM and VW growth. After the formation of a complete two-dimensional layer, the growth of three-dimensional islands takes place. The nature and thickness of the intermediate layer (often called the Stranski-Krastanov layer) depend on the particular situation; for example, the layer might be a sub-monolayer surface phase or a strained film several monolayers in thickness.

The existence of different growth modes can be understood qualitatively in terms of the surface or interface tension $\gamma$. We should recall, that the tension is defined as the work, that must be performed to make a surface (or interface) of the unit area. This arises from a consideration of the free energy change that accompanies the formation of aggregate on the surface of the substrate. If this island is considered to be approximately of circular shape we can estimate the free energy change as (Ohring, 1992)

$$\Delta G = a_1 r^3 \Delta G_v + a_2 r^2 \gamma_F + a_3 r^2 \gamma_{S/F} - a_3 r^2 \gamma_S \tag{4.1}$$

where $r$ refers to the radius of the island, $\Delta G_v$ denotes the change of chemical-free energy for a gas-to-solid transformation, where the second term takes into account the volume of the island. For nucleation to occur $\Delta G_v$ must be negative, which requires that supersaturation of the gas vapor occurs at the surface of the substrate. This term can be expressed as: $\Delta G_v = -(k_B T / \Omega) \ln(1+S)$, where $S = (P_v - P_s)/P_s$, is the vapour supersaturation. We note that $\Omega$ corresponds to the atomic volume. We also have the following; $\gamma_S$ is the surface tension of the substrate surface, $\gamma_F$ the surface tension of the film surface and $\gamma_{S/F}$ is the surface tension of the film/substrate interface.

Bearing in mind, that $\gamma$ can also be interpreted as a force per unit length of the boundary, we should consider the point of contact of the film island and the substrate (see Figure 4.2). If the island whetting angle is $\phi$, the force equilibrium can be written as:

$$\gamma_S = \gamma_{S/F} + \gamma_F \cos \phi \tag{4.2}$$

We can now distinguish between the different growth modes as follows. For the case of layer-by-layer (FM) growth, $\phi = 0$, hence from Eq. 4.2, we obtain the following condition:

$$\gamma_S \geq \gamma_{S/F} + \gamma_F \tag{4.3}$$

This situation corresponds to the case where the depositing film ideally "whets" the substrate. For island growth (VW), $\phi > 0$ and the following condition will hold:

$$\gamma_S < \gamma_{S/F} + \gamma_F \tag{4.4}$$

For the case of layer plus island (SK) growth, the condition given by Eq. (4.3) is initially satisfied, but the formation of the intermediate layer alters

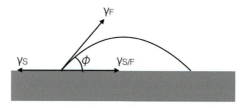

**FIGURE 4.2**   Schematic illustration of a film island on a substrate. $\gamma_S$ is the surface tension for the substrate surface, $\gamma_F$ that for the film surface and $\gamma_{S/F}$ the surface tension at the film -substrate interface. The balance of forces acting along the substrate surface gives rise to Eq. (4.2).

the effective values of $\gamma_S$ and $\gamma_{S/F}$, leading to the condition expressed by Eq. (4.4), for subsequent island growth.

Film growth modes can be identified experimentally by monitoring the Auger electron spectroscopy (AES) signals from the film and substrate during film deposition (see Chapter 5). In practice, it is necessary to stop growth at regular intervals to record the AES spectra and continue growth. Arguably, this could interfere with the normal growth of the layer, since more (diffusion) time is available to the depositing film atoms. However, monitoring RHEED or STM could confirm that the growth is nominally equivalent. Typical plots of the evolution of the Auger peak heights (usually normalized) are shown schematically in the following Figure 4.3. The layer-by-layer growth (a) is manifested by breaks on the gradient of the intensity, which correspond to monolayers. Island growth (c) gives a very slow variation of the Auger peak signals with film deposition. Stranski-Krastanov growth (b) is characterized by an initial linear segment followed by a sharp variation in the gradient of the Auger signal height.

## 4.3   Nucleation and Coalescence

The energy considerations appear to be rather simplistic and a much more complex picture emerges when we consider an atomistic approach. There are a number of factors that are important for the stable formation of clusters and islands, which must be taken into account for a realistic understanding of the complex interplay between the various physical processes at play. Importantly, these combine the energies involved and must also take into account the arrival rate of atoms at a surface, the surface temperature and desorption rate. We can schematically illustrate the main elementary physical processes involved in the formation and growth of films and islands as is illustrated in

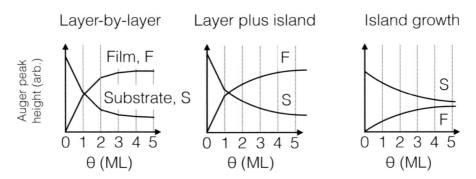

**FIGURE 4.3** Schematic plots for the Auger peak amplitude as a function of coverage, $\Theta$, for the three growth modes: layer-by-layer (FM); layer plus island (SK); and island growth (VW).

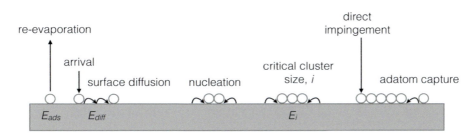

**FIGURE 4.4** Schematic diagram of various atomic processes taking place at the surface involved in the nucleation and growth of films.

Figure 4.4. Here atoms arrive from the gas or vapor phase at a rate $R$ and become accommodated at the surface as adatoms with a binding energy $E_{ads}$. This will produce a population of single adatoms or monomers, $n_1$, on the substrate with $n_0$ adsorption sites per unit area.

As a first approximation, the nucleation rate can be considered as being the product of the concentration of stable nuclei, $n^*$, with the rate of arrival of atoms, expressed as $\omega_a$, multiplied by the area of the critical nuclei $A^*$, such that we can write:

$$\frac{dn}{dt} = n^* \omega_a A^* \tag{4.5}$$

Thermodynamic considerations allow us to express the equilibrium number of nuclei with a critical size per unit area of the substrate. This is given by:

$$n^* = n_s e^{-\Delta G^*/k_B T} \tag{4.6}$$

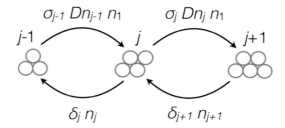

**FIGURE 4.5**    Illustration of fluxes which control the number density of clusters of size $j$.

where $\Delta G^*$ denotes the energy barrier for nucleation. This is defined from the consideration of the critical radius $r^*$ necessary for an island to grow. If it is less than this size it will lose atoms and shrink. The quantity $n_s$ represents the total nucleation site density. Combined with these considerations, we must account for the fact that adatoms migrate over the surface with the diffusion coefficient:

$$D = \frac{v}{4n_0} e^{-\frac{E_{diff}}{k_B T}} \tag{4.7}$$

These adatoms, denoted by the concentration $n_1$, migrate over the surface of the substrate until they are lost by one of the following processes. Firstly, they may be re-evaporated into the vacuum, such as when the substrate temperature is sufficiently high. Such re-evaporation can be characterized by the sojourn time (or residence time) of the adsorbate:

$$\tau_{ads} = v^{-1} e^{\frac{E_{ads}}{k_B T}} \tag{4.8}$$

Secondly, adatoms might be captured by existing clusters or at defect sites, such as steps. Thirdly, adatoms might combine with one another adatom to form a cluster. The small clusters are metastable and can decay back into individual atoms. However, when the clusters grow in size, they become more stable and the probability of its growth exceeds that of its decay. The critical size of an island, $i$, is defined as the minimal size when the addition of just one more atom makes the island stable.

The atomistic processes can be described in quantitative terms as a set of rate equations. The consideration is based on the assessment of the formation and decay rates of clusters. For example, Figure 4.5 illustrates the fluxes controlling the population, $n_j$, of the metastable clusters of size $j < i$, where $i$ is the critical size of the cluster.

From the above analysis we can describe the dynamics of nucleation in terms of the processes by which islands increase and decrease. There are two processes which contribute to the increase of $n_j$:

(1) An additional cluster of size $j$ is formed, when an adatom is attached to a cluster of size $j-1$. The net flux due to this process can be expressed as: $\sigma_{j-1}Dn_{j-1}n_1$.
(2) The detachment of an atom from the cluster of size $j+1$ (i.e., decay of a $(j+1)$-cluster), producing a cluster of size $j$ and an adatom. The net flux of the decay is given as: $\delta_{j+1}n_{j+1}$.

And there are two processes which decrease $n_j$:

(1) The attachment of adatoms to $j$-clusters which transforms them into $(j+1)$-clusters, which have a net rate of: $\sigma_j Dn_j n_1$.
(2) The decay of $j$-clusters to produce $(j-1)$ - clusters, with a net flux of: $\delta_j n_j$.

where, $n_1$ is the number density of adatoms, $D$ is the diffusion coefficient, $\sigma$ the capture number (which describes the capability of islands to capture diffusing adatoms) and:

$$\delta_{j+1} \sim De^{-\frac{\Delta E_j^{j+1}}{k_B T}} \tag{4.9}$$

is the decay rate with $\Delta E_j^{j+1}$ being the energy difference between the $(j+1)$-cluster and the $j$-cluster.

Using a similar approach for the evaluation of the number density of adatoms, $n_1$, and that of stable clusters with $j > i$, denoted by $n_x$, we can write the following rate equations:

$$\frac{dn_1}{dt} = R - \frac{n_1}{t_{ads}} + \left( 2\delta_2 n_2 + \sum_{j=3}^{i} \delta_j n_j - 2\sigma_1 Dn_1^2 - n_1 \sum_{j=2}^{i} \sigma_j Dn_j \right) - n_1 \sigma_x Dn_x \tag{4.10}$$

$$\frac{dn_j}{dt} = n_1 \sigma_{j-1} Dn_{j-1} - \delta_j n_j + \delta_{j+1} n_{j+1} - n_1 \sigma_j Dn_j \tag{4.11}$$

$$\frac{dn_x}{dt} = n_1 \sigma_i Dn_i \tag{4.12}$$

Equation (4.10) describes the time variation of the adatom density $n_1$. Here we note an increase in the concentration of $n_1$ due to deposition, with a flux $R$, and a decrease resulting from desorption at a rate of $n_1/\tau_{ads}$. The terms in the brackets represent the supply and consumption rates due to the formation and decay of subcritical clusters $(j < i)$; $2\delta_2 n_2$ and $2\sigma_1 Dn_1^2$ represent the decay and formation of dimers, (the factor 2 indicates that in each of these processes, adatoms are supplied or consumed in pairs). The $\sum$ terms are for the decay and formation of clusters of size from 3 to $i$. The final term in Eq. (4.10) represents the net capture rate of stable clusters larger than $i$.

Equation (4.11) indicates the density of metastable clusters of size $j$. The terms on the right-hand side are illustrated in Figure 4.5 and have been discussed above. Equation (4.12) describes the growth of the stable cluster density $n_x$ due to attachment of adatoms to critical size clusters. Note that, in these equations, some processes (e.g., coalescence, direct impingement onto clusters and adatoms) are neglected, but can in principle be taken into account by adding the appropriate terms.

Integration of Eqs. (4.10), (4.11), and (4.12) gives the time evolution of island and adatom densities. As an example, Figure 4.6 shows the results of numerical calculations for the case of $i = 1$ (i.e., when a dimer is already a stable cluster) and for sufficiently low temperatures, when adatom evaporation and dimer mobility can be neglected. One can see, that the dynamic behavior of the density of adatoms $(n_1)$ and islands $(n_x)$ can be divided into four different coverage regimes, which we can distinguish in the following manner:

(1) Low coverage nucleation regime (L);
(2) Intermediate coverage regime (I);
(3) Aggregation regime (A); and
(4) Coalescence and percolation regime (C).

In the initial stages of deposition (L), the adatom density is much higher than the island density, $n_1$, so the probability of island nucleation far exceeds the probability of an adatom to becoming incorporated into an existing island. In this regime, $n_1 \propto \theta$, or put more directly $n_1 = Rt$, which comes from the first term in Eq. (4.10) and $n_x \propto \theta^3$. In the course of the deposition, the number density of islands increases until it becomes comparable to that of the adatoms. This can be understood in terms of the probabilities of an adatom encountering another adatom. In this case, the probabilities are comparable. This point corresponds to the onset of the intermediate coverage regime (I). We also note that in the low coverage regime, the adatom density

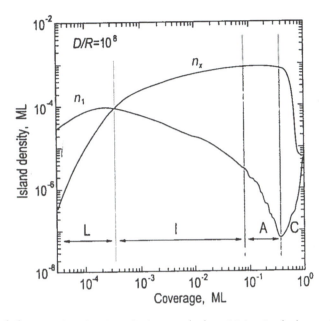

**FIGURE 4.6** Number density of adatoms $(n_1)$ and islands $(n_x)$ as a function of the coverage in the case $i = 1$ and $D/R = 10^8$, illustrating four regimes: Low coverage nucleation regime (L); Intermediate coverage regime (I); Aggregation regime (A) and Coalescence and percolation regime (C). Reprinted figure with permission from J. G. Amar, F. Family & P.-M. Lam, (1994). *Phys. Rev. B, 50,* 8781. Copyright (1994) by the American Physical Society.

peaks and then starts to decrease $(n_1 \propto \theta^{-\frac{1}{3}})$, while the island density is still increasing, though much more slowly $(n_x \propto \theta^{\frac{1}{3}})$. When the density of islands increases, so that the mean island separation is equal to the mean free path of migrating adatoms, the further deposition must result in island growth and the island density will reach a saturation value. This is designated as the aggregation regime (A), and usually occurs at a coverage between 0.1-0.4 ml, see Figure 4.6. Finally, the coalescence and percolation regime (C) will take over, where the islands join together and percolate. This leads to a decrease in the island number density. Simultaneously, this is often accompanied by the growth of a second layer and the adatom density again increases. This scenario corresponds to a layer-by-layer growth mode. One of the key factors here is when the second layer commences. This will clearly be dependent on the rate of arrival of atoms and the diffusion rate. Both factors, as we have discussed are essentially what define the growth mode.

For the case of complete condensation (i.e., with negligible re-evaporation), the saturation density of 2D islands, normalized to the density of adsorption sites, $n_0$, is given by (Venables et al., 1984):

$$\frac{n_x}{n_0} = \eta\,(\theta,i)\left(\frac{R}{Dn_0^2}\right)^{\chi} e^{\frac{E_i}{(i+2)k_BT}} \tag{4.13}$$

This equation shows a fractional power law, depending on the deposition rate, $R$, and an exponential dependence on the temperature. The scaling exponent is $\chi = \frac{i}{i+2}$, $i$ being the critical cluster size, c.f. Stowell (1970), where $i \to i+1$. $E_i$ denotes the binding energy of the critical cluster, given approximately by the number of nearest neighbor atom bonds. $\eta\,(\theta,i)$ is a pre-exponential numerical factor which can vary from $10^{-2}$ up to 10, depending on the regime of condensation, the coverage, and the critical cluster size. In the coverage range of interest, 0.1–0.4 ml, $\eta\,(\theta,i)$ is a weak function of $\theta$ with typical values in the range from 0.1–1; for example for $i = 1 - 5, \eta \simeq 0.2 - 0.3$. We note that in most cases, the nucleation rate occurs under conditions of steady-state. This means that the first two terms of Eq. (4.10) are effectively balanced.

Taking into account the temperature dependence of the diffusion coefficient (Eq. 4.7), Eq. (4.13) can be rewritten for the case of a square lattice as (Müller et al., 1996):

$$\frac{n_x}{n_0} = \eta\,(\theta,i)\left(\frac{4R}{\nu_0 n_0}\right)^{\frac{i}{i+2}} e^{\frac{\left(iE_{diff}+E_i\right)}{(i+2)k_BT}} \tag{4.14}$$

The application of rate theory analysis for the evaluation of experimental data is illustrated in Figure 4.7(a), showing the results of an STM study of the initial stages of Cu epitaxy on Ni(100). Here the temperature dependence of the saturation island density is shown as an Arrhenius plot. Three regions with different slopes can be clearly distinguished. The temperature-independent regime (labeled "post-nucleation") is attributed to the incorporation of adatoms into existing islands after the termination of deposition. The two regimes entered above 160 K and 320 K, respectively, were found to correspond to the different size of the critical cluster, as indicated in the figure. In order to establish these sizes, the rate dependence of island density was measured at 215 and 345 K, located at the center of the corresponding regions in the Arrhenius plot, as shown in Figure 4.7(b). At 215 K, the exponent corresponds to 0.32±0.01, which yields $i = 1$, according to Eq. (4.14).

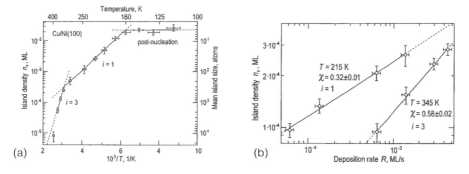

**FIGURE 4.7**    (a) Arrhenius plot of the measured saturation island density of Cu on Ni(100) at a coverage of 0.1 ml and a flux of $1.34 \times 10^{-3}$ ml/s. (b) Log–log plot of the saturation island density versus deposition flux for different growth temperatures. The gradient of the plot is described by the scaling exponent $\chi = i/(i+2)$, which allows for the determination of the critical cluster size, $i$.  Reprinted figure with permission from B. Müller, L. Nedelmann, B. Fischer, H. Brune and Klaus Kern, (1996). *Phys. Rev. B, 54,* 17858. Copyright (1996) by the American Physical Society.

This shows that the critical size corresponds to a monomer and the dimer will be the smallest stable island. At 345 K, the exponent is 0.58±0.02 and has a corresponding value of $i = 3$, i.e., a tetramer. With a knowledge of the sizes of critical clusters, the diffusion activation energy $E_{diff}$, dimer bond energy $E_b$ and attempt frequency $v_0$ can be evaluated by analyzing the Arrhenius plot of the saturation island density (Figure 4.7(b)). They are evaluated as $E_{diff} = 0.351 \pm 0.017$ eV, $E_b = 0.46 \pm 0.19$ eV and $v_0 = 4 \times 10^{11\pm0.3}$ Hz for $i = 1$ data and $5 \times 10^{12\pm2}$ Hz for $i = 3$ data (Müller et al., 1996).

## 4.4   Island Shape

Depending on growth conditions, the shapes of islands can evolve very differently. According to the compactness of their shape, islands can be subdivided into two general classes:

(1) Ramified islands:  fractal-like; dendrite shapes, having rough island edges.
(2) Compact islands:  square, rectangular, triangular or hexagonal shapes, with relatively straight and equi-axial island edges.

The compactness of an island is largely determined by the ability of a captured adatom to diffuse along the island edges and to cross corners where two

20 lattice constants

**FIGURE 4.8**    Diffusion limited aggregate (DLA) with a dendritic-like structure. Reprinted figure with permission from T. A. Witten, & L. M. Sander, (1983). *Phys. Rev. B, 27,* 5686. Copyright (1983) by the American Physical Society.

edges meet. Such processes will again be activation energy-dependent and will be intrinsically dependent on the surface temperature of the substrate.

Typically, the formation of ramified islands will occur at relatively low temperatures, when edge diffusion is slow. In the extreme case, the so-called *hit-and-stick regime,* an adatom sticks to an island and stays immobile at the impact site. This case is described by the classic Diffusion-Limited Aggregation (DLA) model (Witten and Sander, 1983), which predicts that under these conditions fractal islands with average branch thickness of about one atom width will be formed, irrespective of lattice geometry (see Figure 4.8). In real growth, the classic DLA mechanism does not occur, since adatoms reaching an island always walk a certain path to find an energetically more favorable site. The higher the rate of edge diffusion, the greater the branch thickness. The fractal islands observed in STM measurements all have branch thicknesses exceeding one atom width as predicted by the DLA model. Even if the adatom can diffuse along the island edge, it might be unable to cross the corner of the island, since in passing along the corner, the adatom has to lower its coordination, i.e., this forms a barrier to diffusion and will also be temperature-dependent. Without corner crossing, growth also leads to the formation of ramified fractal islands.

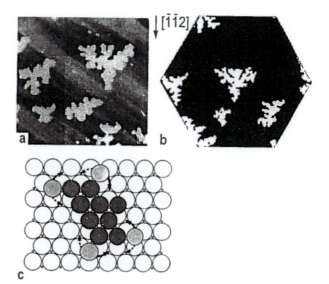

**FIGURE 4.9**    (a) Experimental STM image ($500 \times 500$ Å$^2$) and (b) simulated island shapes in the Pt/Pt(111) system at 245 K. (c) Model of the (111) surface illustrating the mechanism for anisotropic growth. The lightly shaded atoms at the corners are one-fold coordinated. Dashed and full arrows indicate diffusion jumps into two-fold coordinated sites via atop positions (non-preferential path) and via bridge positions (preferential path), respectively. Reprinted figure with permission from M. Hohage, M. Bott, M. Morgenstern, Z. Zhang, T. Michely, & G. Comsa, (1996). *Phys. Rev. Lett., 76*, 2366, Copyright (1996) by the American Physical Society.

Low-temperature Pt growth on Pt(111) gives an example of ramified island formation, as seen in Figure 4.9. The developed fractal islands have a branch thickness of about four atoms, and display clear trigonal symmetry due to the underlying symmetry of the (111) plane. To account for the growth anisotropy, an asymmetric probability of a one-fold coordinated adatom at the corner to jump either side has been suggested. In the center figure (c), for each of the one-fold coordinated atoms (light shades circles), two different paths to adjacent two-fold coordinated sites are indicated. An atom moving along a path indicated by a solid arrow is able to remain at higher coordination with respect to the substrate (passing via a bridge position) than an atom moving along the dashed arrows (via an atop position). Thus, the paths along the solid arrows are preferable. Monte Carlo simulations using these assumptions reproduces well the shape of Pt islands observed experimentally.

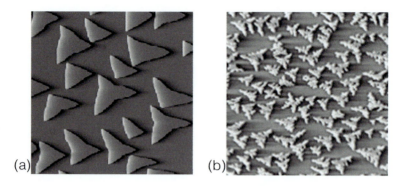

**FIGURE 4.10**  Effect of growth temperature on island shapes in the homoepitaxial growth of Pt on Pt(111). (a) Growth at 400 K results in triangular-shaped islands while (b) growth at 300 K gives rise to ramified island growth. Reprinted figure with permission from T. Michely, M. Hohage, M. L. Bott, & G. Comsa, (1996). *Phys. Rev. Lett., 70,* 3943. Copyright (1996) by the American Physical Society (Michely et al., 1996).

For the case at higher temperatures, where adatoms can easily cross the island corners, the growth is characterized by the formation of compact islands. The change from fractal-like to compact island growth, as we have noted, occurs at higher temperatures. For the Pt/Pt(111) system, this transition occurs between 300 K and 400 K, and leads to the growth of triangular Pt islands (see Figure 4.10).

In general, the compact island shape is controlled by the competition between steps of different orientations in accommodating arriving adatoms. Many rate processes are involved in the growth of islands and the shape of the growing islands might differ from the equilibrium shape due to kinetic limitations. For example, the equilibrium shape of a monolayer Pt island on Pt(111) is not a triangle (as formed in the course of growth, seen in Figure 4.10), but rather a hexagon (as can be obtained on subsequent annealing; see Figure 4.11(a)).

For 3D crystallites, the equilibrium shape reflects the anisotropy of the specific surface free energy. The formation of 3D islands will depend on the ability of adatoms to overcome the energy barrier, at the island edge, to jump up on top of the island. This will have similar energy considerations the one-fold coordination adatom discussed earlier. For 2D islands, the equilibrium shape reflects the anisotropy of the specific step free energy. The 2D Wulff theorem reads as follows:

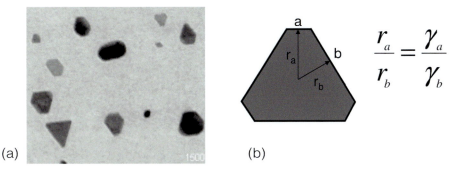

(a)  (b)

**FIGURE 4.11** (a) STM image of Cu islands on Cu (111). (b) Schematic diagram illustrating the 2D Wulff theorem for the determination of the ratio of the step free energies. The shape of islands in (a) conform to this general rule.

In a 2D crystal at equilibrium, the distances of the edges from the crystal center are proportional to their free energy per unit length, see Figure 4.11(b), which shows a similar situation for Cu growth on Cu(111). Hence, the ratio of the specific free energies of the steps, $\frac{\gamma_A}{\gamma_B}$ can be readily evaluated from the aspect ratio $\frac{r_A}{r_B}$. For the case of Pt on Pt(111), the free energy ratio of B and A steps is found to be about 0.87±0.02.

### 4.4.1 The Distribution of Island Sizes

The distribution of nucleated islands is affected by several different parameters, such as the critical island size, coverage, and substrate structure. At relatively high coverages, it is strongly influenced by coarsening phenomena, such as ripening and coalescence of islands.

Effect of Critical Island Size: The island size distribution function $N_s$ gives the density of island size $s$ (where $s$ is the number of atoms in the island). Hence, the total number of stable islands $N$ and coverage, $\Theta$, can be expressed as (Amar et al., 1994):

$$N = \sum_{s>i} N_s \qquad (4.15)$$

$$\Theta = \sum_{s\geq 1} sN_s \qquad (4.16)$$

In these terms, the average island size $\langle s \rangle$ is defined as:

$$\langle s \rangle = \frac{\sum_{s>i} sN_s}{\sum_{s>i} N_s} = \frac{\Theta - \sum_{s\leq i} sN_s}{N} \qquad (4.17)$$

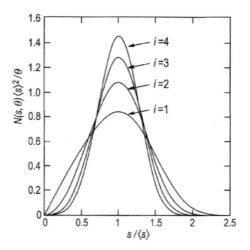

**FIGURE 4.12** Simulated scaling function $f_i$ for the island size distribution for critical sizes of $i$ from 1 to 4.

For a small critical size, such as $i = 1$:

$$\langle s \rangle = \frac{\Theta - N_1}{N} \simeq \frac{\Theta}{N} \tag{4.18}$$

as the relative number of single adatoms $N_1$ is small compared to the total number of islands $N$. Under these assumptions, scaling theory yields the scaling relation as:

$$N_s = \Theta \langle s \rangle^{-2} f_i \left( \frac{s}{\langle s \rangle} \right) \tag{4.19}$$

where $f_i(\frac{s}{\langle s \rangle})$ is the scaling function for the island size distribution for a critical size $i$. Equation (4.19) holds when the ratio $Dn_0^2/R$ is high. A plot of the simulated scaling function for $i$ from 1 to 4 is shown in Figure 4.12. As can be seen, the peak of the island size distribution increases and becomes more sharply peaked with increasing $i$.

By comparing the scaling function obtained experimentally with that from the simulation (Figure 4.12), it is possible to determine the size of the critical island. To obtain the experimental scaling function, we have to scale the measured island size distribution $N(s)$ into $N(s/\langle s \rangle)$, and to plot $N_s(s/\langle s \rangle) \frac{\langle s \rangle^2}{\Theta}$ versus $\frac{s}{\langle s \rangle}$. From Eq. (4.19), it is clear that the plot will yield the scaling function, $f$. The usefulness of this procedure is illustrated in Figure 4.13. This shows the results of the evaluation for Fe island growth on Fe(100). In Figure 4.13(a), the island size distribution obtained from STM images for growth at the relatively low temperatures of 20, 132 and 207°C is shown. In Figure 4.13(b), the island size distribution at relatively high growth temperature is illustrated. In Figure 4.13(c), the scaled island size distribution is shown for data in Figure 4.13(a); this is the experimental scaling function $f$. It can be seen, that after scaling, all three distributions fall onto the same curve, which resembles the simulated $f$ plot for a critical size of $i = 1$. For higher growth temperatures (356°C), the scaled island size distribution fits for simulated $f$ plot for $i = 3$, Figure 4.13(d). Thus, the obtained results indicate that in the temperature range from 20 to 207°C, the critical island size corresponds to a monomer ($i = 1$), but this becomes a tetramer ($i = 3$) at the higher temperature of 356°C.

## 4.4.2  *Coarsening*

The process of increasing the mean island size at the expense of decreasing the number of islands is referred to as *coarsening*. There are two principal mechanisms for coarsening:

(1)  coalescence (i.e., the merging of islands upon contact), and
(2)  ripening (growth of larger islands due to the diffusion flux of adatoms detached from smaller, less stable, islands).

In coalescence, two initially separate islands come into direct contact and transform into a single larger island, as illustrated in Figure 4.14. For monoatomic, or monomer, islands, the area of the new, larger island is the sum of the areas of the initial islands. The shape of the forming islands depends on the edge mobility of atoms. In the case of low edge mobility, the islands attach to each other without much reshaping. However, often the mobility is sufficient to allow the forming island to recover its equilibrium shape, see Figure 4.14.

If the islands are mobile, as is frequently the case, they can encounter each other and coalesce into larger islands. This process is called dynamic coalescence or *Smoluchowski ripening*, named after its discoverer, who formulated the kinetic theory of the coarsening of colloidal particles via dynamic coalescence, in 1916. Static coalescence corresponds to the case when neighboring immobile islands coalesce as their sizes increase due to the deposition process. Dynamic coalescence can take place at low coverages of ~0.1 ml, while static coalescence generally occurs at higher coverages, typically around 0.4–0.5 ml. As coverage increases, the coalescence system reaches a threshold for percolation growth; i.e., for the formation of interconnected structures. The

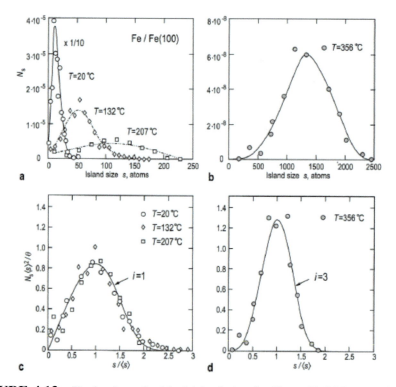

**FIGURE 4.13**   Evaluation of critical island size for Fe on Fe(100) growth. (a) Experimental Fe island distribution at low growth temperatures, 20, 132 and 207°C. The island density per lattice site $N_s$ is plotted against the number of atoms (size) $s$. (b) As in (a), but for higher growth temperature of 356°C. (c) Scaled island size distributions for low growth temperatures, 20, 132 and 207°C, superposed on the simulated scaling function $f$ for critical island size $i = 1$, taken from Figure 4.12. (d) Scaled island size distributions for higher growth temperature of 356°C, superposed on the simulated scaling function $f$ for critical island size $i = 3$. Reprinted figure with permission from J. A. Stroscio, & D. T. Pierce, (1994). *Phys. Rev. B, 49*, 8522. Copyright (1994) by the American Physical Society.

**FIGURE 4.14**   Schematic illustration of the sequential stages of coalescence.

outset of percolation manifests itself by an abrupt change in several physical parameters (e.g., in the conductivity of a metal).

Before coalescence occurs there will be a collection of island structures of various sizes. Over the course of time, the larger islands will tend to grow, or ripen at the expense of the smaller islands. This ripening is often referred to in the literature as Ostwald ripening (named after its discoverer, who in 1900 described the change of the size of granules embedded in a solid matrix by the diffusion flow of particles from one grain to another). In the case of 2D monoatomic islands, ripening is a thermodynamically driven process, which reduces the free energy associated with the island edges. According to the Gibbs-Thompson relation, the chemical potential of a circular island of radius $r$ is given by (Venables et al., 1984):

$$\mu_i(r) = 2A \frac{\gamma_{sl}}{r_i} \qquad (4.20)$$

where $\gamma_{sl}$ is the step line tension and $A$ is the area occupied by one atom. Hence, smaller islands have a higher pressure of adatoms, which results in a net flow of material from smaller to larger islands through the 2D adatom gas, see Figure 4.15(a). Note that ripening occurs only if the net supersaturation of the adatoms is very small and in most experimental studies of island ripening, the post-deposition behavior of islands is monitored, as illustrated in Figures 4.15(b) and (c).

As with most surface processes, the ripening mechanism is thermally driven and is related to the specific island shape, the bonding strength and the thermal energy available which allows atoms to escape from one island and surface diffuse to another. Ripening can thus be understood as the tendency of larger islands to capture more defusing atoms than smaller ones and also to hold on to their atoms since they will on average be held by more bonds.

### 4.4.3 *Magic Islands*

The kinetic stabilization of islands can lead to size distributions, which are multi-peaked. This arises due to the particular stability of certain sized islands where formations of a 2D structure have a specific completion of the bonds between neighbors. In the following, we will illustrate this with some specific examples which for triangular islands

When due to their specific structure, islands of selected sizes demonstrate enhanced stability, one speaks of magic islands or magic clusters. The two examples given below refer to triangular islands, for which the lateral growth of rows is hindered by the barrier for the formation of a new row. This leads to

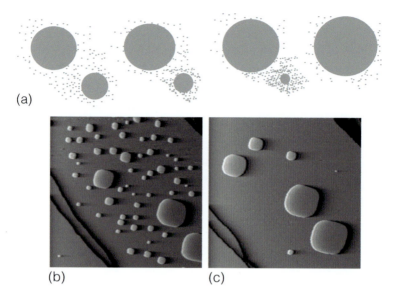

(a)

(b)                         (c)

**FIGURE 4.15**    (a) Schematic illustration of the sequential stages of island ripening. (b) and (c) Two 300 nm × 300 nm STM images, separated in time by 20,000 s, showing island ripening on Cu(001) at 343 K.

a kinetic stabilization of islands of "magic" sizes, which display pronounced peaks in the island size distribution.

The first example we consider is the formation of 2D Ga islands with a deposition coverage of ∼0.15 ML of Ga onto the Si(111) ($\sqrt{3} \times \sqrt{3}$) - Ga surface followed by the annealing to $200 - 500°C$. Scanning tunneling microscopy reveals that most of the formed islands are of triangular shape with specific size preferences, as illustrated in Figure 4.16. The island size distribution clearly shows the existence of *magic clusters* with sizes $N(N + 1)/2$, where $N$ ($= 2, 3, 4$ or $5$) is the number of atoms on each side of the triangular islands. In this particular example, the $N = 4$ island is the most abundant and stable species.

The second example is provided by the growth of epitaxial 2D Si islands with a deposition coverage of 0.14 ml onto substrates of Si(111) ($7 \times 7$) with a temperature of $450°C$. As in the previous example, the Si islands also exhibit a triangular shape, though they are somewhat larger in size. This is due to the size of the ($7 \times 7$) reconstruction, where the island-building block corresponds to one-half of the ($7 \times 7$) unit cell (HUC or half unit cell), and contains 51 atoms. The island size distribution plotted in HUC units is shown

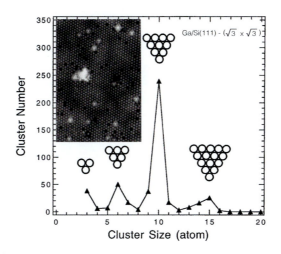

**FIGURE 4.16** Histogram of 2D clusters showing the existence of magic numbers. Inset shows an STM image (size = 31 × 21 nm²) of 2D clusters on the $\sqrt{3} \times \sqrt{3} - R30°$ Ga/Si(111) surface. Reprinted figure with permission from M. Y. Lai, & Y. L. Wang, (1999). *Phys. Rev. B, 60*, 1764. Copyright (1999) by the American Physical Society.

in Figure 4.17. Several peaks corresponding to the magic island sizes $N^2$ ($N = 2, 3, 4, 5, 6$ or 7 HUCs) are observed.

### 4.4.4 Vacancy Islands

In analogy with the formulation of adatom islands in the course of the deposition, the agglomeration of surface vacancies produced by ion bombardment into pits can be visualized as the formation of vacancy islands. In this description, deep pit formation corresponds to the growth of 3D vacancy islands and layer-by-layer sputtering corresponds to layer-by-layer growth. The regularities of the behavior of vacancy and adatom islands are almost identical. For example, the equilibrium shape of both types of the island is controlled by the minimization of the free energy of the step edge bordering the island. As a result, vacancy and adatom islands display similar equilibrium shapes, the only difference being that they are rotated by 180° with respect to each other (see Figure 4.18). We can see, that this is a natural consequence of the fact, that step-down orientation is reversed in the vacancy islands, as compared to the adatom island.

Like that of the adatom island, the evolution of vacancy islands is also governed by processes such as island migration, island ripening, and coa-

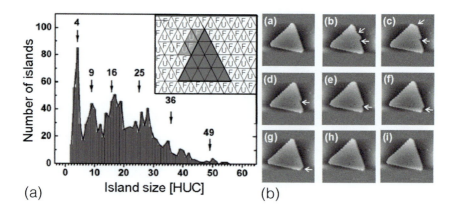

(a)     (b)

**FIGURE 4.17**   (a) Island size distribution for 2D islands epitaxially grown on Si(111) at 450°C. The distribution displays several peaks at *magic sizes*. The island size is expressed units of one-half of the $(7 \times 7)$ unit cell (HUC). (b) A sequence of STM images shows the formation of a new stable magic island as atoms are added along one edge of the triangular island. Reprinted figure with permission from B. Voigtländer, M. Kästner, & P. Smilauer, (1998). *Phys. Rev. Lett., 81*, 858. Copyright (1998) by the American Physical Society.

**FIGURE 4.18**   STM image of Pt islands on the Pt(111) surface. Indicated are an adatom island and a vacancy island, which have characteristic shapes. Reprinted from T. Michely, & G. Comsa, (1991). *Surf. Sci., 256*, 217. Copyright (1991), with permission from Elsevier.

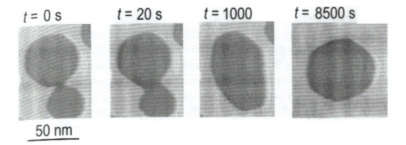

$t = 0\ \text{s}$      $t = 20\ \text{s}$      $t = 1000$      $t = 8500\ \text{s}$

50 nm

**FIGURE 4.19** STM image of vacancy islands on the Ag(111) surface. STM Snapshots are shown at different times, illustrating the coalescence of two vacancy islands.Reprinted by permission from G. Rosenfeld, K. Morgenstern, M. Esser, & G. Comsa, (1999). *Appl. Phys. A, 69,* 489, ©(1999) Springer, Nature.

lescence, etc. As an example, the figure below (Figure 4.19) illustrates the coalescence of two vacancy islands on Ag(111) as observed by STM.

## 4.5 Kinetic Effects in Homoepitaxial Growth

For the case of homoepitaxy, where the substrate and film are of the same chemical species, thermodynamic considerations, which are based on the balancing of the free energies of the film surface, substrate surface, and the substrate/film interface, predict a layer-by-layer Frank-Van der Merwe (FM) growth. However, growth frequently occurs at conditions that are far from equilibrium and kinetic limitations associated with the finite rates of mass transport processes can greatly affect the realities of the deposition process and hence the growth mode. There are two diffusion processes of principal interest:

(1) Intralayer mass transport: diffusion of adatoms on a flat terrace;
(2) Interlayer mass transport: diffusion of adatoms across a step edge.

To characterize interlayer mass transport, we consider the shape of the potential energy curve for an adatom near a step, this is illustrated in Figure 4.20. Here we take into account the periodic potential of the regularly arranged atoms and the interruption to this potential at a step edge. As can be seen, an adatom encountering a step at the lower side would stick to the step, as the lower terrace adsorption site adjacent to the step edge has higher coordination than that for an adatom on the terrace, which gives rise to the higher binding energy. Therefore, upward diffusion is usually neglected. An adatom reaching the step edge from the top of the terrace will encounter an energy

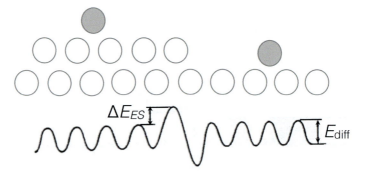

**FIGURE 4.20**  Schematic representation of the potential energy variations associated with a periodic potential of ordered atoms on a surface and its perturbation at a monoatomic step. $E_{ES}$ denotes the Ehrlich-Schwöbel barrier, which an adatom (dark circle) has to overcome in addition to the terrace diffusion barrier, $E_{diff}$, in order to cross the step edge in the downwards direction. Reprinted from R. L. Schwöbel, & E. J. Shipsey, (1966). *J. Appl. Phys., 37*, 3682, with the permission of AIP Publishing.

barrier which is higher than that of the terrace diffusion barrier, expressed as $E_{diff}$. The additional barrier $E_{ES}$, known as the *Ehrlich-Schwöbel barrier*, appears as an adatom reduces its coordination while crossing the step edge. In either case, the atom reacting the step edge, either from able or from below will encounter a barrier to its crossing. However, these barriers are of different strengths, with the atom on the top of the terrace island being lower. The efficiency of interlayer mass transport from the upper to the lower side, can be expressed by the transmission factor:

$$S = e^{-\frac{E_{ES}}{k_B T}} \tag{4.21}$$

This gives the probability for an adatom to cross the step (relative to terrace diffusion and under the assumption that equal prefactors for terrace and step-down diffusion hold). If the Ehrlich-Schwöbel barrier is negligible, $S \simeq 1$, while for an unsurmountable barrier, $S \simeq 0$. As we noted before the process in the opposite direction is considered to be much more unlikely due to the energy barrier been practically insurmountable.

Depending on the relative rate of intra-layer and interlayer mass transport, homoepitaxial growth will proceed in accordance with one of the three possible growth modes, as illustrated in Figure 4.21. These growth modes are labeled:

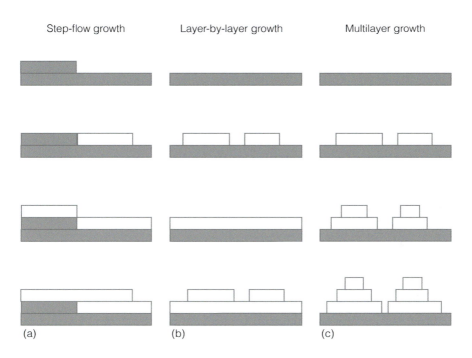

**FIGURE 4.21** Modes of homoepitaxial growth: (a) Step-flow growth, (b) layer-by-layer growth, and (c) multilayer growth.

(1) Step-flow growth;
(2) Layer-by-layer growth; and
(3) Multilayer growth.

In step-flow growth, the layer growth proceeds under conditions close to equilibrium. In this case, the adatom supersaturation is low and the adatom intra-layer mobility is high, hence all of the adatoms reach the step before nucleation on a terrace sets in. As can be seen, interlayer mass transport does not play an essential role in this growth mode.

If the terrace width exceeds the adatom migration length, growth will proceed via the nucleation and growth of adatom islands on the terraces. In this case, depending on the rate of interlayer mass transport, layer-by-layer growth or multilayer growth will occur. Layer-by-layer growth requires a sufficient interlayer mass transport to ensure that all atoms, which are deposited onto the top of the growing island, would reach the island edge and jump down to the lower layer. In the ideal case, we expect a transmission factor of $S = 1$ and a new layer will start to grow only after the previous layer

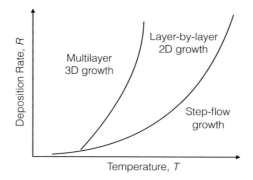

**FIGURE 4.22**    Schematic growth mode phase diagram for the case of a non-zero Ehrlich-Schwöbel barrier.

has been entirely completed. If the interlayer mass transport is suppressed (i.e., $S \simeq 0$), the adatoms cannot escape from the top of the island to the lower terrace and promote the earlier nucleation of a new layer. As a result, the growth will proceed as a multilayer 3D growth.

It is possible to alter the growth mode by changing the deposition rate and/or the substrate temperature. The dependence of the growth mode on these deposition parameters can be summarized in the form of a growth mode diagram, see Figure 4.22. The theoretical consideration of island nucleation assisted by interlayer diffusion shows that the dividing line between layer-by-layer and multilayer growth is defined by the condition:

$$S\lambda = 1 \Longleftrightarrow R = \lambda_0^{\frac{2(i+2)}{i}} e^{-\frac{\left[\frac{E_i}{i} + E_{diff} + \frac{E_{ES}2(i+2)}{i}\right]}{k_BT}} \tag{4.22}$$

while the transition to step flow growth, which depends on the terrace width $L$, is given by the condition:

$$\lambda \cong L \Longleftrightarrow R = \left(\frac{\lambda_0}{L}\right)^{\frac{2(i+2)}{i}} e^{-\frac{\left[\frac{E_i}{i} + E_{diff}\right]}{k_BT}} \tag{4.23}$$

Here $\lambda = N^{-1/2}$ is the average island separation, $R$ is the deposition rate and $i$ is the critical island size. In Eqs. 4.22 and 4.23, the lengths and densities are normalized to the lattice constant and the density of lattice sites, respectively. It can be seen that both transitions obey Arrhenius type relations, with the difference in slope given by the Ehrlich-Schwöbel barrier and the critical island size.

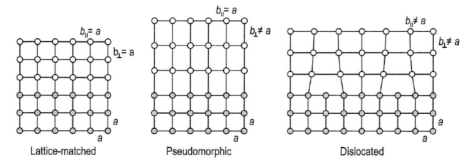

**FIGURE 4.23** Schematic diagram of growth modes (a) lattice-matched growth; (b) strained pseudomorphic growth and (c) relaxed dislocated growth.

## 4.6 Strain Effects in Heteroepitaxy

In heteroepitaxy, a crystalline film of one material is deposited on a crystalline substrate of another material. The alignment of film lattice with respect to the substrate lattice is usually described in terms of parallelism of crystal planes and directions. For example, Al(110)∥Si(100), Al⟨001⟩ ∥Si⟨011⟩ means, that an epitaxial Al film is oriented such, that its (110) plane is parallel to the Si(100) substrate plane and the ⟨001⟩ azimuthal direction of the Al film coincides with the ⟨011⟩ orientation of the Si(100) surface.

Since the substrate and film are of different materials, it is very rare that they have the same lattice constant and ideal lattice-matched, commensurate growth takes place, see Figure 4.23(a). More frequently, the crystal structures of the film and substrate are different. This difference can be in the orientation of the crystalline planes, the crystalline structure as well as the lattice constant. For epitaxy to occur, some accommodation of the structure of the growing film will normally take place. The usual way to measure or quantify the difference in lattice size is referred to as the *misfit*. This quantity is usually defined as the relative difference of their lattice constants and can be expressed in the following form:

$$\varepsilon = \frac{b-a}{a} \tag{4.24}$$

where $a$ and $b$ are respectively the substrate and film lattice constants.

Relatively low misfits can be accommodated by elastic strain; i.e., the deformation of the lattice of the epitaxial film in such a way that the strained film adopts the periodicity of the substrate lattice at the interfacial plane, but can be distorted in the perpendicular direction in order to preserve the volume of the unit cell. This kind of growth is termed *pseudomorphic* growth, as illustrated in Figure 4.23(b). For larger misfits, the elastic strain is relieved by the formation of misfit dislocations at the film-substrate interface, as shown in Figure 4.23(c). It can be seen that the distance between dislocations is regular and given by:

$$d = \frac{ab}{|b-a|} \qquad (4.25)$$

The growth mode that actually occurs in each particular case depends on the relationship between the free energy density associated with strain $(E_\varepsilon)$ and with dislocations $(E_D)$. The general energetics behind the transition from pseudomorphic to relaxed growth is illustrated in Figure 4.24. In Figure 4.24(a), the energy-misfit curve for strained and dislocated films shows the intersection for a certain value or critical misfit, $\varepsilon_c$. Below the critical misfit, a purely strained film will be energetically favorable, while above it the formation of dislocations will be more favorable. In Figure 4.24(b), the effect of film thickness is illustrated. The strain energy increases with film thickness, while the energy of dislocations remains essentially constant. The intersect of these lines yields a critical thickness of $h_c$, which indicates the transition from strained pseudomorphic growth to dislocated relaxed growth. As can be seen in Figure 4.25, the value of the critical thickness changes over several orders of magnitude with a variation of a misfit from a fraction of a percent to a few percents. The example shown here refers to the growth of a $Ge_xSi_{1-x}$ film on Si(100), for which case the misfit value will be controlled by the Ge fraction, $x$.

## 4.7    Deposition Techniques

There is a large variety of thin-film growth techniques available, too many to discuss them all. In the following, we shall give a brief outline of some of the main methods used, which are compatible with an ultra-high vacuum.

### 4.7.1    Molecular Beam Epitaxy (MBE)

In Molecular Beam Epitaxy (MBE), the material for the growing film is delivered to the sample surface by beams of atoms or molecules. During growth, the substrate is typically maintained at an elevated temperature, which is sufficient to ensure that the arriving atoms have the required energy to migrate

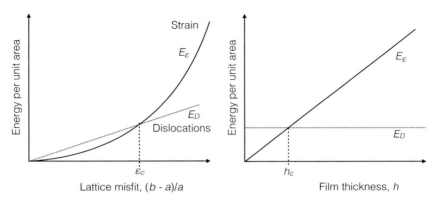

**FIGURE 4.24** Schematic plots of the lattice energy stored at a thin film/substrate interface per unit area (a) as a function of misfit and (b) as a function of the film thickness. Below the critical misfit, $\varepsilon_c$ and below the critical thickness $h_c$, strained pseudomorphous growth is more energetically favorable than relaxed dislocated growth.

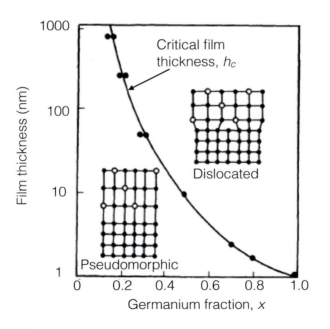

**FIGURE 4.25** Transition between strained pseudomorphic growth and relaxed dislocated growth for $Ge_xSi_{1-x}$ alloy films grown on a Si(100) substrate. The boundary yields the plot of the critical film thickness as a function of the lattice misfit, which is controlled by the Ge content, $x$.

over the surface to lattice sites, while not being too high as to promote interdiffusion between the growing layer and the bulk substrate.

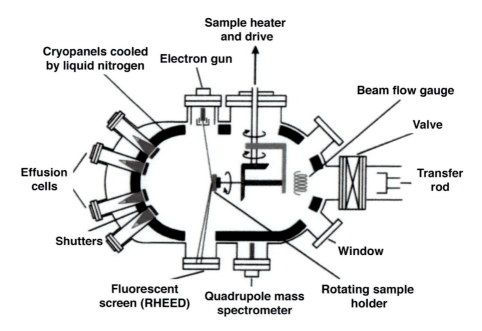

**FIGURE 4.26**   Schematic illustration of the main features of a molecular beam epitaxy (MBE) chamber. The system includes various effusion cells for depositing different materials, an electron gun and fluorescent screen make up the RHEED system for monitoring the sample surface during deposition, all of which is housed in a UHV chamber.

In Figure 4.26, we show a schematic illustration of the MBE growth system. The substrate is mounted on a sample heater with the substrate placed facing various Knudsen cells (or effusion cells), which supply the various materials for deposition. Typically, each source will have a separate shutter, which can be automatically activated via a pre-set growth program and is thus able to grow complex multilayer structures by the controlled/timed shuttering of the various sources. Growth is generally slow in MBE and gives greater control over the deposition process. In-situ monitoring of the film by surface analytical techniques, such as RHEED and AES provide *in-situ* detailed information on the growth of films, as previously discussed.

Molecular beam epitaxy is a versatile thin film deposition technique for the production of well-defined crystalline surfaces, thin films, and multilayers. It is most widely used as a research tool and has extensive use in semiconducting and magnetic metallic thin films and multilayers. While it can be used as a device fabrication technique, it is rather expensive for industrial processing due to the small area coverage.

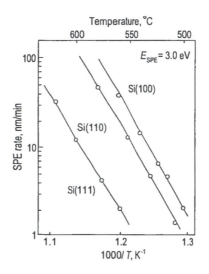

**FIGURE 4.27** Arrhenius plot of the SPE rate for epitaxial crystallization of amorphous Si films deposited in UHV onto Si(100), Si(110) and Si(111) substrates.

## 4.7.2 Solid Phase Epitaxy (SPE)

Solid Phase Epitaxy (SPE) is a specific regime of MBE growth in which an amorphous film is first deposited at lower temperature and is then crystallized by annealing at elevated temperatures. For example, in the case of SPE of Si, the deposition is conducted at room temperature, while annealing treatments take place in the 500–600°C range. The crystallization proceeds by the motion of the amorphous/crystalline interface, from the substrate to the outer surface of the film. Crystallization is a thermally activated process and the speed of the amorphous/crystalline interface motion (SPE rate) is described by an Arrhenius-type expression of the form:

$$V = V_0 e^{-\frac{E_{SPE}}{k_B T}} \tag{4.26}$$

where the activation energy, $E_{SPE}$, for Si-SPE growth is found to be $\sim 3.0$ eV, see Figure 4.27. It is noted that the slopes of the Arrhenius plots are the same for the different substrate orientations, while the pre-exponent factors vary significantly.

Typically, the crystallinity of SPE grown layers are of poorer quality than MBE grown films. However, SPE can be advantageous over MBE in achieving abrupt doping profiles in epitaxial semiconductor films. Many dopants (e.g., Sb, Ga, In) can present difficulties in MBE growth due to surface segregation, which will not normally occur in SPE. Very sharp doping profiles are called delta doping, where dopants are typically confined to a few atomic layers in an intrinsic semiconductor material.

### 4.7.3  *Chemical Beam Epitaxy (CBE)*

Thin-film growth performed by means of surface chemical reactions is defined by the general term Chemical Vapor Deposition (CVD). In this technique, the source material is supplied to the sample surface in the form of gaseous compounds. Precursor gas molecules decompose at the hot surface, leaving behind the desired species, while waste fragments of molecules are desorbed from the surface and pumped away. The group IV and V components are usually supplied as hydrides such as $SiH_4$, $GeH_4$, $AsH_3$, $PH_3$, etc., while group III components can be supplied as metal-organic compounds such as trimethyl-gallium [$Ga(CH_3)_3$, TMGa], triethyl-gallium [$Ga(C_2H_5)_3$, TEGa], triethyl-indium [$In(C_2H_5)_3$, TEIn], etc. Conventionally, the term Metal-Organic CVD (MOCVD) refers to growth conducted at relatively high pressures ($\sim$1-760 Torr). If the growth is performed at UHV pressures, the technique is usually referred to as Metal-Organic MBE (MOMBE) or Chemical Beam Epitaxy (CBE). Due to the large mean free path of the gas molecules at low pressures, CBE growth is not affected by reactions in the gas phase and only surface chemistry determines the growth process.

For CBE growth, a UHV chamber is used, much as in conventional MBE, but the reacting components are supplied by means of a refined gas inlet system, which is an important component in the CBE apparatus. Material flux control is provided by adjusting the input pressure of the gas injection capillary. Sometimes it is required to crack (decompose) the gaseous compound already before it enters the growth chamber. The so-called cracker cell is composed of a heated filament or foil in the capillary stage. In Figure 4.28, we illustrate the growth rate of GaAs film, which increases linearly with the TMGa beam pressure (i.e., the Ga supply), but saturates at a value which strongly depends on the cracking efficiency of the $AsH_3$ inlet capillary, i.e., on the elemental As supply.

In comparison to MBE, CBE growth is far more complex and conventionally requires higher temperatures. However, it provides higher growth rates with the conservation of good film crystallinity. An additional advantage of CBE lies in the ability to supply materials which are inconvenient for evaporation from effusion cells, either due to the extremely low vapor pressure even at very high temperatures (e.g., W, B, Nb) or due to extremely high vapor pressure already at low temperatures (e.g., P). Note that MBE

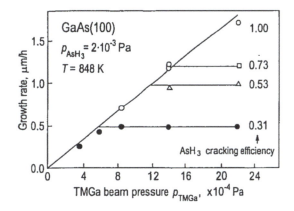

**FIGURE 4.28** Dependence of chemical beam epitaxy (CBE) growth rate for GaAs films on the beam pressure of trimethyl-gallium $Ga(CH_3)_3$, TMGa, for various cracking efficiencies of $AsH_3$.

and CBE techniques can be combined in the same growth process, when one component is supplied in the form of a gaseous compound, while the other is deposited from an effusion cell.

CBE has a major disadvantage in that many of the gases used (such as $AsH_3$) are highly toxic and require expensive gas handling systems, such as scrubbers.

### 4.7.4 Pulsed Laser Deposition (PLD)

Pulsed Laser Deposition (PLD) is a deposition technique where a high power laser (Excimer or Nd:YAG) is focused onto a target of material to be deposited. The material is vaporized or ablated onto a substrate which faces the target material. The deposition typically takes place in a UHV chamber, or in the case of the deposition of oxide materials, in the presence of oxygen gas.

While the basic setup of the deposition technique is relatively simple, the physical phenomena of the laser-target interaction and film deposition can be highly complex. When the laser pulse is absorbed by the target, energy is first converted to electronic excitation and then into thermal, chemical and mechanical energy, resulting in evaporation, ablation, plasma formation. The ejected species from the target expand into the vacuum in the form of a plume, containing many energetic species, including atoms, molecules, electrons, ions, clusters, particulates, etc, before being deposited onto a typically hot substrate, see Figure 4.29.

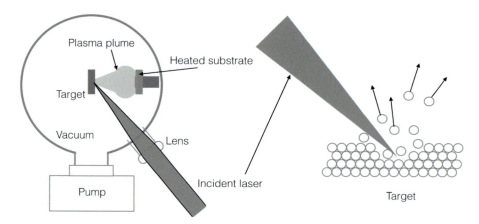

**FIGURE 4.29**   Schematic illustration of the main features of a PLD growth system. Also illustrated (right) is a representation of the target surface under illumination from the incident laser beam, which provokes the ablation of the target surface.

This technique is quite commonly used in the deposition of oxide and ceramic thin films, due to the ability to produce stoichiometric layers of the desired materials. Oxide materials are difficult to produce in conventional deposition processes, such as MBE. Such films typically require very high deposition temperatures, up to 1000°C. Recent developments in this technique have allowed better control of the deposition of thin films, where the use of in-situ analytical tools will allow for the production of high-quality epitaxial layers.

### 4.7.5   Sputter Deposition

Sputter Deposition is another example of a physical vapor deposition; i.e., ejection of a material from a target, which is then deposited onto a closely placed substrate. Sputtered atoms ejected from a target typically have a broad energy distribution, generally in the range from very low energies up to tens of eV. The sputtered ions (typically only a small fraction, $\sim 1\%$, of the ejected particles is ionized) can ballistically fly from the target in a straight line and impact energetically on the substrate (or vacuum chamber wall) and cause re-sputtering or, at higher pressures, collide with gas atoms which will act as a moderator and move diffusively, reaching the substrates and condensing after undergoing the normal random walk migration on the substrate surface. The entire range, from high-energy ballistic impact to low-energy thermalized motion is accessible by changing the background gas pressure.

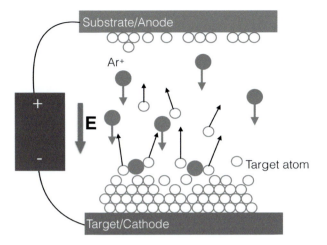

**FIGURE 4.30** Schematic representation of the principal features of sputter deposition. An electric field produced by a bias potential accelerates inert gas ions (Ar) towards the surface of the target (cathode). The ejected target atoms are then emitted from the surface and deposit on the substrate (anode).

The sputtering gas is often an inert gas, such as argon. For efficient momentum transfer, the atomic weight of the sputtering gas should be close to the atomic weight of the target material so, for example, the sputtering of light elements, Ne gas is preferable, while for heavier elements Kr or Xe are used. Reactive gases can also be used to sputter compounds. The compound can be formed on the target surface, in-flight or on the substrate, depending on the process parameters. The availability of many parameters that control sputter deposition makes it a complex process, but also allow experts a large degree of control over the growth and microstructure of the film.

Sputtering sources are usually magnetrons which use strong electric and magnetic fields to trap electrons close to the surface of the magnetron, which contains the target material. The electrons follow helical paths under the action of a magnetic field and therefore undergo more ionizing collisions with the inert gas near the target surface than would otherwise occur. The sputter gas, typically a noble gas such as Ar, is ionized and then accelerated onto the target. This causes the atoms to be sputtered from the target and deposited onto a substrate, placed in close proximity. The sputtered atoms are neutrally charged and will be unaffected by the magnetic trap. Charge build-up on insulating targets can be avoided by the use of RF sputtering, where the sign of the anode-cathode bias is varied at a high (RF) frequency. RF sputtering works well in the production of highly insulating oxide film but only with the added expense of RF power and impedance matching networks. Stray magnetic fields, which leak from ferromagnetic targets may disturb the sputtering process and specially designed sputter guns, which have strong permanent magnets to compensate that of the target, are thus required.

Sputtering has become a commonly used industrial deposition procedure, used in various applications such as the semiconductor industry for IC processing and in antireflection coatings for optical applications. Sputtering is also used for metalizing plastics and in the production of CDs and DVDs.

A variant of the sputtering process is Ion Beam Deposition (IBD) in which an independent ion source is generated and then directed, using an electric field, onto a target for sputter deposition. This has an advantage over conventional sputtering in that the energy and flux of the ions can be controlled independently. Since the flux that strikes the target is composed of neutral atoms, both insulating and conducting targets can be sputtered. IBD has found applications in the manufacture of thin-film heads for disk drives.

## 4.8   Surfactant-Mediated Growth

With the deliberate introduction of a specific impurity, it is possible to alter the growth mode in the desired way. For example, it is possible to change island growth to step-flow or layer-by-layer growth. The surface-active impurity (typically in monolayer or submonolayer amounts) is called a *surfactant*. The classical definition states that a surfactant is a substance that will lower the surface tension, thereby increasing spreading and whetting properties. To be a surfactant, an impurity should satisfy the following experimental requirements:

(1)  Promote 2D growth under conditions where normally 3D growth occurs.
(2)  Be immiscible in the film, such that no appreciable amount of surfactant atoms become incorporated in the bulk of the depositing film itself.

This second requirement can be satisfied in two ways: firstly, the surfactant can be segregated to the film surface (i.e., due to a continuous exchange with the adsorbing atoms, the surfactant atoms always remain on top of the growing film). Secondly, surfactant atoms can be trapped at the buried film/substrate interface. The latter kind of surfactant is sometimes referred to as an *interfactant*.

As an example of interfactant-mediated growth, we can consider the growth of an Ag film on a hydrogen-terminated Si(111) surface, as determined by Time-Of-Flight-Impact-Collision Ion Scattering Spectroscopy

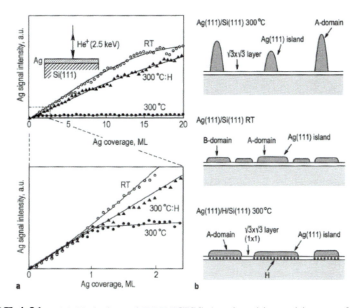

**FIGURE 4.31**   (a) Variation of TOF-ICISS Ag signal intensities as a function of Ag coverage for three deposition conditions. Data points indicated as RT and 300°C are for Ag deposited onto Si(111)-(7 × 7) at room temperature and 300°C, respectively. Those labeled as 300°C:H are for deposition onto H-terminated Si(111) at 300°C. (b) Schematic illustration of the growth modes derived for deposition of Ag onto Si(111)-(7 × 7) surface at 300°C (top), at RT (middle) and onto H-terminated Si(111) at 300°C (bottom). Reprinted figure with permission from K. Sumitomo, T. Kobayashi, F. Shoji, K. Oura, & I. Katayama, (1991). *Phys. Rev. Lett., 66,* 1193. Copyright (1991) by the American Physical Society.

(TOF-ICISS), as illustrated in Figure 4.31. On the clean Si(111)-(7 × 7) surface at 300°C, the growth of an Ag film proceeds according to the Stranski-Krastanov mode (after completion of the $(\sqrt{3} \times \sqrt{3})$ surface phase at 1 ml, rather thick Ag islands develop). The room temperature growth on the Si(111)-(7 × 7) proceeds in a layer-by-layer fashion, but the film contains rotational disorder (A and B type domains). In the case of Ag film growth on the H-terminated Si(111) surface kept at 300°C, the growth proceeds according to the layer-by-layer mode and in a single domain orientation; i.e., it is epitaxial.

Another well-known example of growth with a segregating surfactant is the Sb-mediated epitaxy of Ge on Si(111), see Figure 4.32. Growth of Ge on the bare Si(111)-(7 × 7) surface proceeds in the SK mode, in which 3D islands are formed on top of a 3 ml thick pseudomorphic (5 × 5) reconstructed Ge-Si layer. With ∼ 1 ml of Sb as a surfactant, the formation of 3D islands

Ge/Si(111) no surfactant          Ge/Si(111) Sb surfactant

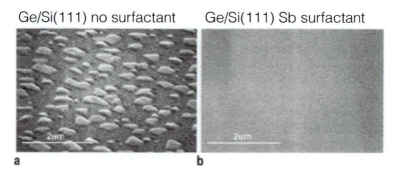

a                                                          b

**FIGURE 4.32**    SEM images of 50 ml thick Ge films grown on Si(111) (a) without surfactant and (b) with a Sb surfactant. Reprinted by permission from P. Zahl, P. Kury, & M. Horn von Hoegen, (1999). *Appl. Phys. A, 69*, 481, ©(1999), Springer, Nature.

is suppressed and continuous Ge film is grown in a layer-by-layer fashion. The surface of the film displays a well ordered Ge(111)-$(2 \times 1)$-Sb(1 ml) reconstruction.

Depending on the particular case, various atomic mechanisms might be responsible for the surfactant effect. The surfactant-induced enhancement of adatom diffusion along the terrace leads to the earlier achievement of step-flow growth. However, in the case of layer-by-layer growth, the reduction of adatom mobility might also lead to improving 2D growth for the following reasons.

Firstly, lower mobility results in an increase in the island density. (Note that the process of site exchange between adatoms and surfactant atoms also leads to a reduction in the adatom migration length.) Secondly, the increase of the surface diffusion barrier leads to an effective decrease of the Ehrlich-Schwöbel barrier, see Figure 4.33, and hence to the enhancement of inter-layer mass transport. Another possibility is a direct decrease of the Ehrlich-Schwöbel barrier by surfactant atoms being incorporated at the step edge (see Figure 4.33). The reduction of the edge atom mobility inducing the ramified island shape also favors smoother growth.

It is worth noting that none of the above mechanisms is related to the classical definition of surfactant in the sense that it reduces the surface free energy. Actually, the surfactant affects the film growth by a modification of the surface kinetics.

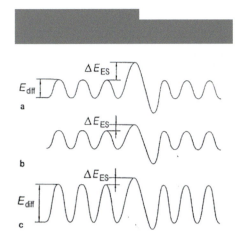

**FIGURE 4.33** Possible surfactant induced modification of the surface potential at a step-edge with the Ehrlich-Schwöbel barrier. The Ehrlich-Schwöbel barrier for a clean surface (a) can be reduced by either locally decreasing the total barrier height (b), or globally increasing the surface diffusion barrier at the terrace (c).

## 4.9 Summary

Thin films perform a large number of tasks in a wide range of applications and are of interest in the areas of semiconductors, magnetism and optical materials. To understand the relation between the physical properties of solid films it is crucial to control the growth of thin films. The nature of the application will determine how much control is required of the film quality, by which we mean the crystalline properties, such as the epitaxy or grain size for polycrystalline films. We are also interested in the nature of the surface and importantly the surface roughness.

Our understanding of the nature of the growth mechanism for a thin film is based on the consideration of the surface energies, relating to both that of the substrate and the overlayer. For good quality flat surfaces, where the roughness is on the level of a few atomic layers, we require that the surface energy of the film be inferior to that of the substrate surface and interface. This means that the whetting condition is satisfied. This type of growth is referred to as Frank-van der Merwe growth. On the other end of the scale, for the case where the surface energy is lower than that of the film and the film-substrate interface, a three-dimensional growth will be favored. This is known as the Volmer-Weber growth mode. An intermediate growth mode is a mixture of the two, where an initial monolayer is deposited and is then followed by 3D growth and is referred to as Stranski-Krastanov growth.

To fully comprehend these mechanisms it is necessary to consider the surface energies as well as the diffusion behavior of adatoms on a surface. This approach provides the main ingredients for understanding the physics involved in the early stages of deposition. When atoms arrive at a surface they can adhere and diffuse across the surface. The probability of meeting other atoms and forming clusters is an important consideration in the nucleation process. This will be fundamentally linked to the rate of arrival of depositing atoms and the surface temperature of the substrate. This will determine whether the islands thus formed will be of a dendritic nature or of a more compact shape. Simulations based on the rate equations allow a good understanding of these processes. Once islands are formed on a surface, the atoms are still mobile, allowing for the coarsening and ripening processes, which occur over time. Specific structures can lead to stable island shapes and are related to the surface energies involved. Steps and kinks on the surface can also play an important role in the process since they will produce a break in the symmetry of the surface potential and can form barriers to the diffusion process.

There are a large number of deposition techniques available. Most are based on good control of the ambient conditions. Frequently this will involve the use of vacuum chambers in which the base pressure is maintained to limit contamination of the film during growth. This is particularly important for methods such as molecular beam epitaxy which use very slow deposition rates. The sources are frequently solid though gas sources are also quite common. The transfer of the material to the substrate will depend on the nature of the materials used. For example, metals can be evaporated by heating, oxide materials require the transfer to maintain stoichiometry and commonly require the use of oxygen atmospheres. This is the case for methods such as pulsed laser deposition. Semiconductors can be prepared using MBE or MOCVD methods.

Once the material has been deposited it will need to be characterized, both to determine its crystalline and morphological properties. These can be performed in-situ using electron diffraction techniques and scanning probe microscopies. We also frequently require a chemical analysis to determine purity and the level of contamination. This can be analyzed using electron spectroscopies. In the next chapter, we will discuss these and a number of other important analytical tools used in the study and characterization of thin films and surfaces.

## 4.10 Problems

(1) The activation energy of surface self-diffusion for a clean Ag(111) surface, 100 meV, is doubled by the presence of a Sb surfactant. Estimate how the diffusion constant, $D$, and the saturation Ag island density, $n_x$, will be changed at room temperature. Assume that the critical island size is $i = 1$. What happens if the critical island size increases to $i = 2$?

(2) An Arrhenius plot of saturation island density, measured for Ag growth on Pt (111), shows two regimes with slopes of 56 and 122 meV for critical islands of $i = 1$ and $i = 2$, respectively. Calculate the Ag−Ag dimer bond energy and migration energy for Ag on Pt (111).

(3) The homoepitaxy of Si on Si(111)-(7×7) was studied using STM. 2D islands are observed with triangular shapes and size distribution as illustrated below. Interpret this result. (N.B. HUC = half unit cell.)

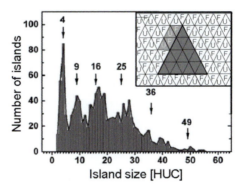

(4) The hopping rate of a N atom on Fe (100) surface is $10^{-3}$ s$^{-1}$ at 300 K and $3 \times 10^{-3}$ s$^{-1}$ at 330 K. Estimate the diffusion coefficient and calculate the activation energy. Take into account that Fe is a bcc crystal with a lattice parameter of 2.87 Å.

(5) After the deposition of equal amounts of aluminum at the same deposition rate onto a silicon surface, the number density of Al islands was found to be $10^{10}$ cm$^{-2}$ at 350°C and $10^{12}$ cm$^{-2}$ at 80°C. Estimate the activation energy for the surface diffusion of Al adatoms on this Si surface. Assume that the critical cluster density is $i = 3$. What does the activation energy for the surface diffusion depend on in physical terms?

(6) Consider the deposition of epitaxial Fe on a GaAs (001) substrate. By considering the misfit of the two crystalline systems, suggest the most probable epitaxial relationship between these materials. You will need to look up the crystalline structures (if you do not already know them) and their lattice parameters.

(7) Random walk diffusion of Ag atoms occurs on the Si(111)-($\sqrt{3} \times \sqrt{3}$) surface. Estimate the mean displacement of the atom in a time of 1 sec, 1 min and 1 hour at 450°C. $D_0 = 10^{-3}$ cm$^2$ s$^{-1}$ and $E_{diff} = 0.33$ eV. The lattice parameter for Si is 5.43 Å.

(8) Describe the physical parameters that determine the island shape in the early stage of film growth. Justify your answer.

(9) Explain what is meant by heteroepitaxy and what are the consequences of a film grown in this manner?

(10) Compare and contrast the PLD and MBE film deposition techniques.

(11) The process of nucleation of clusters can be described in terms of the density of islands of a certain dimension. Express the variation of densities of islands of just one atom with time. Explain the physical significance of each term in the expression.

(12) Explain when RF, as opposed to DC sputtering, should be used.

(13) The surface concentration of a particular GaAs dopant is temperature independent, has an activation energy of 3.5 eV and value of diffusion coefficient at 700°C of $10^{-16}$ cm$^2$ s$^{-1}$. The diffusion length of all processing should be $10^{-6}$ cm. Calculate:

(a) the process time at 700°C,
(b) the area density of impurities in the diffused layer at the same temperature,
(c) the change in process time if an elevated temperature of 800°C is used,
(d) the flux of atoms after 30 minutes at 800°C.

## References and Further Reading

Amar, J. G., & Family, F. (1994). *Mechanisms of Thin Film Evolution*, Yalisove, S. M., Thompson, C. V., & Eaglesham, D. J. (Eds.), Materials Research Society, Pittsburgh.

Amar, J. G., Family, F., & Lam, P.-M. (1994). *Phys. Rev. B, 50*, 8781.

Amar, J. G., & Family, F., (1995). *Phys. Rev. Lett., 74*, 2066.

Antczak, G., & Ehrlich, G., (2010). *Surface Diffusion: Metals, Metal Atoms, and Clusters*, Cambridge University Press, Cambridge, U.K.

Brune, H., H. Röder, Boragno, C., & Kern, K., (1994). *Phys. Rev. Lett., 73,* 1955.

Esch, S., Hohage, M., Michely, T., & Comsa, G., (1994). *Phys. Rev. Lett., 72,* 518.

Frank, C., Novák, J. Banerjee, R., Gerlach, A., Schreiber, F., Vorobiev, A., & Kowarik, S., (2014). *Phys. Rev. B, 90,* 045410.

Hohage, M., Bott, M., Morgenstern, M., Zhang, Z., Michely, T., & Comsa, G., (1996). *Phys. Rev. Lett., 76,* 2366.

Jnawali, G., Hattab, H., Bobisch, C. A., Bernhart, A., Zubkov, E., Möller, R., & Horn-von Hoegen, M. (2008). *Phys. Rev. B 78,* 035321.

Lai, M. Y., & Wang, Y. L., (1999). *Phys. Rev. B, 60,* 1764.

Michely, T., Hohage, M., Bott, M. L. & Comsa, G., (1996). *Phys. Rev. Lett., 70,* 3943.

Michely, T., & Comsa, G., (1991). *Surf. Sci., 256,* 217.

Morgenstern, K., Rosenfeld, G., Poelsema, B., & Comsa, G., (1995). *Phys. Rev. Lett. 74,* 2058.

B. Müller, Nedelmann, L., Fischer, B., H. (1996). Brune and Klaus Kern, *Phys. Rev. B, 54,* 17858.

People, R., & Bean, J. C., (1985). *Appl. Phys. Lett., 47,* 322.

Ohring, M., (1992). *The Materials Science of Thin Films*, Academic Press, San Diego.

Oura, K., V.Lifshits, G., A.Saranin, A., A.Zotov, V., Katayama, M., (2003). *Surface Science: An Introduction*, Springer Berlin–Heidelberg.

Prutton, M., (1998). *Introduction to Surface Physics*, Oxford University Press.

Rosenfeld, G., Morgenstern, K., Esser, M., & Comsa, G., (1999). *Appl. Phys. A, 69,* 489.

Schwöbel, R. L., & Shipsey, E. J., (1966). *J. Appl. Phys., 37,* 3682.

Stowell, M. J., (1970). *Phil. Mag., 21,* 125.

Stroscio, J. A., & Pierce, D. T., (1994). *Phys. Rev. B, 49,* 8522.

Sumitomo, K., Kobayashi, T., Shoji, F., Oura, K., & Katayama, I., (1991). *Phys. Rev. Lett., 66,* 1193.

Venables, J. A., Spiller, G. D. T., & Hanbücken, M. (1984). *Rep. Prog. Phys., 47,* 399.

Venables, J. A., (2000). *Introduction to Surface and Thin Film Processes*, Cambridge University Press, Cambridge, U.K.

Voigtländer, B., Kästner, M., & Smilauer, P., (1998). *Phys. Rev. Lett., 81,* 858.

Witten, T. A., & Sander, L. M., (1983). *Phys. Rev. B, 27,* 5686.

Woodruff, D. P., & Delchar, T. A., (1994). *Modern Techniques of Surface Science,* Second Edition, Cambridge University Press.

Zahl, P., Kury, P., & Horn von Hoegen, M. (1999). *Appl. Phys. A, 69,* 481.

Zangwill, A., (1988). *Physics at Surfaces,* Cambridge University Press.

Zhang, Z., Chen, X., & Lagally, M. G., (1994). *Phys. Rev. Lett., 73,* 1829.

# Chapter 5

# Techniques for Surface and Nanostructure Analysis

## 5.1 Surface Diffraction Techniques

### 5.1.1 *Introduction*

Diffraction techniques using electrons or X-rays are widely used to characterize the structure of surfaces. The structural information is obtained from the analysis of the particles/waves scattered elastically by the crystal. The intensities of diffracted beams contain information on the atomic arrangement within a unit cell. The spatial distribution of the diffracted beams tells us about the crystal lattice and can be interpreted via the reciprocal lattice construction along with the Ewald sphere (We have already introduced the concept of the reciprocal lattice in Chapter 3). We can define the diffraction condition, also referred to as the von Laue condition, with respect to this using the vector notation:

$$\mathbf{k} - \mathbf{k}_0 = \mathbf{G}_{hkl} \Longleftrightarrow \mathbf{k}_0 + \mathbf{G}_{hkl} = \mathbf{k} \tag{5.1}$$

where $\mathbf{k}_0, \mathbf{k}$ are the incident and scattered wave vectors and $\mathbf{G}_{hkl}$ is the reciprocal lattice vector. Since the scattering is elastic:

$$|\mathbf{k}| = |\mathbf{k}_0| \tag{5.2}$$

Eqs. (5.1) and (5.2) express the conservation of momentum and energy, respectively. We will use the Ewald construction to represent the diffraction phenomenon as follows:

(1) Construct the reciprocal lattice of the crystal.
(2) Draw the incident wave vector $\mathbf{k}_0$ with the origin chosen such that $\mathbf{k}_0$ terminates at a reciprocal lattice point.
(3) Draw a sphere of radius $k = \frac{2\pi}{\lambda}$ centered at the origin of $\mathbf{k}_0$. This sphere is referred to as the *Ewald sphere*.

(4) Find the reciprocal lattice points which lie on the surface of the sphere and draw the scattered vectors $\underline{k}$ to these points.

From the construction (see Figure 5.1) we can see that the scattered wave vectors **k** obtained in this way will satisfy Eqs. (5.1) and (5.2).

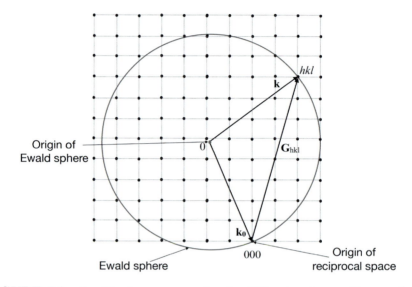

**FIGURE 5.1**    Ewald sphere construction for diffraction from a 3D crystal. We note that the origin (center, 0) of the Ewald sphere corresponds to the origin of the incident electron wave-vector, while the origin of reciprocal space (at reciprocal lattice point 000) is at the end of the incident electron wave-vector. As such, the radius of the Ewald sphere is $|\mathbf{k_0}| = 2\pi/\lambda$. Diffraction conditions, as expressed by the von Laue condition, will be met for all reciprocal lattice points that coincide with the Ewald sphere, see Eq. (5.1).

To demonstrate the use of diffraction at the surface of crystals, we will now consider the most commonly used experimental techniques: LEED and RHEED, both of which are based on the use of monochromatic electron beams. Before doing this, it is interesting to point out that the diffraction condition expressed in Eq. (5.1) can be shown to be equivalent to the Bragg law, expressed as $n\lambda = 2d_{hkl} \sin \theta_{hkl}$. This will be left as an exercise for the student, see the problem section at the end of the chapter.

### 5.1.2    *Low-Energy Electron Diffraction (LEED)*

The use of low energy electrons for surface analysis stems from the following:

The de Broglie wavelength of an electron is:

$$\lambda = \frac{h}{\sqrt{2mE}} \quad ; \quad \lambda\,[\text{Å}] = \sqrt{\frac{150}{E\,[\text{eV}]}} \qquad (5.3)$$

In the typical range of energies used in LEED (30-200 eV), the electrons have an associated wavelength of 1–2 Å, which satisfies the atomic diffraction condition; i.e., $\lambda$ is of the order of interatomic distances.

The mean free path of the low energy electrons is very short, of the order of a few atomic layers. Therefore, most elastic collisions occur within the uppermost atomic layers of the sample.

Therefore LEED provides principally information regarding the surface 2D structure and is said to be sensitive to the surface.

The corresponding 2D diffraction condition will be modified since the crystal periodicity in the normal direction to the surface is missing, and Eq. (5.1) becomes:

$$\mathbf{k}'' - \mathbf{k}_0'' = \mathbf{G}_{hk} \qquad (5.4)$$

The law of conservation of momentum concerns only the wave-vector components parallel to the surface and implies that the scattering components $\mathbf{k}'' - \mathbf{k}_0''$ must be equal to the vector of the 2D reciprocal lattice $\mathbf{G}_{hk}$. The wave vector components normal to the surface are not conserved in this process.

The Ewald construction modified for 2D diffraction is shown below (Figure 5.2). In contrast to the 3D reciprocal lattice points, the reciprocal lattice is made up of rods which lie perpendicular to the surface and are attributed to every 2D reciprocal lattice point—a 2D lattice can be conceived as a 3D lattice with infinite periodicity in the normal direction $|\mathbf{c}| \to \infty \Rightarrow |\mathbf{c}^*| = 0$; i.e., the reciprocal lattice points along the normal direction are infinitely dense, thus forming rods. The incident wave vector $\underline{k}_0$ terminates at the reciprocal rod, 00. The intercepts of the rods with the Ewald sphere define the scattered wave vectors $\mathbf{k}$ for each of the diffracted beams.

The basic LEED experiment allows direct observation of the diffraction pattern, as illustrated in Figure 5.3.

The main components are electron gun, sample holder and a hemispherical fluorescent screen.

The electrons emitted by the cathode (at $-V$) are accelerated to an energy of eV within the gun and propagate and scatter at the sample surface

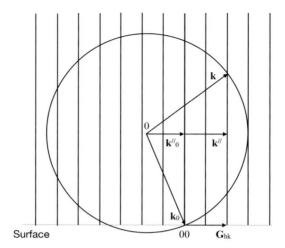

**FIGURE 5.2** Ewald sphere construction for diffraction from a 2D crystal or surface lattice. The origin of the Ewald sphere remains the origin of the incident electron wave-vector, while that for the reciprocal space corresponds to its endpoint, at reciprocal lattice rod 00. The diffraction condition is now satisfied for all reciprocal lattice rods, as expressed by Eq. (5.4).

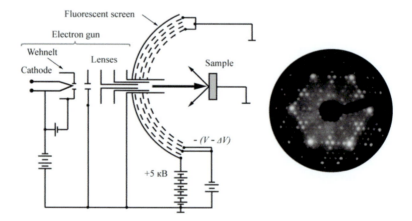

**FIGURE 5.3** Schematic diagram of the standard four-grid LEED apparatus and a LEED pattern of the $Si(111) - (7 \times 7)$ surface reconstruction. The LEED geometry, in the reverse-view arrangement shown here, means that a shadow is cast of the electron gun.

in the field-free space. Second and third grids are used to reject inelastically scattered electrons. The potential of the second and third grids is close to that of the cathode, but lower in magnitude (at $-(V - \Delta V)$). The greater

$\Delta V$, the brighter the LEED pattern, but the higher its background. So the retarding voltage is adjusted to get a LEED pattern with the highest spot-to-background contrast. The fourth grid is at earth and screens other grids from the field at the screen which is biased to $\sim +5$ kV, so after the elastically scattered diffracted electrons pass the retarding grids, they are then accelerated to high energy to cause fluorescence at the screen where the diffraction pattern can be observed.

Due to the geometry of the LEED set-up and the Ewald construction, the diffraction pattern observed at the screen is essentially a view of the surface reciprocal lattice. The observed spots are indexed as of the reciprocal lattice itself, *hk*, where the specular spot is taken as (0,0). The number of spots that are observed in a given LEED pattern depends on the size of the Ewald sphere. By increasing electron energy, we will decrease the electron wavelength, thus increasing the radius of the Ewald sphere. Consequently, the diffracted spots move closer together and more spots will be observed.

In general, spot sharpness will give an indication of the structural perfection of the sample surface. Sharp features are associated with a well-defined surface, while the presence of structural defects and crystallographic imperfections will lead to a broadening and weakening of spot intensity and a corresponding increase in the overall background intensity. The absence of any diffraction spots is indicative of a disordered or amorphous surface, which could be coated with residual gas species.

However, the principal interpretation of the LEED pattern concerns the geometrical distribution of the spots. In the following figures (Figures 5.4–5.6), we show some examples of LEED patterns for various real space lattice superstructures.

Additional information can be extracted by probing the spot profile; intensity distribution across the width of the spot. Such information will typically concern surface imperfections, since any deviation from an ideal 2D periodicity disrupts the sharp spot profile.

It should be noted that even for a perfect periodic structure, the spot will have a finite width due to (i) finite coherence length ($\sim 100$ Å) and (ii) instrumental limitations—finite energy and angular spread of the primary electron beam.

Examples of surface imperfections and the corresponding reciprocal space and diffraction peak profiles are illustrated in Figure 5.7.

The expected spot profile will be obtained by the intersection of the Ewald sphere by the modified reciprocal rod shape.

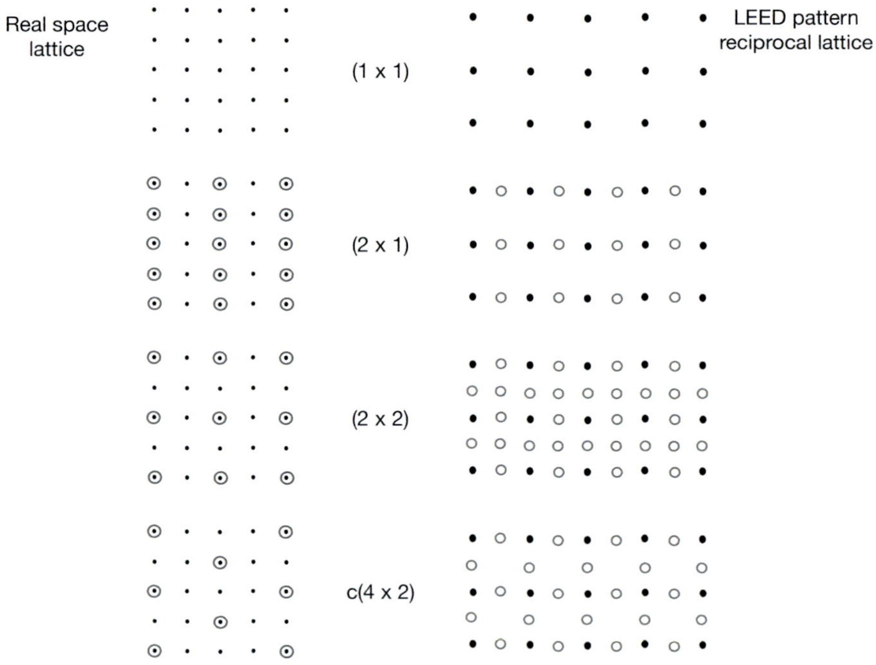

**FIGURE 5.4**  Examples of real space superlattices on a substrate with a square lattice and corresponding LEED patterns for (1 × 1), (2 × 1), (2 × 2) and c(4 × 2) surface reconstructions. In real space, substrate lattice points are represented by black dots and the superstructure by open circles. In the LEED patterns, the main spots are shown as black dots, extra spots are shown as open circles.

The use of spot position or profile does not give information about the atomic arrangement within the surface unit cell. In order to do this, it is necessary to perform an analysis of spot intensity ($I$) as a function of the primary electron energy ($\propto V$)—a so-called I–V curve. The analysis is rather complex and requires the use of the dynamic theory of scattering, which also accounts for multiple scattering events of low energy electrons. A comparison between theory and experiment is required to make any reasonable assessment of the atomic arrangement within the unit cell. We will not extend our analysis here, but should be aware of this important aspect of LEED surface structure analysis.

## 5.1.3  *Reflection High-Energy Electron Diffraction*

Reflection high-energy electron diffraction (RHEED) provides an alternative electron diffraction technique and is widely used in UHV for MBE growth

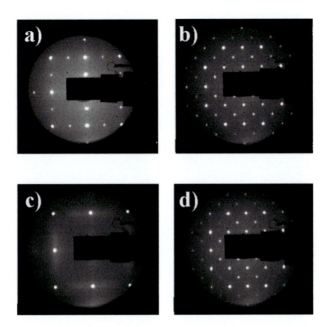

**FIGURE 5.5**   LEED patterns obtained at 56 eV from various surfaces: (a) clean Si(001)-(2 × 1); (b) Si(001)c(4 × 4) by exposing to 900 l ethylene at T = 600°C; (c) c(4 × 4) converted to a (1 × 1) pattern by exposing to 22,500 l hydrogen at room temperature; (d) preceding surface after a further 20 min annealing at T = 600°C.

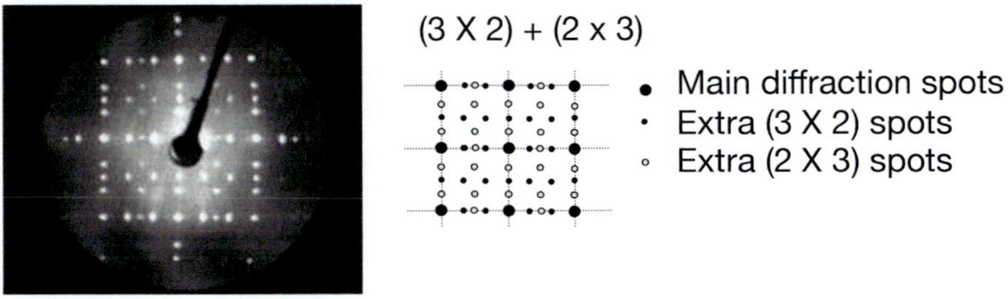

(3 X 2) + (2 x 3)

● Main diffraction spots
· Extra (3 X 2) spots
○ Extra (2 X 3) spots

**FIGURE 5.6**   LEED pattern for a cubic (100) surface exhibiting both (3 × 2) and (2 × 3) surface reconstructions. The interpretation of the pattern is illustrated on the right of the LEED image.

of ultra-thin films and multilayer systems. Despite using high-energy electrons, the RHEED technique is a surface-sensitive technique using a grazing angle $(1-3°)$ incidence of the primary electron beam, such that the detection must be made also at grazing incidence to observe the diffracted beams. The high energy of the incident electrons means that the mean free path is

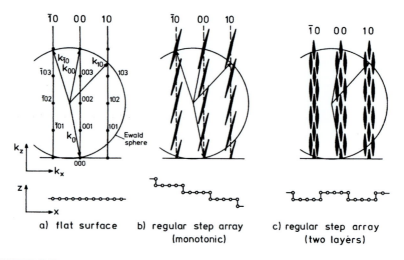

a) flat surface    b) regular step array    c) regular step array
(monotonic)    (two layers)

**FIGURE 5.7**    Reciprocal space and Ewald sphere construction for (a) a flat surface, (b) a regular stepped or terraced surface, and (c) a regular step array (After M. Henzler, (1977)).

correspondingly high. However, the grazing incidence maintains a low penetration depth and hence its surface sensitivity. For 50–100 keV electrons, the mean free path is of the order of ~ 1000 Å, which at a grazing angle of incidence of $1°$ gives a penetration depth of about 10 Å.

The Ewald construction for the RHEED experiment is shown in Figure 5.8.

The diffraction condition for RHEED is the same as that for LEED, as given by Eq. (5.4). It will be noticed that since the electron primary energy is much higher in RHEED than LEED, the radius of the Ewald sphere will be relatively larger and will thus sample a greater area of the reciprocal lattice. Also, the grazing incidence will give an appreciable elongation or streaking of the diffraction spots since both the Ewald sphere and the reciprocal lattice rods have a finite thickness due to instrumental limitations and surface imperfections.

The experimental set-up for the RHEED experiment is shown in Figure 5.9. Typically the electron gun will produce a monochromatic beam of energies in the range 5–20 keV, which is directed at grazing angles of 1–5° onto the sample surface. The energy and wavelength are related, and a small correction is usually taken into account for the high energies, and Eq. (5.3) can be written in the form:

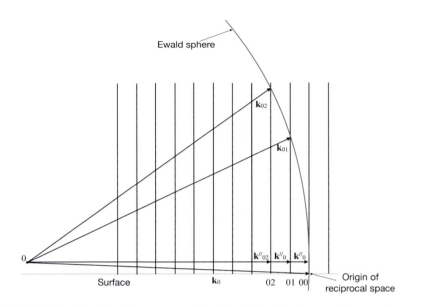

**FIGURE 5.8** Ewald sphere construction for diffraction from a 2D crystal or surface lattice in the RHEED geometry. The origin of the Ewald sphere remains the origin of the incident electron wave-vector, while that for the reciprocal space corresponds to its endpoint, at reciprocal lattice rod 00. The diffraction condition is satisfied for all reciprocal lattice rods, as expressed by Eq. (5.4).

$$\lambda = \frac{h}{\sqrt{2meV\left(1 + \frac{eV}{mc^2}\right)}} \qquad (5.5)$$

where $V$ is the potential of the electron gun.

The sample is mounted on a rotatable manipulator to allow the observation of diffraction patterns along various crystallographic axes. The fluorescent screen is placed diametrically opposite to the electron gun, at a distance L from the sample. The primary electron energy is sufficient that no further acceleration is required (as in LEED). Furthermore, no energy filtering is necessary since the intensity of the diffracted beams is much higher than that of the background of inelastically scattered electrons.

Due to the difference in scattering geometry, the RHEED diffraction pattern differs in various aspects to that of the LEED pattern. In the following Figure 5.10, we show the RHEED patterns for the Si(111) $(7 \times 7)$ surface, recorded at the azimuthal direction : $[\bar{1}2\bar{1}]$. Also shown is the sketch of the 2D reciprocal lattice and the observed RHEED pattern, to aid interpretation.

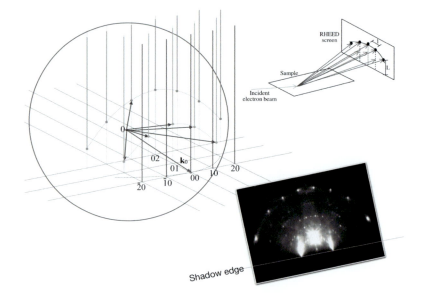

**FIGURE 5.9**    Ewald sphere construction for diffraction from a 2D crystal or surface lattice in the RHEED geometry. We also illustrate schematically the arrangement for the RHEED experiment.

In Figure 5.11, we show the RHEED patterns for the Si(111) $(\sqrt{19} \times \sqrt{19})$ surface.

One of the most striking differences in the appearance of the RHEED pattern is the existence of the ring structure of the diffracted spots. These are called Laue zones. In cases where the surface is flat, the appearance of Laue zones is common and provides a good signature of surface perfection. The spots are more elongated for the lower Laue zones due to the geometrical construction of the reciprocal lattice rods and Ewald sphere. Higher-order Laue zones show a more round spot structure. In addition, a RHEED experiment will require patterns recorded at least two azimuthal orientations to establish the surface structure. In addition to the spot structure, we see several bright bands and lines. This is called the Kikuchi structure and arises from multiple scattering inside the bulk of the crystal, and is typical for high purity substrates.

RHEED can be used for quantitative structural analysis; i.e., checking models of surface arrangement. The analog of the LEED I-V curves is RHEED rocking curves, where the variation of spot intensity is measured as a function of the primary electron beam angle of incidence. Dynamic scattering theory is again required for a detailed analysis.

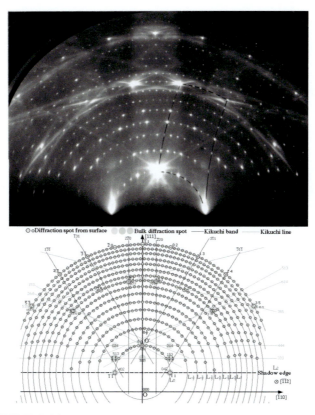

**FIGURE 5.10**    RHEED image for the Si(111) – (7 × 7) surface.

One of the major advantages of RHEED over LEED is its suitability of measurement during the thin film deposition process. The fact that both electron gun and fluorescent screens are placed on either side of the sample means that the deposition process is not impeded and allows in-situ analysis during deposition. By monitoring the RHEED pattern during deposition, changes in crystalline periodicity and structure can be observed as well as diffraction (or specular) spot intensity as a function of growth time. In the figure below (Figure 5.11), oscillations of the RHEED beam (specular) spot intensity are shown as a function of the overlayer thickness. One oscillation of the intensity corresponds to the formation of a single atomic layer. The observation of such RHEED oscillations can be used to calibrate the thickness of thin-film growth. This is often used in MBE growth.

When a surface or overlayer has a rough appearance then the RHEED pattern will alter since the electron beam will penetrate essentially into a 3D

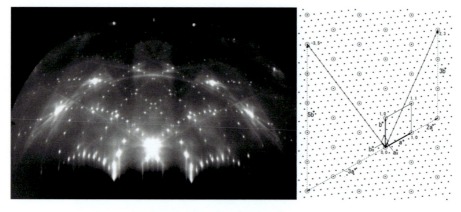

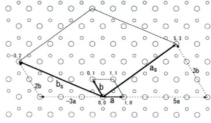

**FIGURE 5.11**     RHEED image for the $Si(111) - (\sqrt{19} \times \sqrt{19})$ surface.

structure giving rise to a 3D diffraction pattern; i.e., one from the transmission. These can still be used for structural analysis of samples.

### 5.1.4   *Grazing-Incidence X-Ray Diffraction*

While other diffraction techniques are available, such as grazing incidence X-ray diffraction (GIXRD), they are broadly related to the electron diffraction techniques discussed above, with some specific differences in experimental set-up and analysis. However, in practice LEED and RHEED are by far the most extensively used. GIXRD is mainly an ex-situ method and is used to study the crystalline properties of thin-film structures.

## 5.2   Electron Spectroscopies

### 5.2.1   *Introduction*

Electron spectroscopies are probes of the electronic structure of the surface through the analysis of the energy spectra of secondary electrons emitted

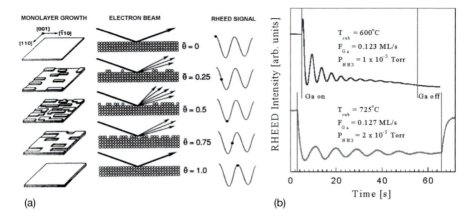

MONOLAYER GROWTH    ELECTRON BEAM    RHEED SIGNAL

(a)                                      (b)

**FIGURE 5.12** (a) Schematic illustration of the intensity of the specular spot as a function of the coverage for a growing thin film. (b) RHEED intensity oscillations were recorded using the specular $<\bar{2}110>$ beam. These oscillations were observed only under excess $NH_3$ (Ga-limited) conditions, indicating growth by island nucleation. After closing the Ga shutter the RHEED intensity recovers above approx. 700°C. Continued growth at low temperatures eventually leads to a rough surface, but the samples can be annealed to their initial RHEED intensity.

from the sample. The secondary electrons are generally created by bombarding the surface with electrons or photons. The typical energies of secondary electrons analyzed in surface electron spectroscopy usually fall in the range of 5–2000 eV. The surface sensitivity of electron spectroscopy stems from the fact that electrons within this range are strongly scattered in solids. The electron inelastic mean free path as a function of kinetic energy is shown below (Figure 5.13). Although the data are energy and material dependent, the magnitude of the inelastic mean free path in the energy range shown is of the order of several tens of Å and in a favorable region (20–200 eV) is less than 10 Å.

The main techniques of electron spectroscopy are Auger Electron Spectroscopy (AES), Electron Energy Loss Spectroscopy (EELS) and Photoelectron Spectroscopy (PES). There are many variations of these techniques, but the basic principles are common. In PES, secondary electrons are generated by irradiation of the sample surface by photons (UV $\rightarrow$ UPS, X-rays $\rightarrow$ XPS). In AES and EELS, the surface is bombarded with electrons. To specify the difference between these techniques, we must consider the secondary electron energy distribution.

If a solid is bombarded by mono-energetic electrons of energy $E_p$, a typical secondary electron spectrum $N(E)$ shows several features:

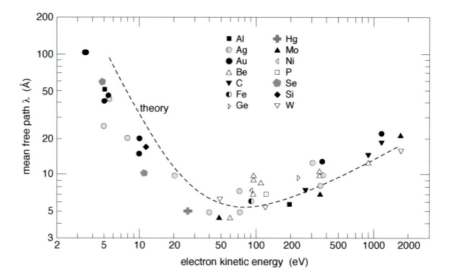

**FIGURE 5.13**   Inelastic mean free path of electrons as a function of their kinetic energy. The compilation of data for a range of materials illustrates the quasi-universal nature of this curve. This is a result of the fact that the principal interaction mechanisms between the electrons and the solid are the excitation of plasmon waves whose energy is determined by the electron density in the solid.

(1)  A sharp elastic peak at the primary electron energy;
(2)  A broad structureless peak near $E = 0$, which extends as a weak tail up to $E_p$;
(3)  A long region with relatively few electrons between $E_p$ and $E = 0$ with a number of small peaks. The nature of these small peaks falls into two categories:

Related to the elastic peak, so that if $E_p$ is raised by $\Delta E_p$, all peaks in this group move by $\Delta E_p$ – these are called loss peaks, due to the primary electrons that have lost discrete amounts of energy, say for discrete energy level ionization or plasmon excitation. These are analyzed as electron energy loss spectroscopy (EELS).

The energy position of the peaks is fixed and independent of the primary energy of the excitation source. The most important of these are due to Auger electrons which are of interest in Auger electron spectroscopy (AES). Figure 5.14 illustrates the points made above.

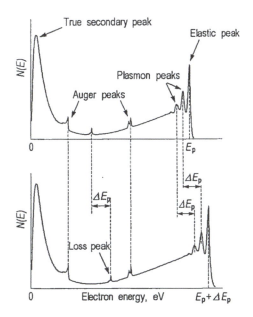

**FIGURE 5.14** Schematic illustration of the electron energy spectrum, showing different types of electron backscattering mechanisms from a solid.

## 5.2.2 *Electron Energy Analyzers*

The detection of electrons is central to the problem of recording the secondary electron energy for spectroscopies. Depending on the task, the spectra are recorded in the form of $N(E)$ – number of electrons vs energy – and also in the first and second derivatives; $\frac{dN(E)}{dE}$ and $\frac{d^2N(E)}{dE^2}$. The device used for the recording of electron spectra is called the electron energy analyzer. The objective is to separate out the entire secondary electron flux so only electrons with definite energy called the pass energy are detected. The pass energy is controlled by the voltages applied to the electrodes of the analyzer. To restore the whole spectrum, the electrode voltages are varied (swept) and the electron current is recorded as a function of the pass energy.

Most analyzers used for electron spectroscopy utilize electrostatic forces for electron energy separation and can be divided into two main classes:

(1) Retarding field analyzers; and
(2) Deflection analyzers.

The retarding field analyzer (RFA) works by repelling electrons with an energy less than $E_0 = eV_0$, where $V_0$ is the voltage applied to the grids. Therefore, the collector receives an electron current:

$$I(E) \propto \int_{E_0}^{+\infty} N(E)dE \qquad (5.6)$$

This corresponds to the shaded area illustrated in Figure 5.15(a).

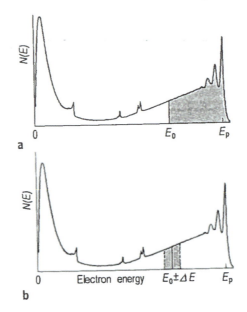

**FIGURE 5.15**   Schematic illustration of the electron energy spectrum.  The shaded area of the secondary electron spectrum corresponds to electrons selected by (a) a retardation field analyzer (all electrons with energies greater than $E_0$) and (b) deflection analyzers (electrons in the energy window $E_0 \pm \Delta E$).

The most common RFA is a four-grid analyzer, which uses conventional four-grid LEED optics as shown below (Figure 5.16). In this case, it is usually referred to as a LEED-AES device. The second and third grids of the analyzer are used as retarding grids for energy analysis. The fluorescent screen is used as a collector of electrons emitted by the sample over a wide solid angle.

In the deflection type analyzers, electrons in a narrow energy window are collected, see Figure 5.15(b). The electron separation arises from the use of geometries in which only the electrons with the desired energy move along a specific trajectory leading to the collector. This is ensured by an electro-

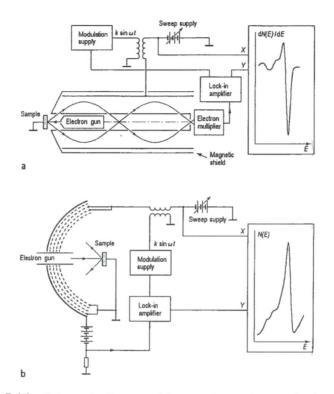

**FIGURE 5.16**   Schematic diagram of the experimental set-up for Auger electron spectroscopy with the most widely used analyzers: (a) cylindrical mirror analyzer (the CMA illustrated is of the double-stage type, with an electron gun integrated inside the inner cylinder). and (b) four-grid retarding field analyzer (LEED-AES). The synchronous detection of the signal at the frequency $\omega$ of the modulation supply results in recording the spectra in the $dN(E)/dE$ mode in the case of the CMA and the $N(E)$ mode in the case of the four-grid RFA. To obtain the $dN(E)/dE$ spectra in the latter case one uses detection at the frequency $2\omega$.

static field crosswise to the direction of electron motion. To enhance analyzer efficiency, they are designed so that electrons having the same energy but entering the input aperture at different angles are all focused at the output aperture. The most commonly used deflection type analyzers are:

(1)  cylindrical mirror analyzer (CMA);
(2)  concentric hemisphere analyzer (CHA); and
(3)  27°-angle analyzer.

In the CMA, electrons leaving the target enter the region between the two concentric cylinders through a conical annulus. For a negative voltage $V_a$

applied to the outer cylinder with the inner cylinder grounded, electrons are deflected in the region between the cylinders and only electrons with a certain energy $E_0$ pass through the output aperture (see Figure 5.17(a)) and are collected by an electron multiplier. The energy $E_0$ is proportional to $V_a$ and is determined by the analyzer geometry (the ratio $\frac{V_a}{E_0}$ usually lies between 1 and 2). To ensure focusing of the electrons, the location of the sample and the windows in the cylindrical electrode guarantee the electron entrance angle of $42°19'$. The CMA is characterized by high sensitivity but moderate energy resolution. To enhance resolution a double stage CMA is used with two successive analyzer units. CMAs are widely used in AES with an electron gun integrated into the CMA, see Figure 5.17(a).

A CHA is shown below (Figure 5.17(b)). The main elements of a CHA are two metallic concentric hemispheres. The outer is biased negatively with respect to the inner hemisphere to produce an electrostatic field which balances the centrifugal force of the electrons on their trajectory. The entrance and exit apertures are circular holes. The efficiency of the CHA arises from the focusing condition for electrons deflected through an angle of $180°$. CHA is widely applied to PES and AES, especially when the angular resolution is required.

The $127°$-angle analyzer (see Figure 5.17(c)), also known as cylindrical sector analyzer, uses an operational principle similar to that of the CHA. As electrodes, it uses two concentric cylinder sectors with an angle of $127°17'$, which satisfies the focusing condition. The $127°$-angle deflector is characterized by high energy resolution, but relatively poor transmission. It is mainly used in high-resolution EELS (HREELS) both as a monochromator and as an energy analyzer.

As for the energy resolution, deflection type analyzers can be used in two modes:

(1)  constant $\frac{\Delta E}{E}$ mode; and
(2)  constant $\Delta E$ mode.

The constant $\frac{\Delta E}{E}$ mode is used when the electron pass energy $E_0$ is swept by the variation of the voltage applied to the electrodes. In this case, the energy window $\Delta E$ grows continuously with electron energy, leaving the ratio $\frac{\Delta E}{E}$ constant. Its value is mainly determined by the angular dimensions of the input and output apertures. The electron current collected in constant $\frac{\Delta E}{E}$ mode is proportional to $EN(E)$; i.e., $I(E) \propto EN(E)$.

In the constant $\Delta E$ mode, the electron pass energy $E_0$ is held constant, thus preserving the constant resolution $\Delta E$. In this case, the electron spec-

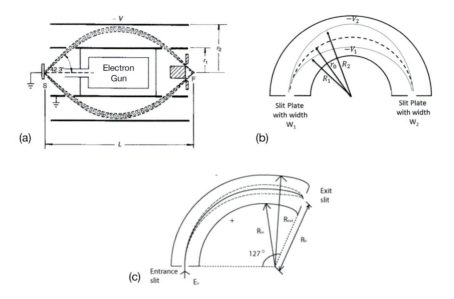

**FIGURE 5.17**   Illustration of the principal deflection analyzers: (a) Cylindrical mirror analyzer (CMA); (b) concentric hemisphere analyzer (CHA) and (c) 127°-angle cylindrical sector analyzer. In all these analyzers, the outer electrode is biased negatively with respect to the inner electrodes.

trum is constantly shifted through a fixed energy window $\Delta E$ by varying the acceleration or deceleration voltage in front of the analyzer.

### 5.2.3   Auger Electron Spectroscopy

Auger electron spectroscopy (AES) is one of the most commonly used techniques for the analysis of surface chemical composition by measuring the characteristic energies of Auger electrons. It was developed in the late 1960s, deriving its name from the effect first observed by Pierre Victor Auger (1899–1993), in the mid-1920s.

The principle of the Auger process is illustrated in Figure 5.18. The primary electron, with energy typically in the range 2–10 keV, creates a core hole; i.e., the atom is ionized, in the example given this corresponds to an electron from the K-shell. This hole is then filled by a less tightly bound level, in this case, an $L_1$ electron, and a further de-excitation can be performed by a relaxation of the energy by either:

(1)  Auger emission (non-radiative transition); and
(2)  X-ray fluorescence (radiative transition).

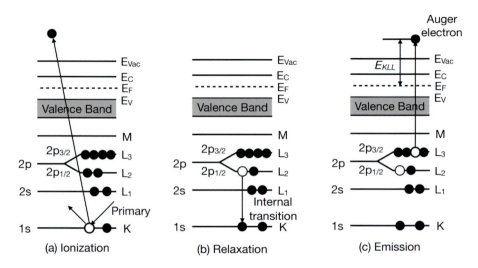

**FIGURE 5.18**   Schematic illustration of the Auger process, showing (a) the ionization of the atom via the primary electron beam; (b) the partial relaxation of the atom from its excited stated via the de-excitation with the decay of an $L_2$ electron to the $K$ shell, and (c) emission of an Auger electron from the $L_3$ level.

For low energy transition ($\leq$ 500 eV), especially in the case of lighter elements, X-ray fluorescence becomes negligible and consequently, Auger emission is favored. Only at around 2 keV does X-ray production become comparable to the Auger efficiency.

The Auger process involves various electronic states; the initial-one-hole state and the final-two-holes states are both excited states. All three electrons are involved in the transition and the Auger process can take place in all elements, besides H and He.

In the Auger process, after the initial ionization (K-level in Figure 5.18(a)), the $L_1$ electron de-excites the atom partially (relaxation, Figure 5.18(b)). In this process, a further electron is released, in the example an $L_3$ electron (emission, Figure 5.18(c)).

This shake-up electron is emitted and is the Auger electron of interest, see Figure 5.19(a). The labeling of this process is $KL_1L_{2,3}$, or simply KLL, since these levels are involved in the emission process. If the Auger process occurs in the solid and the valence band electrons are involved, the process will use the V notation. The example is given in Figure 5.19(c) of the previous figure shows the $L_{2,3}$VV, or LVV, Auger transition. The excitation of the atom via the primary electron beam can also give rise to X-ray fluorescence,

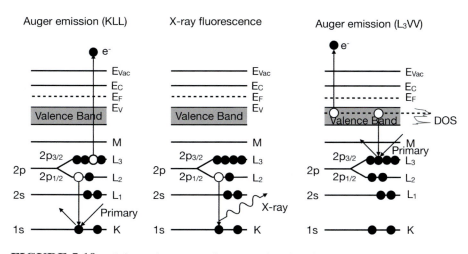

**FIGURE 5.19**   Schematic energy diagrams showing the competing paths (a) and (b) for energy dissipation with Si as an example. The $KL_2L_3$ Auger electron is about 1591 eV ($E_{KL_2L_3} \simeq E_K - E_{L_2} - E_{L_3}$) and the X-ray photon energy is 1690 eV ($\hbar\omega = E_K - E_{L_2}$). (c) Diagram for the $L_3VV$ Auger transition.

as illustrated in Figure 5.19(b), and is a competing de-excitation process with the Auger emission mechanism.

An Auger transition is primarily characterized by the location of the initial hole and the final two holes. Thus the kinetic energy $E_{KL_1L_{2,3}}$ of the ejected Auger electron—example (a)—can be estimated from the binding energies of the levels involved:

$$E_{KL_1L_{2,3}} \simeq E_K - E_{L_1} - E_{L_{2,3}} - \phi \qquad (5.7)$$

where $\phi = E_{\text{vac}} - E_F$ is the work function of the material.

It should be noted, that Eq. (5.7) is only an approximation, which does not take into account that the final emission occurs from an ion and not a neutral atom. This produces a shift of the final energy levels downwards and will thus affect the energy of the Auger electron. This will be due to the change of electron screening effects from the nucleus. Also, the hole-hole interaction in the final state configuration will depend on whether the holes are both in core shells, one in a core-shell and one in a band, or both in a band. A better approximation for non-band transitions would be; ABC transition:

$$E_{ABC}(Z) \simeq E_A(Z) - \frac{1}{2}[E_B(Z) + E_B(Z+1) + E_C(Z) + E_C(Z+1)] \qquad (5.8)$$

The use of binding energies for combinations of the atomic number $Z$ and $(Z+1)$ makes a crude allowance for the hole-hole interaction. For band-state transitions, Eq. (5.7) is a reasonable approximation.

The experimental set-up for the AES is shown in Figure 5.16 and consists of an electron gun, an energy analyzer, and data processing electronics. The electron gun provides the primary electron beam with typical energies of 1–5 keV. The most commonly used energy analyzers for AES are a cylindrical mirror, hemispherical and four-grid analyzers.

The small Auger peaks (see Figure 5.20) are superposed on a high background of true secondary electrons—electrons that undergo multiple losses of energy. To improve signal-to-noise and suppress the large background, Auger spectra are frequently recorded in derivative mode $\frac{dN(E)}{dE}$. Differentiation is performed using a modulation of the analysis energy—perturbation voltage $\Delta u \propto \sin \omega t$ (see Figure 5.20).

The detection current in this mode is given by:

$$I(E_0 + k\sin \omega t) \simeq I_0 + \frac{dI}{dE}k\sin \omega t - \frac{d^2I}{dE^2}\frac{k^2}{4}\cos 2\omega t \qquad (5.9)$$

So $\frac{dN(E)}{dE}$ can be obtained with CMA detecting output signal at $\omega$ while for LEED-AES the detection is at frequency of $2\omega$.

Since the emitted Auger electron has well-defined kinetic energy directly related to the nature of the atomic emitter, elemental identification is possible from the positions of the Auger peaks, providing a chemical signature of the emitter species. Values of Auger energies are widely available for all the elements of the periodic table (excluding H and He) and are shown in Figure 5.21.

As seen, each element typically has a series of peaks which have characteristic energies. In addition, the Auger spectrum of a particular element may contain information on the chemical bonding states of atoms. The change in the chemical environment of a certain atom changes the electronic binding energies—a so-called chemical shift—and leads to a redistribution of the electron density of states in the valence band. These changes are reflected in the changes in the Auger peak position and/or peak shape. However, a quantitative analysis of such data is compounded by the fact that three electrons are involved in the Auger process.

Other applications of AES include Auger mapping and Auger compositional depth profile analysis. In Auger mapping—also known as Scanning Auger Microscopy (SAM) or Multichannel SAM (MULSAM)—is the spatial distribution of an element across the sample surface can be determined.

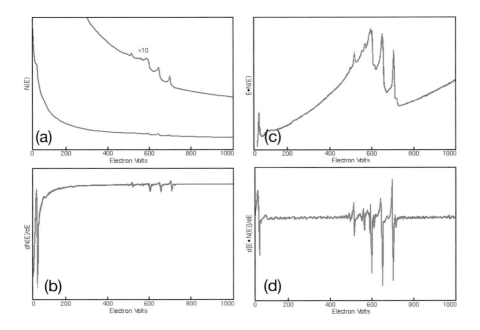

**FIGURE 5.20**     (a) Auger spectrum (of iron) plotted as total electron signal, $N(E)$, versus electron energy.     The Auger peaks are obscure even after expanding the vertical scale.   (b) Plotting the spectrum as the differential of the electron signal, $dN(E)/dE$, clarifies some of the spectral details. (c) Both of the first two plots under emphasize the high energy end of the spectrum.  Multiplying the total electron signal by the electron energy, $E \times N(E)$, accentuates the high energies. (d) Finally, plotting the differential, $d[E \times N(E)]/dE$, of the above function provides for a clear display of the features in an Auger electron spectrum. This $d[E \times N(E)]/dE$ format is the most common mode for presenting Auger data.

To acquire such an Auger map, the intensity of a particular Auger peak is monitored as a function of the electron beam position as it is scanned across the sample surface. The advantage in using the primary electron beam is that it can be focused down to give a good spatial resolution of the order of tens of nanometers. The variation of sample composition with depth can also be studied using AES in conjunction with ion beam sputtering. The intensity of a particular Auger peak can be monitored as a function of sputter time and, if correctly calibrated, can be used to measure the depth and consequently thickness of nanostructures-films.

   AES also provides the potential for quantitative analysis, i.e., the number of atoms of a certain species present on the sample surface can be calculated based on the measured Auger intensity.  The general formula for the

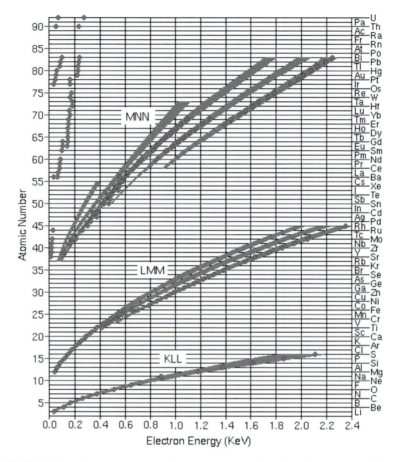

**FIGURE 5.21**  Principal Auger electron energies for the elements used in quantitative analysis. Three main series, KLL, LMM, and MNN are shown. The points denote the strongest and most characteristic Auger peaks while the bands indicate the rough structure of less intense peaks.

Auger current, $I_i$, from an atom of species i can be written in the general form (e.g., for a KLM Auger transition):

$$I_i = I_p \sigma_i \gamma_i (1 + r_i) T \frac{1}{4\pi} \int n_i(z) e^{-\frac{z}{\lambda_i \cos\theta}} \sin\theta \, d\theta \, d\phi \, dz \qquad (5.10)$$

where:

(1)  $I_p$ is the intensity of primary electron beam of energy $E_p$.
(2)  $\sigma_i(E_K, E_p)$ is the ionization cross-section of core level K by electrons of energy $E_p$.

(3)  $\gamma_i(KLM)$ is the probability of relaxation by the KLM Auger transition.

(4)  $(1+r_i)$ is the backscattering factor, which accounts for the backscattered primary electrons (ionization can be produced by backscattered or secondary electrons, which results an increase of the total number of core holes by a factor of $[1+r_i(E_K, E_p, \theta_0)]$, where $\theta_0$ is the angle of incidence).

(5)  $n_i(z)$ is the number of atoms $i$ as a function of depth $z$, known as depth profile function.

(6)  $e^{-\frac{z}{\lambda_i \cos \theta}}$ is the probability of no-loss escape of Auger electrons from depth $z$. $\lambda_i(KLM)$ is the inelastic mean free path of the electrons, otherwise known as the electron attenuation length, and $\theta$ the escape angle of Auger electrons with respect to the surface normal. $\lambda_i$ depends on the electron energy and material and is shorter than the inelastic mean free path and accounts for the role of elastic scattering.

(7)  $T$ characterizes the transmission of the energy analyzer.

The integration in Eq. (5.10) is over azimuthal $\phi$ and polar $\theta$ angles and the depth $z$. We note that the inelastic mean free path can be expressed in an empirical form as (Seah and Dench, 1979):

$$\lambda = \frac{A_i}{E^2} + B_i \sqrt{E} \qquad (5.11)$$

here $A_i$ and $B_i$ are constants that depend on the material and $E$ is the energy of the electron. This equation attempts to provide the form of the universal curve shown in Figure 5.13. This relation can also be expressed as (Seah, 1984):

$$\lambda = \frac{538}{E^2} a_A + 0.41 a_A \sqrt{a_A E} \qquad (5.12)$$

where $E$ is expressed in eV and $\lambda$ and $a_A$ are in nm. Here $a_A$ is the atomic size deduced from the relation $\rho n N_V a^3 = M$, where $\rho$ is the density, $N_V$ is Avogadro's number, $M$ is the molecular weight and $n$ the number of atoms in the molecule.

Eq. (5.10) shows the considerations necessary when accounting for quantitative analysis of the AES signal. However, its use is hindered since: (i) exact values of factors (e.g., $\sigma_i, \gamma_i$) for a given system are frequently not available; (ii) the formula concerns the current measurement which is not convenient in practice; (iii) in general, the distribution of atomic species in the sample, $n_i(z)$, is not known and integration of $z$ cannot be performed.

The first two problems can be overcome to a certain extent if we measure the Auger spectrum in arbitrary intensity units and use a normalization procedure. To overcome the third point it is necessary to make some reasonable assumptions about the character of profile $n_i(z)$ and conduct an evaluation within a chosen model. We will consider the following two commonly occurring cases: a homogeneous binary material and a uniform layer on the substrate.

**Case 1: Homogeneous binary material AB**

In this case, $n_i(z)$ $(i = A, B)$ is constant and integration over $z$ is possible in Eq. (5.10). Then by using the so-called elemental sensitivity factors $I_A^\infty$ and $I_B^\infty$ for elements A and B (i.e., the intensity of AES signals from semi-infinite bulk samples, which can be found in the literature), the Auger ratio of peak intensities can be written as:

$$\frac{I_A/I_A^\infty}{I_B/I_B^\infty} = \frac{[1 + r_{AB}(E_A)]\,\lambda_{AB}(E_A)\,[1 + r_B(E_B)]\,\lambda_B(E_B)\,X_A}{[1 + r_{AB}(E_B)]\,\lambda_{AB}(E_B)\,[1 + r_A(E_A)]\,\lambda_A(E_A)\,X_B}\left(\frac{a_A}{a_B}\right)^3 \quad (5.13)$$

where $a_{A,B}^3$ are the atomic volumes of elements A and B and $X_{A,B}$ are the atomic fractions of the two elements. The greatest source of uncertainty comes from the attenuation lengths. A rough estimate of the surface composition, where the backscattering is neglected, from this Eq. (5.13) can be simplified to:

$$X_A = \frac{I_A/I_A^\infty}{I_A/I_A^\infty + I_B/I_B^\infty} \quad (5.14)$$

where $X_A + X_B = 1$. A more general form can be written for the case where other elements are present:

$$X_A = \frac{I_A/I_A^\infty}{\sum_i I_i/I_i^\infty} \quad (5.15)$$

However, since the backscattering contribution to the Auger intensity and the inelastic mean free path are matrix sensitive, a better approximation is given by (Seah, 1984):

$$X_A = \frac{I_A/I_A^\infty}{\sum_i F_i^A I_i/I_i^\infty} \quad (5.16)$$

where $F_i^A$ is a calculable Auger matrix factor and have been evaluated by Hall and Morabito (1979).

**Case 2: Layer of material A on a substrate of material B**

If a uniform layer has a thickness $d_A$, it means that:

$$X_A(z) = \begin{cases} 1 & ,0 \leq z \leq d_A \\ 0 & ,z > d_A \end{cases} \tag{5.17}$$

$$X_B(z) = \begin{cases} 0 & ,0 \leq z \leq d_A \\ 1 & ,z > d_A \end{cases} \tag{5.18}$$

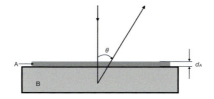

**FIGURE 5.22** Experimental geometry for the measurement of under-layer/overlayer AES signal for a continuous layer.

After integration of Eq. (5.13) over $z$ under these conditions and some calculations, we obtain:

$$I_A = I_A^\infty \frac{1 + r_B(E_A)}{1 + r_A(E_A)} \left[ 1 - e^{-\frac{d_A}{\lambda_A(E_A)\cos\theta}} \right] \tag{5.19}$$

$$I_B = I_B^\infty e^{-\frac{d_A}{\lambda_A(E_B)\cos\theta}} \tag{5.20}$$

If the layer of material A is not continuous, but covers a surface fraction $\phi_A$, as shown in Figure 5.23, the exponential terms in Eqs. (5.19) and (5.20) change respectively to:

$$I_A = \phi_A \left[ 1 - e^{-\frac{d_A}{\lambda_A(E_A)\cos\theta}} \right] \tag{5.21}$$

$$I_B = \left[ 1 - \phi_A e^{-\frac{d_A}{\lambda_A(E_B)\cos\theta}} \right] \tag{5.22}$$

A typical application of these equations is to distinguish between different growth modes; layer-by-layer or clusters, etc. We will briefly indicate these results in a later section when we discuss thin film growth.

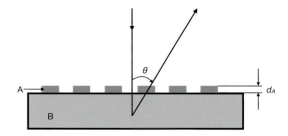

**FIGURE 5.23** Experimental geometry for the measurement of under-layer/overlayer AES signal for a discontinuous layer.

### 5.2.4  *Electron Energy-Loss Spectroscopy*

Inelastically scattered electrons, which have lost well-defined energies in the course of interaction with the solid surface, are considered in electron-energy loss spectroscopy (EELS). These losses cover a broad range of energies, from $10^{-3} - 10^4$ eV and can originate from various scattering processes:

(1)  Core level excitations: $100 - 10^4$ eV;

(2)  Excitation of plasmons and electronic inter-band transitions: 1–100 eV; and

(3)  Excitation of vibrations of surface atoms and adsorbates: $10^{-3} - 1$ eV.

The study of the first group of energy losses is specifically labeled core level electron energy loss spectroscopy (CLEELS). Depending on the core levels studied, it requires excitation sources with a relatively high primary energy of several keV or higher. As a result, the contribution of the bulk to the CLEELS signal can be very significant.

Conventional EELS deals with the second group of losses related to plasmon excitation and inter-band transitions. These losses are studied using electrons with primary energy from 100 eV up to a few keV. The resulting EELS spectra contain both bulk and surface components.

When low energy ($E_p \lesssim 20$ eV) EELS is performed with high resolution, the designated spectra are called high-resolution electron energy loss spectroscopy (HREELS). Such spectroscopy is used to study surface phonons and vibrational modes of adsorbed atoms and molecules.

The EELS and CLEELS techniques employ conventional energy analyzers such as CMA or CHA, while HREELS requires more sophisticated apparatus with cylindrical sector deflectors used both as monochromators for the primary electron beam and as energy analyzers for the secondary electrons.

**CLEELS**

An electron passing through a sample can lose some of its kinetic energy to induce an electron transition from a core level to an empty state. For metals, these empty states are those above the Fermi level which, in semiconductors or insulators, are located above the bandgap in the conduction band. If the primary electron excites a transition, say of a K electron to a final state $\varepsilon_c$ in the conduction band (see Figure 5.24), the energy loss, $\Delta E$, can be written as:

$$\Delta E = E_K + \varepsilon_c \tag{5.23}$$

and an inelastically scattered primary electron will have an energy of:

$$E_s = E_p - \Delta E = E_p - E_K - \varepsilon_c \tag{5.24}$$

The energy does not depend on the material work function since the primary electron passes through the surface barrier twice.

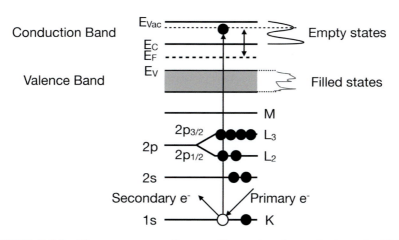

Core level excitation

**FIGURE 5.24** Electron energy diagram illustrating the excitation of a K-level electron in a semiconductor, via the energy transfer from a primary electron, to an empty state in the conduction band.

In Figure 5.25, we show an example of a CLEELS spectrum of oxidized silicon. The CLEELS signal is typically quite weak and the second derivative spectrum $\frac{d^2}{dE^2} N(E)$ is often used to improve its visibility.

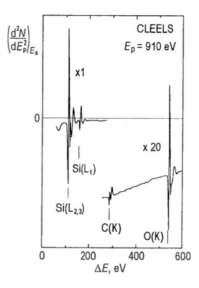

**FIGURE 5.25**    Core-level electron energy loss spectrum of oxidized silicon. In the measurement, the pass energy of the analyzer, $E_S$, was fixed, while the primary energy, $E_S$, was scanned. To enhance the visibility of the CLEELS peaks, the spectrum is recorded as the second derivative. The loss peaks due to excitation of K-levels of oxygen (O) and carbon (C) as well as the $L_1$ and  $L_{2,3}$ levels of Si are observed.

We can see that the loss energy identifies the transition and as such EELS can be used for elemental analysis. Quantification of the data is possible since the intensity is proportional to the elemental concentration. As the transition probability of the electron depends on the density of the final (empty) states, the fine structure of the CLEELS spectrum provides information on the energy distribution of the density of empty states.

**(Conventional) EELS**

The term electron energy loss spectroscopy (EELS) has a double meaning. Firstly, it is used as a general term to characterize all the techniques dealing with electron energy losses (including CLEELS and HREELS). On the other hand, it is used as a particular term to characterize the electron energy loss technique that studies only the losses in the range from a few eV to several dozen eV. The losses in this range are originated primarily from plasmon excitation and electronic inter-band transitions, see Figure 5.26.

It should be noted that it is not a simple task to interpret conclusively the origin of a given loss peak.

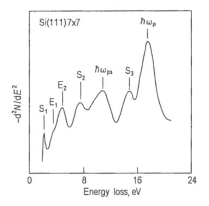

**FIGURE 5.26**    Electron energy loss spectrum of a clean $Si(111) - (7 \times 7)$ surface. The loss peaks in the spectrum are due to the excitation of bulk plasmons $(\hbar\omega_p)$, surface plasmons, $(\hbar\omega_p s)$, bulk inter-band transitions, $(E_1, E_2)$ and transitions from occupied surface states, $(S_1, S_2, S_3)$. The primary electron energy is 100 eV.

An electronic inter-band transition in a semiconductor or insulator involves the excitation of an electron from an occupied state in the valence band or occupied surface state to the normally empty state in the conduction band. Thus, the process is similar to that in CLEELS, but the loss energies are essentially lower.

A primary electron can lose some energy to induce oscillations of the electron density (or plasma oscillation) in a solid. The quantum of plasma oscillations is called a plasmon. A calculation using a classical Maxwell equation approach can be used to show that the eigenfrequency of the bulk oscillations of a homogeneous electron gas with respect to a positively charged ionic skeleton is given by:

$$\omega_p = \sqrt{\frac{4\pi n e^2}{m}} \qquad (5.25)$$

This is known as the Langmuir formula, where $e$ and $m$ are the electronic charge and mass, respectively, and $n$ is the electron concentration.

Accordingly, $\Delta E$ is the energy of a bulk plasmon.

$$\Delta E = \hbar\omega_p \qquad (5.26)$$

The Langmuir formula is obtained for a homogeneous electron gas, which seems only suitable for simple metals. However, the formula appears to be a reasonable approximation for many metals and some semiconductors (e.g., Si, Ge, InSb) and insulators (SiC, SiO$_2$). In the latter, valence electrons participate in the plasma excitations.

The presence of a surface manifests itself by a surface plasmon. This type of oscillation is localized at the surface and its amplitude rapidly decays with depth. In the classical case of an abrupt interface between a homogeneous electron gas and a vacuum, the surface plasmon frequency is related to the bulk plasmon frequency by the relation:

$$\omega_{sp} = \frac{\omega_p}{\sqrt{2}} \tag{5.27}$$

In Figure 5.27, we show an EELS spectrum for Al. Surface and bulk plasmon peaks are indicated at energies of 10.3 eV and 15.3 eV. Other peaks are due to losses arising from multiple plasmon excitations.

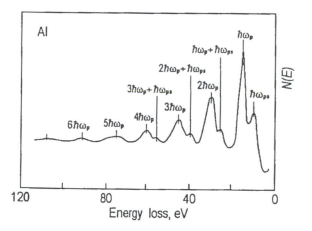

**FIGURE 5.27**   Electron energy loss spectrum of aluminum. Losses due to bulk (15.3 eV) and surface (10.3 eV) plasmons are shown. Additionally, losses due to multiple plasmon excitations are also clearly observed.

The use of EELS in the study of surfaces can be summarized as follows:

(1) Electron density determination: Using Eq. (5.25), it is possible to evaluate the density of electrons participating in plasma oscillations. However, it must be borne in mind that for such applications we require that the effective mass of the electron in the solid be close to that of the free electron and that the electron density is a constant throughout the whole probing depth.

(2) Chemical analysis: Since the electron density is characteristic of a given solid, plasmon energies can be used to identify surface species. Fig-

ure 5.28 shows the distribution between Si with a bulk plasmon at 17 eV and $SiO_2$ with a bulk plasmon at 22 eV.

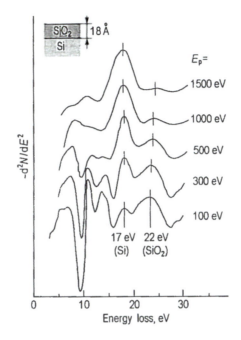

**FIGURE 5.28** Electron energy loss spectrum from a $SiO_2$ film (18 Å) on a Si(111) substrate recorded at a series of primary electron energies (100, 300, 500, 1000 and 1500 eV). The loss peaks due to bulk plasmons in Si (17 eV) and bulk plasmons in $SiO_2$ (22 eV) are indicated.

(3) Analysis of depth distribution of species: The probing depth in EELS is related to the energy of the primary electrons. By varying the primary electron energy, it is possible to obtain information about the depth distribution of a particular elemental species. Figure 5.28 shows a set of EELS spectra from a sample with an 18 Å film of $SiO_2$ on Si(111) recorded at different primary electron energies, from 100–1500 eV. It is clear that by increasing the primary electron energy, the contribution from the Si substrate increases while that from the $SiO_2$ film decreases. In principle, it is possible to quantify such data to evaluate the depth profile of the film.

## HREELS

High-resolution electron energy loss spectroscopy (HREELS) concerns losses due to the excitation of vibrational modes of surface atoms and adsorbates. Since such losses are typically a fraction of an eV, the energy separation between the elastic peak and loss features is very small and thus requires high energy resolution. This is achieved using the following experimental set-up (Figure 5.29).

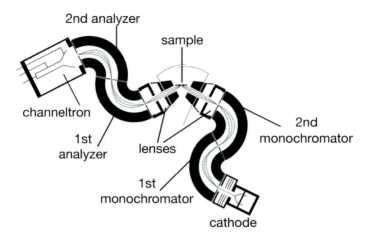

**FIGURE 5.29** Typical experimental arrangement for high-resolution EELS. Both analyzer and monochromator systems employ 127°-angle cylindrical sector deflectors as energy dispersive elements.

The electrons from a cathode system pass into the (energy) monochromator which, after focusing, impinge on the sample. A second lens system collects the secondary electrons into the analyzer. The 127°-angle cylindrical sector deflectors select electrons with a narrow window of energies ($\sim 1 - 10$ meV down to a fraction of a meV in sophisticated systems). The monochromator and analyzer systems are very similar in appearance and function, both performing energy bandpasses of narrow linewidth.

HREELS is a powerful tool in the study in adsorption of atoms and molecules on surfaces, allowing species identification and providing information on bonding geometries. Comparison is usually based on known vibrational modes of molecules in the gas phase using IR absorption or Raman spectroscopy. Energies are frequently expressed in terms of wavenumber (e.g., $100 \text{ cm}^{-1} = 12.41 \text{ meV}$).

The applications of HREELS include the following:

(1) Identification of adsorbates: As each molecule is characterized by a specific set of modes, identification of adsorbed species is possible. An example is shown below (Figure 5.30). Atomic O and molecular $O_2$ adsorption are distinguished.

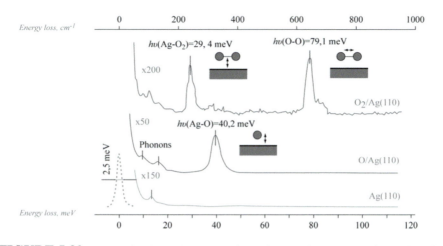

**FIGURE 5.30** Investigation of oxygen adsorption on the silver surface using the HREELS technique. In case of adsorption in normal room temperature conditions (medium spectrum), a typical peak is observed at 40.2 meV that corresponds to the vibrational mode O−Ag, which reflects dissociation of $O_2$ molecules. In case of specimen's exposure to $O_2$ at 100 K (upper spectrum), molecular adsorption of $O_2$ is observed; typical peaks at 29.4 meV and 79.1 meV correspond to vibrational modes Ag−$O_2$ and O−O, respectively. Ag(110) clean surface spectrum (lower curve) is provided for benchmarking.

(2) Identification of adsorption sites: This can be evaluated from the observation of a certain mode of vibration of adsorbate-substrate; e.g., As-H stretching modes indicate that surface As atoms in GaAs is bonding sites.

(3) Identification of spatial orientation of adsorbed molecules: This is based on the consideration of dipole selection rules in HREELS: in specular geometry (see Figure 5.31), only dipoles oriented normal to the surface give rise to significant losses. Dipoles oriented parallel to the surface can be detected only in the off specular geometry. Therefore, one can only discriminate between the chemical bonds oriented normal and parallel to the surface.

## 5.2.5 Photoelectron Spectroscopies

Photoelectron spectroscopy (PES) is the most commonly used analytical technique for probing the electronic structure of occupied states at the surface and near-surface region. It is broadly based on the photoelectric effect, where an electron, initially in a bound state $E_i$, absorbs a photon of energy $\hbar\omega$ and leaves the solid with a kinetic energy of:

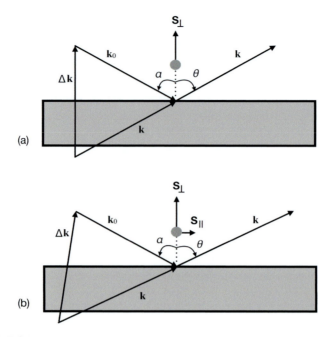

**FIGURE 5.31**    (a) Specular ($\alpha = \theta$) and (b) off-specular ($\alpha \neq \theta$) geometries in HREELS measurements.  The energy loss, $\hbar\omega$, is assumed to be small when compared to the primary energy, $E$, i.e., $|\mathbf{k}| = |\mathbf{k_0}|$.  In (a), the scattering vector $\Delta\mathbf{k}$ is normal to the sample surface, therefore the atom vibration $\mathbf{S}_{||}$, parallel to the surface is not detectable and vibrations, $\mathbf{S}_\perp$, will be normal to the surface. In (b), $\Delta\mathbf{k}$ has a component parallel and normal to the surface, thus both vibrations $\mathbf{S}_\perp$ and $\mathbf{S}_{||}$ can be studied.

$$E_{kin} = \hbar\omega - E_i - \phi \tag{5.28}$$

where $\phi = E_{\text{vac}} - E_F$ is the work function of the solid in question (see Figure 5.32). The necessary conditions for detecting the escaping elastic electron are as follows:

(1)  The energy of the photon is sufficient to allow the electron to escape from the solid; $\hbar\omega \geq E_i + \phi$
(2)  The electron velocity is directed towards the outer surface.
(3)  The electron does not lose energy in collisions with other electrons on its way to the surface.

Depending on the energy (wavelength) of the photons used for electron excitation, PES is considered as:

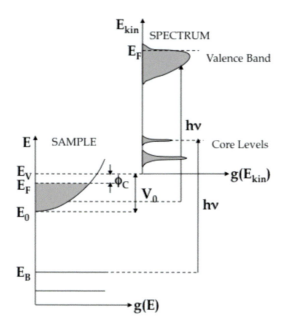

**FIGURE 5.32** Scheme of the photoemission process in a generic metal. On the left, the energy level scheme inside the sample is shown; on the right, a typical photoemission spectrum is shown. A spectrum is constituted by the distribution of the electrons as a function of their kinetic energy as a result of the interaction with the incident radiation.

(1) X-ray photoelectron spectroscopy (XPS) – or electron spectroscopy for chemical analysis (ESCA), when X-ray radiation is used with the photon energy in the range 10 eV–10 keV ($\lambda \sim 100 - 1$ Å). As a consequence of the high energies of the X-rays, XPS probes deep core levels.
(2) Ultraviolet photoelectron spectroscopy (UPS) – when the photons are in the UV range, 10–50 eV ($\lambda \sim 1000 - 250$ Å). As a result, UPS is used for studying valence and conduction bands.

Frequently, a synchrotron is used as a source of radiation since a continuum of energies can be obtained from soft ultraviolet to hard X-rays. The division, as such, can be a little arbitrary, since the physical experiment is essentially the same for both XPS and UPS.

The experimental set-up for photoemission experiments, shown in Figure 5.33, includes a monochromatic source of photons, the sample which is maintained in UHV and an electron energy analyzer to record the spectra of photoelectrons. The photon source can be laboratory sources which are spe-

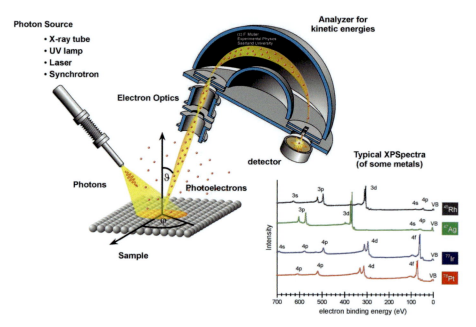

**FIGURE 5.33** Schematic diagram for the photoemission experiment, which includes photon source, sample, and concentric hemispherical analyzer. Only electrons with the appropriate energy and an acceptable entry angle will reach the detector. Biasing of the whole CHA with respect to ground can be used to decelerate the electrons as they enter.

cific for UPS and XPS, or a synchrotron source, which covers all ranges of PES.

### 5.2.5.1   *X-Ray Photoelectron Spectroscopy (XPS)*

Referring to the schematic diagram of Figure 5.32 for the photoemission process for XPS, one can see that the photoemission spectrum, $I(E_k)$, is a fingerprint of the density of occupied states, $D(E_i)$, in the probed material. However, in reality, things are rarely so straightforward. Besides the peaks due to elastic photoelectrons, there is a number of additional features in the XPS spectrum, like the continuous background of inelastic secondary electrons, Auger peaks and peaks due to plasmon losses (on the low-energy side of each photoemission peak), see Figure 5.34. It is also important to bear in mind that the cross-section for excitation is different for different electron levels, which can greatly affect the spectral lineshape. However, the valence band features in the XPS spectrum are very weak due to the low photoelectric cross-section

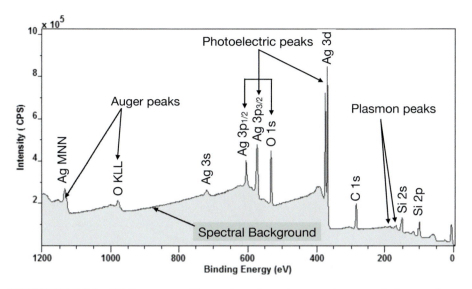

**FIGURE 5.34** XPS spectrum illustrating photoelectron, Auger and plasmon loss peaks. The background in XPS is non-trivial in nature and results from all those electrons with initial energy greater than the measurement energy for which scattering events cause energy losses prior to emission from the sample.

for shallow valence band levels at typical XPS photon energies. In general, photoemission has a maximum probability at photon energies close to the threshold and drops off rapidly when the photon energy exceeds sufficiently the electron binding energy. Therefore XPS is a useful tool to probe mainly deep core levels.

The core levels show up in XPS spectra as sharp peaks, whose locations are defined by the electron binding energies, which is characteristic of atomic species; i.e., the presence of peaks at certain energies can be treated as a sign of the presence of a particular element at the surface region, and thus the XPS spectrum contains information on the surface composition.

On the qualitative level, one can distinguish what atomic species are present by comparing peak energies from the spectrum with tabulated values of the elements (see Figure 5.35).

On the quantitative level, one can evaluate the concentration of the atomic species on a surface layer from the XPS peak heights. By analogy with AES, we can write a general equation for the intensity of the photoelectrons from a core level of an arbitrary species $i$ as:

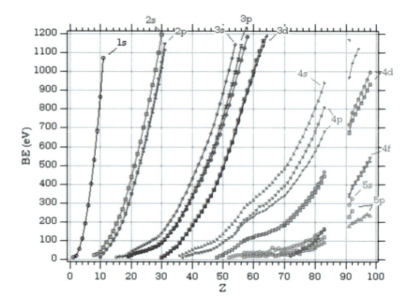

**FIGURE 5.35**   Electronic binding energies of core filled levels of the elements.

$$I_i = J_0 \sigma_i (h\nu) T \int_{\gamma=0}^{\pi} \int_{\phi=0}^{2\pi} L_i(\gamma) \int_{z=0}^{+\infty} n_i(z) e^{-\frac{z}{\lambda_i \cos\theta}} \, dz \, d\phi \, d\gamma \qquad (5.29)$$

where $J_0$ is the intensity of the primary X-ray beam; $\gamma$ is the angle between the incident X-ray beam and the direction of the ejected photoelectron; $L_i(\gamma)$ is the angular dependence of photoemission; $\sigma_i(h\nu)$ is the photo-ionization cross-section of the core level by a photon of energy $h\nu$. All other symbols agree with those of AES, Eq. (5.10).

In contrast with the analysis for AES, there is no background factor in Eq. (5.29), but the angular anisotropy of photoemission is added. The corresponding simple cases given for AES of a homogeneous binary material and a uniform layer on a substrate can be expressed as follows:

(1)  Homogeneous binary material AB:

$$\frac{\frac{I_A}{I_A^\infty}}{\frac{I_B}{I_B^\infty}} = \frac{\lambda_{AB}(E_A)}{\lambda_{AB}(E_B)} \frac{\lambda_B(E_B)}{\lambda_A(E_A)} \frac{X_A}{X_B} \left(\frac{a_A}{a_B}\right)^3 \qquad (5.30)$$

(2)  Layer A of thickness $d_A$ on substrate B:

$$I_A = I_A^\infty \left( 1 - e^{-d_A/\lambda_A (E_A) \cos \theta} \right) \qquad (5.31)$$

$$I_B = I_B^\infty e^{-d_A/\lambda_A (E_B) \cos \theta} \qquad (5.32)$$

It will be noted that, compared to the analysis for AES, the corresponding PES equations do not include backscattering and, as such, quantitative PES analysis is somewhat more accurate than that for AES.

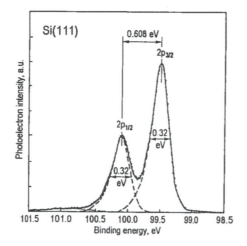

**FIGURE 5.36** The spin-orbit splitting of the Si 2p core level is observed with the partner $2p_{1/2}$ and $2p_{3/2}$ lines. The splitting of 0.608 eV and the intensity ratio of 1:2 are atomic properties that are practically independent of chemical environment.

Using XPS with high resolution, it is possible to visualize the fine structure of core levels. In particular, the spin-orbit splitting is well resolved as shown in Figure 5.36. Accurate measurements of the core level peaks for a given element reveal that they vary, depending on its chemical environment, as can be seen in the Figure 5.37. These are so-called chemical shifts. Their typical values are in the range from 1 to 10 eV. The origin of chemical shifts can be understood as follows:

(1) The binding energy of the electron at a given level is defined by the interplay between the Coulomb attraction to the nucleus and the screening of this attraction by other electrons in the atom.

(2) Formation of the chemical bond involves electron transfer, thus the charge density on the atom is changed, which results in turn in changing the electron binding energy itself.

(3)  Electron charge transfer to a given atom enhances the electron screening and hence weakens the electron binding energy—the corresponding peak shifts to shallower binding energies relative to the Fermi energy. Conversely, electron charge transfer from a given atom weakens the electron screening and enhances the electron binding to the nucleus—the peak shifts to the deeper binding energies.

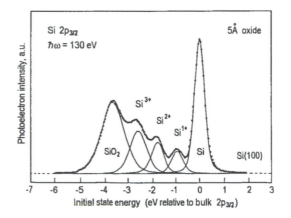

**FIGURE 5.37**   Si $2p_{3/2}$ core-level spectrum from a thin $SiO_2$ film on Si(100). In addition to the Si peak, also observed are the different oxide states, showing a shift to lower energies with increasing oxidation state.

The ability of XPS to provide information on chemical composition and chemical bonding states warrants its other name: Electron Spectroscopy for Chemical Analysis (ESCA).

The atomic environment of atoms at the surface differs from that in the bulk even in the case of an adsorbate free atomically clean surface. This difference is reflected in the shifts of the core levels. In XPS spectra, the surface components are superposed with bulk components due to the finite penetration depth of X-rays and the electron mean free path. To enhance the surface sensitivity of XPS, grazing incidence radiation (refraction limits the penetration depth) and collection of photo-emitted electrons also at grazing angles (which reduces the effective mean free path normal to the surface) are typically used. In addition, by changing the polar angle of the detector, the fraction of the signal coming from various depths can be altered for analyzing the depth distribution of a given atomic species.

### 5.2.5.2   Ultraviolet Photoelectron Spectroscopy

Since ultraviolet photoelectron spectroscopy (UPS) utilizes relatively low photon energies (typically less than $\sim 50$ eV), only the valence levels become excited in the photoelectric process. Note that, besides levels, which correspond to the occupied band states of a solid surface, these are also the filled bonding orbitals of adsorbed molecules. If one takes into account the large photoemission cross-section of the valence states under UPS excitation energies, it is clear, why UPS has proven to be a powerful tool for studying the valence band structure of surfaces and its modification during various surface processes such as adsorption, thin-film growth, surface chemical reactions.

Depending on the task, UPS is employed conventionally in one of two regimes: (i) angle integrated UPS and (ii) angle-resolved UPS.

In case (i), angle integrated UPS, the electrons are collected, in the ideal scenario, over the whole half-space above the sample surface. The usage of the retarding field (LEED) analyzer is a good approximation. The obtained data are used to evaluate the density of states within the surface valence band.

In angle-resolved UPS (ARUPS), the photo-emitted electrons are collected only in a chosen direction. The hemispherical and $127°$-angle detectors are well suited to such measurements. In this case, one deals with not only the electron energy, but also with the corresponding wave vector, which provides access to the dispersion of the surface states.

The kinetic energy of photoemission electrons can be expressed as:

$$E_{kin} = \frac{\hbar^2 \left[ (k_\perp^{ex})^2 + (k_\parallel^{ex})^2 \right]}{2m} \tag{5.33}$$

where $k_\perp^{ex}$ and $k_\parallel^{ex}$ are the perpendicular and parallel components of the wave vector, where $ex$ denotes external, i.e., a free electron which has escaped the solid. If $\mathbf{k}^{ex}$ makes an angle $\theta$ with the surface normal, we can write:

$$k_\parallel^{ex} = k^{ex} \sin \theta = \sqrt{\frac{2mE_{kin}}{\hbar^2}} \sin \theta \tag{5.34}$$

To consider the wave vector of an electron inside the solid, $\mathbf{k}^{in}$, it is important to remember that, when passing the solid-vacuum interface, only the parallel component of the electron momentum is conserved:

$$\mathbf{k}_\parallel^{ex} = \mathbf{k}_\parallel^{in} + \mathbf{G}_{hk} \tag{5.35}$$

where $\mathbf{G}_{hk}$ is a vector of the 2D surface reciprocal lattice. In contrast, the perpendicular component $\mathbf{k}_{\perp}^{ex}$ is not preserved and thus does not bear any particular relationship to $\mathbf{k}_{\perp}^{in}$.

Experimentally, to restore the dispersion relation $E_i(\mathbf{k}_{\parallel}^{in})$ of the surface states along a certain surface direction, the photoemission spectra are recorded as a function of the polar angle, with azimuth being fixed. For each polar angle $\theta$, the binding energy $E_i$ and the parallel component of the wave vector $\mathbf{k}_{\parallel}^{in}$ are extracted from the UPS spectra using Eqs.(5.28), (5.34) and (5.35), respectively. In Figure 5.38, we see an example of the experimental determination of the energy dispersion relation for the surface states of Cu(111).

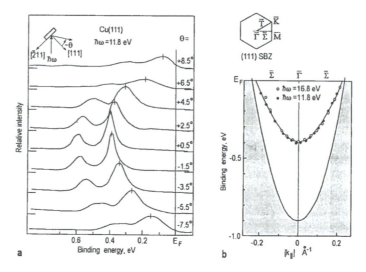

**FIGURE 5.38**    ARUPS determination of the dispersion for the Cu(111) sp surface states. (a) Experimental photoemission energy distribution curves from Cu(111) for several angles near the normal direction. (b) Evaluated dispersion of Cu(111) surface states plotted with a projection of bulk continuum of states (shaded area).

It should be born in mind, that UPS is a surface-sensitive technique, but not a surface-specific one. This means that some effort is required to distinguish between the surface and bulk contributions to the photoemission spectrum. There are some tests to help verify whether particular peaks are actually due to surface states:

(1) The surface state dispersion curve is one and the same at different photon energies used for excitation (see Figure 5.38(b)). This is a consequence of the fact that for surfaces states only the parallel component of the wave

vector is essential, without regard for the value of the wave vector itself. This will not be the case for bulk bands.

(2) Surface states reside in the bulk band-gap.

(3) Surface states are more sensitive to surface treatment. For example, if the peaks in a spectrum from a clean surface vanish with gas adsorption, they are likely to correspond to surface states.

The electronic structure of graphene has been intensively studied, both, experimentally and theoretically. In Figure 5.39, we show the valence band structure of graphene, shown in the form of ARUPS band maps and Fermi surface maps, which we compare to available experimental data from the literature.

## 5.3    Microscopies at the Nanoscale

### 5.3.1    *Introduction*

Microscopy techniques are used to produce real space magnified images of a surface or object. We should more correctly speak of *Nanoscopy*, but this does not yet exist as a recognized word. Typically, microscopy information regards surface crystallography, surface morphology, and surface composition. The operational principles vary greatly from one technique to another, including the following:

- Electron beam transmission–transmission electron microscopy (TEM);
- Electron beam reflection–reflection electron microscopy (REM), Low-Energy electron microscopy (LEEM), scanning electron microscopy (SEM);
- Field emission of electrons–field emission microscopy (FEM), scanning tunneling microscopy (STM);
- Field emission of ions–field ion microscopy (FIM)
- Scanning the surface by a tip (also referred to generically as scanning probe microscopies)–STM, atomic force microscopy (AFM), magnetic force microscopy (MFM);
- canning the surface with a probing electron beam–SEM;
- canning the surface with an optical probe–SNOM.

Most techniques of microscopy used in surface science ensure resolution on the nanometer scale, while some, such as FIM, STM, AFM (under specific conditions) and high-resolution TEM (HRTEM) allow resolution at the

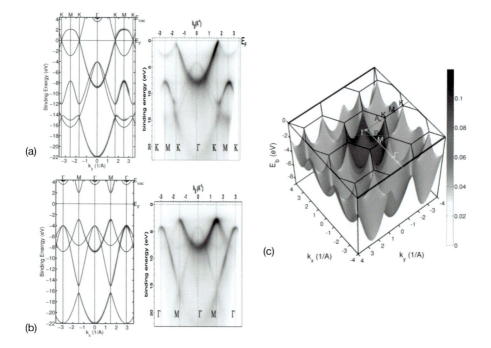

**FIGURE 5.39**    (a) Band structure (blue lines) and simulated photoemission intensities (colormap) of graphene along the $k_x$ - direction, i.e., along $\Gamma - K - M - K - \Gamma$ for 50 eV photons with in-plane polarization and 60° off-normal incidence. Right panel: experimental ARPES data of epitaxial graphene/SiC (b) Same as (a) but for the $k_x$-direction, i.e., along $\Gamma - M - \Gamma$. (c) Brillouin zones of graphene (honeycombs) with special points labeled in black. The red dashed lines indicate the directions for which the photoemission cross-section has been simulated. In addition, points P and A display the one-dimensional Brillouin zone boundaries of poly-para-phenylene ($\infty$P) and poly-acene ($\infty$A), respectively. The surface depicts the occupied $\pi$-band of graphene in an extended zone scheme, and the color code denotes the photoemission intensity with 35 eV incident photons.

atomic scale. The last item in this list sticks out a little since it is the only one, which relies directly on the optical imaging of a nanoscale object.

## 5.3.2    *Field Emission Microscopy*

Field emission microscopy (FEM) was invented in 1931 by Erwin Müller. The design consists of a metallic sample made up of a sharp tip and a conducting fluorescent screen inside an evacuated chamber, as shown in Figure 5.40(a).

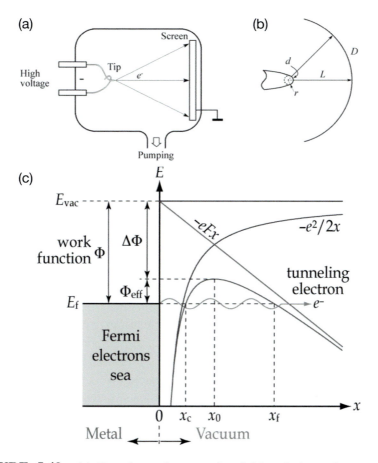

**FIGURE 5.40** (a) Experimental set-up for field emission microscopy. (b) Schematic illustration of the FEM optics. An object with linear dimension $d$ on the tip surface will be magnified by a factor of $L/r$ and will appear on the screen with a size $D$. (c) Potential energy diagram for an electron near a metal surface in the presence of an applied electric field $F$.

A large negative potential is applied to the tip, with respect to the screen, typically of $\sim 1 - 10$ keV. The radius of curvature of the tip is of the order of $r \sim 1000$ Å and hence the electric field near the tip apex is of the order of 1 V/Å. At such high fields, the field emission of electrons takes place (see Figure 5.40(c)).

The electrons emerge from the tip with a current given by the Fowler-Nordheim equation as:

$$j = \frac{AF^2}{\Phi t^2(\xi)} \exp\left(-\frac{Bf(\xi)\Phi^{\frac{3}{2}}}{F}\right) \tag{5.36}$$

where $F$ is the electric field (V cm$^{-1}$), $\Phi$ is the work function of the metal (eV), $t(\xi)$ and $f(\xi)$ are slow varying functions of dimensionless parameter $\xi$ and $A$ and $B$ are constants $(A = 1.54 \times 10^{-6}; B = 6.83 \times 10^7)$. The electrons will diverge radially along lines of force and the magnification of the microscope is given as:

$$M = \frac{L}{r} \tag{5.37}$$

where $r$ is the tip apex radius and $L$ the tip-screen distance. Linear magnifications of around $10^5 - 10^6$ are attainable. The resolution limit of the instrument is about 20 Åand is determined by the electron tangential velocity, which is of the order of the Fermi velocity of the electron in the metal.

Figure 5.41 shows an FEM image of a clean tungsten tip, with the wire axis perpendicular to the (110) plane. The emission current varies strongly with the local work function, as shown by the Fowler-Nordheim equation, Eq. (5.36). The FEM image displays the projected work function map of the emitter surface. The closely packed faces of the structure $\{110\}, \{211\}$ and $\{100\}$ have higher work functions than atomically rough regions and thus they show up in the image as dark spots on the brighter background.

Application of FEM is limited by materials, which can be fabricated in the form of a sharp tip, can be cleaned in a UHV environment and can withstand high electrostatic fields. For these reasons, refractory metals with high melting temperatures (e.g., W, Mo, Pt, Ir) are conventional objects for FEM experiments. FEM allows the measurement of the work function for various crystallographic planes on a single sample. In cases when the adsorbate affects the work function, the technique is applicable for studying adsorption-desorption kinetics and surface diffusion.

### 5.3.3   *Field Ion Microscopy*

The field ion microscopy (FIM) was also invented by Erwin Müller (1951). It is an offshoot of the FEM, in an effort to improve resolution by using the field desorption technique. The FIM apparatus is very similar to that used in

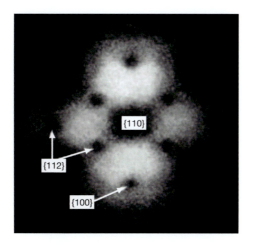

**FIGURE 5.41**    Field emission microscope image of tungsten clean surface with orientation (110) showing locations of different crystal planes.

the FEM, with a sharp tip and a fluorescent screen, which can be replaced with a multichannel plate, see Figure 5.42(a).

The field strength at the tip apex is a few V/Å. However, there are some essential differences in the two techniques which are summarized as follows:

(1) The tip potential is positive.
(2) The chamber is filled with an imaging gas, typically He or Ne at $10^{-5} - 10^{-3}$ Torr.
(3) The tip is cooled to low temperatures ($\sim 20 - 80$ K).

The principles of the FIM image formation are illustrated in Figure 5.42(b).

The imaging gas atoms in the vicinity of the tip become polarized by the electric field and, as the field is non-uniform, the polarized atoms are attracted towards the tip surface. Reaching the surface, they lose their kinetic energy through a series of hops and accommodate to the tip temperature. Eventually, the atoms are ionized by tunneling electrons into the surface and the resulting ions are accelerated to the screen to form a field ion image of the emitter surface. The resolution of the FIM is determined by the velocity of the imaging ion. Effective cooling of the tip to low temperatures can achieve a resolution of $\sim 1$ Å; i.e., atomic resolution. Figure 5.43 shows images obtained from a tungsten tip.

Limitations of FIM are similar to those of FEM, hence FIM experiments are performed using metals, which are prepared by electropolishing thin

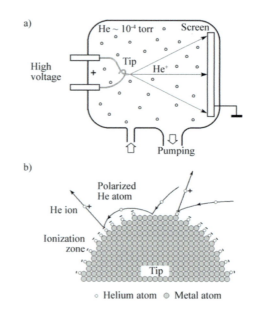

**FIGURE 5.42**     (a) Schematic illustration of the field ion microscope set-up. (b) Illustration of the image formation for FIM. Due to the large electric field at the terrace edges, ionized He atoms (imaging gas) are attracted preferentially to these regions. Other hopping He atoms may be field ionized above the adsorbed He atoms and accelerated as positive ions to the screen to produce the FIM image.

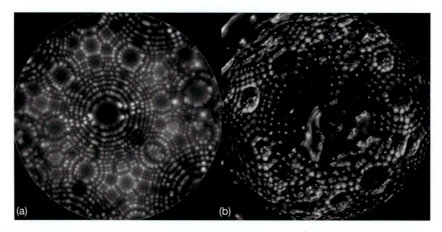

**FIGURE 5.43**     FIM images of a W tip of radius 120 Å acquired at 21 K using a He - $H_2$ mixture as the imaging gas.

wires. However, these tips usually contain asperities. The final preparation resides in the in-situ removal of these asperities by field evaporation by raising the tip voltage. For most materials, field evaporation (desorption of surface atoms in the form of ions) occurs in the range 2–5 V/Å. This process is self-regulating as the most protruding atoms desorb first, making the surface smooth. The tips used in FIM are sharper (tip radius $\sim 100 - 300$ Å) compared to those used in FEM, where the radius is about 1000 Å.

The most spectacular results obtained by FIM are associated with investigations of the dynamical behavior of surfaces and the behavior of adatoms on the surface. The elongated features observed in Figure 5.43(b) are traces captured as atoms move during the imaging process. Typical problems of the investigation will be adsorption-desorption phenomena, surface diffusion of adatoms and clusters, adatom-adatom interactions, step motion, equilibrium crystal shapes. It should, however, be taken into account that experimental results could be affected by the limited surface area; i.e., edge effects and the presence of the large electric field.

### 5.3.4 *Transmission Electron Microscopy*

In transmission electron microscopy (TEM), the image is formed by electrons passing through the sample. The principle of operation of the TEM is very similar to that of the optical microscope, using magnetic lenses instead of glass lenses and electrons (with an associated de Broglie wavelength) instead of photons. A beam of electrons emitted by an electron gun is focused by a condenser lens into a small spot ($\sim 2$–3 m) on the sample and, after passing through the sample (in the diffraction plane), is then focused by the objective lens to project the magnified image onto the screen (see Figure 5.44). An essential element is an aperture located at the back focal plane of the objective lens; this will determine the image contrast and resolution limit of the microscope. Note that this simple scheme illustrates only the principle of image formation in the TEM set-up, which is more sophisticated.

Due to the limited penetration depth of electrons in solids, the samples need to be very thin; the acceptable thickness is $\sim 100 - 1000$ Å for conventional microscopes with accelerating voltages of 50–200 keV and a few thousand Å for high voltage microscopes with accelerating energies of up to 3 MeV. Of course, the required sample thickness depends on the sample material: the larger the atomic number, the greater the electron scattering and hence the thinner the sample needs to be.

The diffraction limit for TEM resolution can be estimated from:

$$\Delta = \frac{\lambda}{2\sin\alpha} \tag{5.38}$$

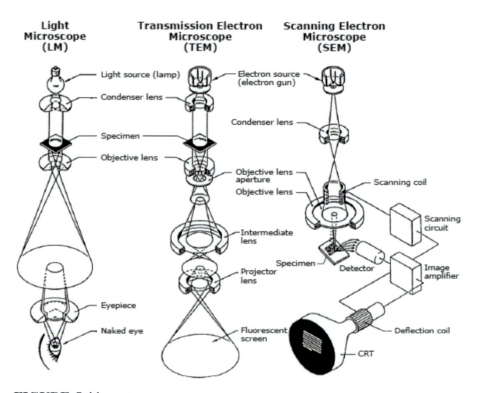

**FIGURE 5.44**  Schematic illustration of the principles of transmission electron microscopy (TEM). Also illustrated for comparison are the corresponding components for optical microscopy (left). The scanning electron microscope (SEM) differs from the transmission microscope in the imaging methodologies, as will be discussed in Section 5.3.5, this is shown on the right hand side.

where $\lambda$ is the electron wavelength and $\alpha$ is half the angular aperture, which in turn can be approximated by the ratio of the objective diaphragm radius to the objective focal length. For example, for a voltage of 100 kV ($\lambda$ = 0.037 Å), diaphragm radius of 20 m and a focal length of 2 mm, the estimation yields $\Delta \simeq 2$ Å. In practice, the resolution is usually worse due to non-idealities in the electron optic system.

The formation of the TEM image contrast may be understood in the following way. When passing through a sample, the electron flux loses part of its intensity due to scattering. This part is greater for thicker regions or regions with species of a higher atomic number. If the objective aperture effectively cuts-off the scattered electrons, these thicker regions or regions of higher atomic numbers appear darker. The smaller aperture enhances the contrast, but leads to the loss of resolution, as seen from Eq. (5.38). In crystals, the elastic scattering of electrons results in the appearance of diffraction contrast.

In experiments studying surface phenomena, metals can be deposited onto alkali halide surfaces cleaved in a vacuum. At metal thicknesses of around 10 Å, a continuous film is not formed, but instead, a large number of small islands form on the surface. A thin film of carbon can then be deposited onto the surface to fix the metal nuclei. The substrate—alkali halide—can be dissolved leaving the carbon film with metal nuclei embedded. Subsequent TEM observations can then be made to study mechanisms of island nucleation, growth, and coalescence. In addition, a step structure of alkali halide surfaces can be made using step decoration, where preferential island nucleation along step edges can be used. Today, such studies will be typically made using other methods, such as SEM, STM or AFM.

The TEM provides an image based on the amplitude of the interference diffraction pattern. More recent developments have employed techniques to utilize the phase information or phase-contrast, which provides much better resolution. This technique is labeled by high-resolution TEM (HRTEM). The resolution of modern microscopes is about 0.8 Å. With such high resolution, it is possible to image crystalline structures and defects at the atomic scale and HRTEM is an invaluable tool in the study of nanoscale properties.

Due to the inability to measure the phase of the electron waves, we usually rely on the amplitude from the interference pattern. However, the phase of the electron wave still carries information about the sample and generates contrast in the image—thus the name phase-contrast imaging. This will only be the case if the sample is sufficiently thin so that amplitude variations only slightly affect the image.

The interaction of the electron wave with the crystallographic structure is not entirely understood, though a qualitative idea can be obtained. Above the sample, the wave of an electron can be approximated as a plane wave incident on the sample surface. As it penetrates the sample, it is attracted by the positive atomic potentials of the atom cores and channels along crystallographic planes. At the same time, the interaction between the electron wave in different atom columns leads to Bragg diffraction. This complex situation can be simulated to interpret the images obtained is HRTEM.

As a result of the interaction with the sample, the exit wave of the electrons below the sample, $\phi_e(x, u)$, is a function of position, $x$, being a superposition of a plane wave and a multitude of diffracted beams with different in-plane spatial frequencies, $u$. The phase change of $\phi_e(x, u)$ compared to the incident wave peaks at the location of the atom columns. The exit wave now passes through the imaging system of the microscope where it undergoes a further phase change and interferences in the image plane. The image that is recorded is not a direct representation of the samples' crystallographic structure, for example, an intensity peak may or may not indicate the presence of an atom column; see Figure 5.45 of exit wave simulation and recorded images.

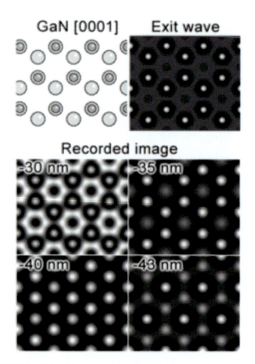

**FIGURE 5.45**     Simulated HREM images for GaN[0001].

The relationship between the exit wave and image wave is highly nonlinear and is a function of the aberrations of the microscope. The relation is described by the contrast transfer function (CTF):

$$CTF(u) = A(u)E(u)\sin[\chi(u)] \qquad (5.39)$$

where $A(u)$ is the aperture function, $E(u)$ describes the attenuation of the wave for higher spatial frequency and is called the envelope function and $\chi(u)$ is a function of the aberrations of the electron optic system.

The basis of the interpretation of HRTEM is the so-called exit-wave reconstruction of $\phi_e(x, u)$ in the image plane. If all properties of the microscope are known, it is possible to reconstruct the real exit wave with very high accuracy. Some representative HRTEM images of crystalline interfaces and nanoparticles showing atomic resolution are illustrated in Figure 5.46.

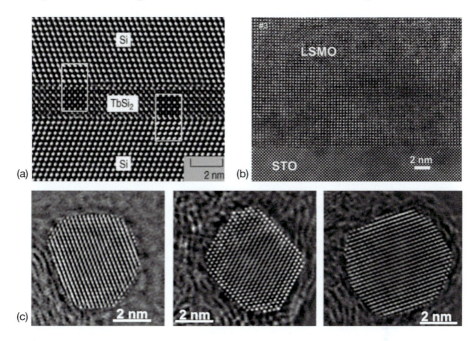

**FIGURE 5.46** (a) An atomic resolution TEM image of a Si/TbSi$_2$/Si heterostructure with simulated images pasted for direct comparison (in the white boxes). (b) Cross-section HRTEM image of a PZT/LSMO interface and corresponding GPA image showing the strain field distribution perpendicular to the interface direction. (c) 5–6 nm Pt nanoparticles supported on graphitic C, obtained by applying exit wavefunction restoration to defocus the series of high resolution TEM images.

In (a), we see an atomic resolution TEM image of a Si/TbSi$_2$/Si heterostructure, while in (b), a sharp interface between a STO (SrTiO$_3$) crystal and a La$_{0.7}$Sr$_{0.3}$MnO$_3$ (LSMO) film grown by PLD (Pulsed Laser Deposition) is shown. In (c), we can see some high definition atomic resolution images of Pt nanoparticles.

The techniques of reflection electron microscopy (REM) and low-energy electron microscopy (LEEM) are broadly extensions based on the RHEED and LEED techniques and find some applications in surface physics and surface analysis. However, the technique of scanning electron microscopy (SEM) is one of the more popular for the study of surfaces.

### 5.3.5   Scanning Electron Microscopy

The experimental set-up of the SEM technique is illustrated below (Figure 5.47). An electron beam with primary beam energy in the range 1–10 keV is focused by a lens system into a spot of about 1–10 nm in diameter onto the surface of a sample. The focused beam is scanned in a raster across the sample surface by a deflection coil system in synchronism with an electron beam of a video tube, which is used as an optical display. Both beams are controlled by the same scan generator and the magnification is just the size ratio of the display and scanned area on the sample surface. A variety of signals can be detected, including secondary electrons, backscattered electrons, X-rays, cathodoluminescence, and sample current. The 2D map of the signal yields the SEM image.

The main applications of SEM concern the visualization of the surface topography and elemental mapping. Figure 5.48 shows a typical SEM image for a film surface structure with crystallite formations. To consider the nature

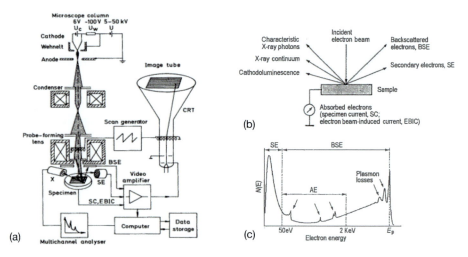

**FIGURE 5.47**   (a) Schematic illustration of a scanning electron microscope. (b) The types of signal that can be generated by irradiating the sample with a primary electron beam. (c) Energy spectrum of electrons emitted from the sample under illumination from a primary electron beam of energy $E_p$. The energy regions corresponding to secondary electrons (SE), backscattered electrons (BSE) and Auger electrons (AE) are shown.

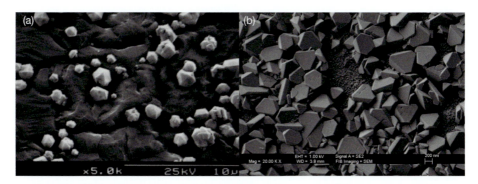

**FIGURE 5.48**   (a) Surface of a diamond film. (b) Manganese oxide film.

of the contrast in various SEM modes, it should be noted that the energy spectrum of the electrons emitted from the sample surface which is irradiated by an electron beam, of energy $E_0$, will produce various peaks depending on the material. Besides the peak of elastically scattered electrons at $E_0$, the spectrum will have secondary electrons in the region of 0–50 eV (SE). The region of inelastic backscattered electrons (BSE) will range from 50 eV–$E_0$. The Auger and loss peaks fall into this region. By appropriate choice of detector, the signal of the electrons from the desired energy range can be monitored.

In secondary electron (SE) mode, electrons with an energy of a few eV are collected by a directional detector. Due to the angular dependence of the SE yield and shadowing effect, the SE image shows the surface topography. The contrast due to the chemical composition is not so essential. Most of the secondary electrons originate from a shallow ($\sim$ nm) layer within the primary beam. This allows a resolution of the order of 5–20 nm. However, some secondary electrons are excited by the backscattered electrons (for which the information region exceeds the primary spot), hence typical BCE contrast is superposed on all SE micrographs.

**Backscattered electron mode (BSE)**: For collecting backscattered electrons of relatively high energy, detectors with a large acceptance angle are used. The most essential contrast in BSE mode is due to the dependence of the backscattering coefficient on the atomic number of the species. The higher the atomic number, the greater the backscattering and consequently the brighter the corresponding area in the image. Furthermore, if using energy filtering, one can select an Auger peak or characteristic loss peak for a given atomic species, and a map of the spatial distribution of this species on the surface can be obtained.

**Sample current (SE)**: The sample current is essentially the primary beam current minus the emission currents of SE and BSE. If the SE emission is suppressed by a positive bias applied to the sample, the map of the sample current will correspond to the map of the backscattered coefficient in the inverted contrast. It will be noted that this mode does not require an additional detector.

**X-rays**: Besides the secondary and backscattered electrons, the primary electron beam induces the emission of X-rays, whose signal can also be employed for surface imaging. With the detector turned to the characteristic X-ray energy, a map of the spatial distribution of a given element can be obtained. This is qualitatively similar to elemental mapping using the AES signal. The difference is the greater information depth of the X-ray probe (0.1–10 m), compared to that of the AES signal (a few nm). As a consequence, X-ray probe analysis will have a worse spatial resolution, but offers better depth analysis. In contrast to AES, X-ray probing does not require UHV conditions.

**Electron-beam-induced current (EBIC)**: The electron irradiation of a semiconductor generates a lot of electron-hole pairs (a few thousand per incident electron). The generated charge carriers are separated in the depletion layers of the p-n junction. This process results in an electron-beam-induced current (EBIC) which can be amplified and used for the inspection of semiconductor devices (for example, to image p-n junctions, to localize avalanche breakdowns, to visualize the electrically active defects).

**Cathodo-luminescence (CL)**: Cathodo-luminescence, i.e., the emission of ultraviolet or visible-light-induced by an electron bombardment, is another source of analytical information. CL signals are typically of low intensity, hence a sensitive detector with a large collection solid angle is required. CL is often used in combination with EBIC for the characterization of semiconductor devices and, in particular, for the imaging of lattice defects which affect the recombination rate of charge carriers.

SEM can be applied to many types of specimens, from biological materials to metals and mechanical parts. SEM plays an enormously important role in materials science, biology, physics, and engineering sciences. However, sample preparation is required, where the surface must be of a conducting nature, since charging in insulating samples can severely hamper or distort the SEM image.

A further variation of the SEM technique can be obtained by performing a polarizing analysis on the secondary electrons, which are emitted from the sample surface. Scanning electron microscopy with polarization analysis (SEMPA) provides a powerful analytical tool in the study of ferromagnetic materials. The spin polarization of the emitted secondary electrons provides a measure of the surface magnetization. In Figure 5.49(a), the apparatus is illustrated and the surface state magnetization can be used for magnetic domain imaging in magnetic thin films and multilayers. In addition, Figure 5.49(b) shows the magnetization oscillations observed in magnetic multilayer systems as a function of the interlayer metallic non-magnetic spacer (wedge).

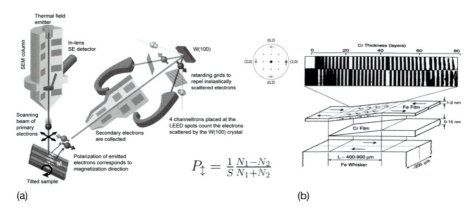

(a)  (b)

**FIGURE 5.49** (a) Schematic illustration of the SEMPA apparatus. An unpolarized focused electron beam is scanned across the surface of a ferromagnetic material and the spin polarization of the emitted secondary electrons is determined using a spin analyzer. (b) The difference in magnetic coupling is illustrated for a Fe/Cr/Fe sandwich structure where the Cr interlayer has a wedge form. The variation of this interlayer thickness across the sandwich effectively varies the magnetic interlayer coupling between ferromagnetic and antiferromagnetic. (Such effects will be considered in more detail in Chapter 4 of Volume 2.) Reprinted from Pierce, D. T., Unguris, J., Celotta, R. J., Stiles, M. D. (1999). *J. Magn. Magn. Mater., 200*, 290, © (1999), with permission from Elsevier.

### 5.3.6 *Scanning Tunneling Microscopy*

The STM was developed in 1981 by Gerd Binnig and Heinrich Rohrer, who were subsequently awarded the Nobel prize in Physics for their invention in 1986. A schematic illustration of the STM apparatus is shown below in Figure 5.50. The main components of the STM are given as follows:

(1) An atomically sharp tip. STM tips are typically made from metal wires (e.g., W, Pt-Ir, Au). The preparation procedure of the atomically sharp

tip includes a preliminary ex-situ treatment, such as mechanical grinding, cleavage, electrochemical etching and later in-situ treatments such as field emission/evaporation or even soft-crashing of the tip by touching the surface of a sample.

(2) A scanning stage or raster to scan the tip over the area of interest on the sample surface. Piezoelectric ceramics are used in scanners as electromechanical transducers, as they can convert electrical signals of 1 mV to 1 kV into mechanical motion, in the range from a fraction of an Ångström to a few m.

(3) Feedback electronics are required to control the tip—sample separation.

(4) A computer system to control the tip position, to acquire data and to convert the data into an image of the surface.

(5) A course positioning system is required to bring the tip to within the tunneling distance of the sample and, when required (for changing the sample), to retract it back to a sufficient distance—a few mm.

(6) Vibration isolation: For stable operation of the STM, changes of tip-sample separation caused by vibrations should be kept to within $\sim 0.01$ Å. The required vibration damping is achieved by suspending the inner STM stage with the tip and sample on very soft springs and by using the interaction between the eddy current induced in copper plates attached to the inner stage and the magnetic field of the permanent magnets mounted on the outer stage.

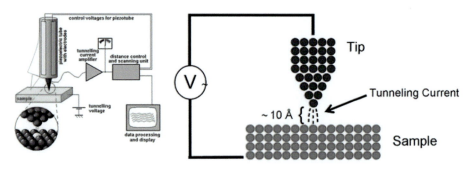

**FIGURE 5.50**   Schematic illustration of the STM apparatus.

The operating principles of the STM are as follows: The very sharp tip of the microscope is placed so close to the probed surface that the wave functions of the electrons in the closest tip atom and the surface atom overlap with one another. This takes place at tip-sample separations of around 5–10 Å. If one applies a bias voltage $V$ between the tip and the sample, a tunnel current can flow through the gap. In a simplified form, the tunnel current density, $j$, can be expressed in the form:

$$j = D(V)\frac{V}{d}e^{-A\phi_B^{0.5}d} \tag{5.40}$$

where $d$ is the effective tunneling gap, $D(V)$ reflects the electron state densities, $A$ is a constant and $\phi_B$ is the effective barrier height of the junction (see Figure 5.51).

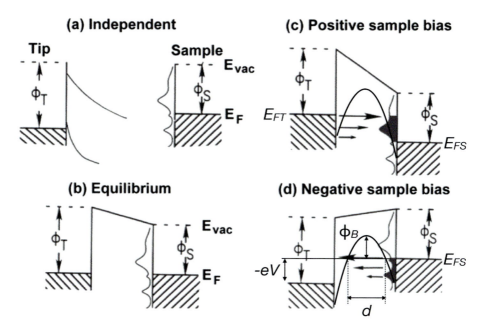

**FIGURE 5.51** Energy diagram of a metallic sample and tip in an STM experiment. (a) For tip and sample well separated and independent. (b) When tip and surface are in close contact, but with no potential applied between them. The Fermi levels for the two metals will be at the same potential and no tunneling will occur. (c) When a forward bias $V$ is applied between tip and surface, empty states are open for tunneling to take place from tip to surface. (d) With reverse bias, the opposite is true. We note, that when there is a difference of potential between tip and surface, the tunneling probability will be sensitive to the empty (or filled) states of the sample, depending on the forward (or reverse) bias that is applied.

The sharp dependence of the tunnel current on the gap width determines the extremely high vertical resolution of the STM. Typically, a change of the gap by $\Delta d = 1$ Å results in a change in the tunnel current by an order of magnitude or, if the current is kept constant to within 2%, the gap width remains constant to within 0.01 Å. As for the lateral resolution of the STM, this is determined by the fact, that up to 90% of the tunnel current flows through the gap between the last atom of the tip and the atom of the surface, which is closest to it. Surface atoms with an atomic separation of less than 2 Å can be resolved.

By scanning the tip along the surface, one obtains the pattern of the surface topography. However, one should bear in mind that STM is not primarily sensitive to atomic positions, but rather to the local density of electronic states. When the tip bias voltage is positive with respect to the sample, the STM image will correspond to the surface map of the filled electronic states. With a negative tip bias voltage, the empty-state STM image will be obtained. Hence, the maximum in the STM image might correspond to both the topographical protrusions on the surface and the increased local density of states.

There are five main variable parameters in STM. These are the lateral coordinates ($x$ and $y$), the height above the sample ($z$), the bias voltage ($V$) and the tunnel current ($I$). Three main STM modes of operation are defined, depending on the manner in which these parameters are varied:

(1) Constant current mode: In this mode, $I$ and $V$ are kept constant, $x$ and $y$ are varied by rastering the tip and the height $z$ is measured.
(2) Constant height mode, also called current imaging: In this mode, $z$ and $V$ are kept constant, $x$ and $y$ are varied by rastering the tip and the tunnel current $I$ is measured.
(3) Scanning tunneling spectroscopy (STS): This is a series of various modes in which $V$ is varied.

We will now consider each of these modes in greater detail.

**1. Constant Current Mode**: This is the most widely used technique for the acquisition of STM images. In this mode, the tip is scanned across the surface at constant voltage and current. To maintain the tunneling current at a preset value, a servo system continuously adjusts the vertical position of the tip by a variation of the feedback voltage $V_z$ on the $z$-piezoelectric driver. In ideal circumstances of an electronically homogeneous surface, the constant current will essentially mean a constant gap; i.e., the scanning tip tracks all the features of the surface topography, see Figure 5.52(a). The height of the surface features is derived from $V_z$. Thus, the surface relief height $z(x,y)$ is acquired as a function of the tip position at the surface. The advantages of the constant-current mode reside in the possibility of probing surfaces, which are not necessarily atomically flat and in the ability to determine the surface height quantitatively from $V_z$ and the sensibility of the piezoelectric driver. The disadvantage is the limited scan speed due to the finite response time of the servo system.

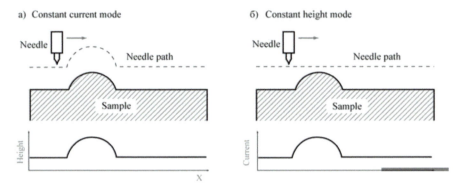

**FIGURE 5.52** Schematic illustration of STM imaging in (a) constant current mode and (b) constant height mode.

**2. Constant Height Mode**: In this mode, the tip is scanned across the surface with $V_z$ kept constant and the variations of tunnel current are recorded as a function of the tip position, see Figure 5.52(b). The bias voltage is fixed and the feedback is slowed or switched off. The rastering of the tip can be done at a greater speed than in the constant-current mode, since the servo system does not have to respond to the surface features which pass under the tip. This ability is especially valuable for studying real-time dynamical processes and, in particular, for recording the STM video. The disadvantages are that the technique is applicable only for relatively flat surfaces; and quantitative determination of the topographic heights from the variation of the tunneling current is not a simple process since a separate determination of $\phi_B^{0.5}$ is required to calibrate the z-response.

**3. Scanning Tunneling Spectroscopy (STS)**: Since the tunneling current is determined by the summation over electron states in the energy interval determined by the bias voltage $V$, by varying $V$ it is possible to obtain information on the local density of states as a function of energy. One way is to acquire a set of constant current STM images of the same surface area at various values of $V$ and both polarities. Another way is to conduct, at each pixel of the scan, a measurement of the tunneling current $I$ versus the bias voltage $V$ at constant tip-sample separation. From this measured $I - V$ curves it is possible to calculate $\frac{dI/dV}{I/V}$, which closely corresponds to the sample density of states. Thus, the spatial distribution of particular states can be mapped. This technique is referred to as current-imaging tunneling spectroscopy (CITS).

STS allows the probing of the electronic properties of a very local pre-elected area, even of an individual adatom on the surface (see Figure 5.53). This provides the significant possibility of distinguishing surface atoms of different chemical natures. In general, spectroscopic information is valuable for consideration of the energy gap, band bending, and chemical bonding.

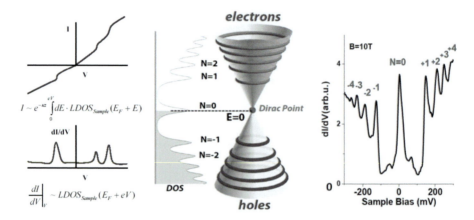

**FIGURE 5.53**     Schematic illustration of STS between tip and surface for $I - V$ and it is differential. We note that the latter helps to accentuate the features of the $I - V$ curve. In the center, we show the schematic representation of the quantization of states due to an applied magnetic field (the so-called Landau levels) in graphene. On the right-hand side, we see the corresponding d$I$/d$V$ STS spectrum for an applied magnetic field of 10 T.

However, in general, there is a lack of a sure and simple way for the interpretation of STS data. For meaningful consideration, a thorough theoretical calculation is required and, even in this case, one should be cautious of conclusions. For example, it is usually assumed that the tip is featureless and its electron density of states resembles that of the free electron gas in the ideal bulk metal. In fact, the validity of the assumption for a real tip requires specific testing. Another problem is associated with the fact that the tunneling current depends on the tunneling transmission probability density. As a result, the large density of states of the sample will not be accessible in measurements if these states do not overlap with those of the tip.

The atomic resolution of the STM technique makes it an almost unique surface analytical tool. Figure 5.54, we show some typical STM images.

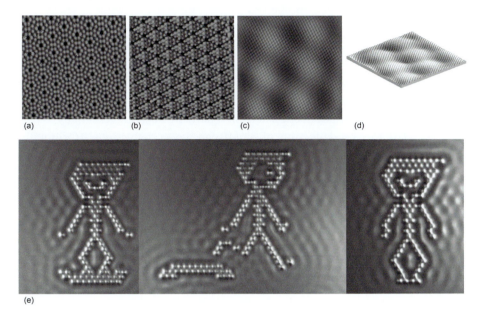

(a)        (b)        (c)        (d)

(e)

**FIGURE 5.54** Si(111)-(7x7) surface. Various topography imaging of (a) empty and (b) occupied states (STM, 20 nm × 20 nm). (c) Superstructure on HOPG surface (8.4 nm × 8.4 nm). (d) Perspective view. (e) Frames from a STM animation, which were made by moving atoms one at a time using a scanning tunneling microscope. The animation can be seen at: http://newatlas.com/smallest-animation-ibm-atom/27346/ (accessed on 30 March 2020).

## 5.3.7 *Atomic Force Microscopy*

The enormous success of the STM technique have inspired the development of a set of novel scanning probe microscopy (SPM) methods; for example, atomic force microscopy, lateral force microscopy, magnetic force microscopy, ballistic electron-emission microscopy, scanning ion conductance microscopy, near-field scanning optical microscopy, to name the most important. As in the STM technique, the SPM methods are based on the use of piezoelectric transducers that provide the ability to control the spatial position of the probing tip relative to the surface of the sample with great accuracy and thus to map the measured surface property on an atomic or nanometer scale.

Among the various SPM techniques, atomic force microscopy has found the broadest applicability.

The atomic force microscope (AFM) was invented by Binnig, Quate, and Gerber in 1986. AFM measures the forces between the sample and the tip. The principle of the AFM operation is illustrated below (Figure 5.55). A sharp tip, a few microns in length, is located at the free end of a cantilever (usually 100–200 m long). The interatomic forces between the tip and the sample surface atoms cause the cantilever to deflect. The cantilever displacement is measured by a deflection sensor. There are several techniques, which can be used for the deflection of small cantilever deflections. Originally, Binnig et al. used an STM as the deflection sensor and measured the tunnel current between the tip and the conductive rear side of the lever (see Figure 5.56(a)). Other sensors utilize optical interferometry (Figure 5.56(b)), a reflection of the laser beam from the rear side of the cantilever (Figure 5.56(c)) or measurement of the capacitance between the cantilever and the electrode located close to the rear side of the cantilever (Figure 5.56(d)). Typically, sensors can detect deflections of $10^{-2}$ Å. Measuring the deflection of the cantilever while the tip is scanned over the surface of the sample (or as the sample is scanned under the tip) allows the surface topography to be mapped. The evident advantage of AFM is that it is applicable for the study of all types of surfaces: conducting, semiconducting and insulating. The latter being impossible in conventional STM due to surface charging problems.

When considering the tip-surface interaction, we refer to the interatomic force versus distance curve, as illustrated in Figure 5.57. When the tip-to-sample separation is relatively large—right-hand side of the curve—the cantilever is weakly attracted to the sample. With decreasing distance, this attraction increases until the separation becomes so small that the electron clouds of the tip and sample surface atoms begin to repel each other electrostatically. The net force goes to zero at a distance on the order of the length of a chemical bond (a few Å) and at closer distances, the repulsive force will dominate, as shown. The range of tip-to-sample separation used in AFM imaging defines the mode of operation: contact mode, non-contact mode, and tapping mode.

### 5.3.7.1   AFM Contact Mode

In contact mode, the tip-sample separation is on the order of a few Å. Thus, an AFM tip is in soft physical contact with the sample and is subjected to repulsive forces. To avoid damaging the probed surface, the cantilever should not

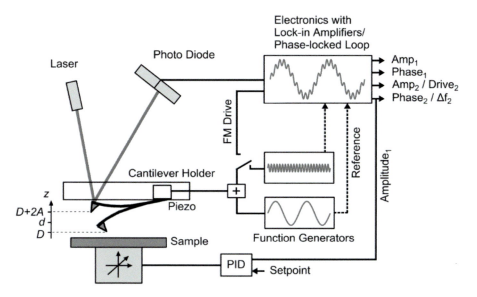

**FIGURE 5.55** Schematic illustration of the AFM apparatus and operation.

be too stiff; i.e., the cantilever spring constant should be lower than the effective spring constant of the sample atomic bonding. Therefore, the tip-sample interaction causes the cantilever to bend upwards, following the changes in surface topography. Topographic AFM images are acquired typically in one of two modes: constant-height mode or constant-force mode:

(1) In the constant-height mode, the scanner height will be fixed and the cantilever deflection is monitored to generate the topographic image. A constant-height mode is preferred for the acquisition of atomic-scale images for atomically flat surfaces; i.e., where cantilever deflections are small and for fast recording of real-time images of changing surfaces; the case where rapid scanning is required, for dynamical surface studies.

(2) In constant-force mode, the cantilever deflection is fixed (which means that the net force applied to the cantilever is constant) through the continuous adjustment of the scanner height by a servo system. The image is generated via the scanner motion. Constant-force mode is the most widely used mode, as the net force is well controlled and the data set is easy to interpret. The disadvantage of this mode is the limited scan speed that can be achieved due to the finite response time of the feedback circuit.

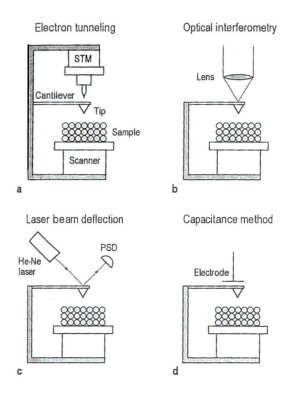

**FIGURE 5.56**    Schematic illustration of the AFM techniques for monitoring the cantilever deflection. (a) electron tunneling; (b) optical interferometry; (c) laser beam deflection; and (d) capacitance method.

### 5.3.7.2    AFM Non-Contact Mode

The tip-sample separation in non-contact mode is conventionally on the order of 10-100s of Å. As such, the cantilever is affected by weak attractive forces. In this mode of operation of AFM, the stiff cantilever is kept vibrating near its resonance frequency. Typical frequencies are from 100-400 kHz with a typical oscillation amplitude of a few 10s of Å. Due to the interaction with the sample, the cantilever resonance frequency, $f_1$, changes according to:

$$f_1 \propto \sqrt{c - F'} \qquad (5.41)$$

where $c$ is the cantilever spring constant and $F'$ is the force gradient. If the resonance frequency (or the vibrational amplitude) is kept constant by a feedback system, which controls the scanner height, the probing tip traces

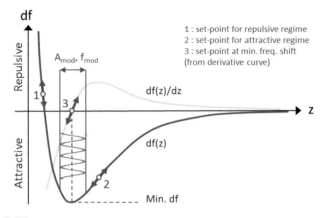

**FIGURE 5.57** The variation of the van der Waals force as a function of the distance between tip and surface is responsible for the interaction necessary for the atomic force microscope (AFM). Illustrated are the different regions of operation, from contact to non-contact modes, as well as the intermittent or tapping mode of operation between the two limits. In this latter the AFM is sensitive to the derivative of the force as a function of separation.

lines of the constant gradient. The motion of the scanner is used to generate the full data set. This mode is also referred to as constant-gradient mode. The efficiency of this detection scheme has been proved by reaching atomic resolution in AFM images acquired in the non-contact mode. An example is shown in Figure 5.58 for a Si(100) clean surface indicating a $(2 \times 1)$ surface reconstruction.

**FIGURE 5.58** Non-contact image of the Si(100) clean $(2 \times 1)$ surface.

### 5.3.7.3 AFM Tapping Mode

This AFM mode of operation is similar to the non-contact mode. The only difference being that the cantilever tip at the bottom point of its oscillation barely touches (taps) the sample surface. Tapping mode does not provide atomic resolution, but appears to be advantageous for imaging rough surfaces with high topographical corrugations.

### 5.3.8   *Magnetic Force Microscopy*

A variation of AFM, magnetic force microscopy (MFM) images the spatial variation of magnetic stray forces on a sample surface. For MFM the tip will be coated with a ferromagnetic film. The magnetic force between the sample and the tip is given by:

$$\mathbf{F}_m = (\mathbf{m} \cdot \nabla) \mathbf{H} \tag{5.42}$$

where $\mathbf{m}$ is the magnetic moment of the tip and $\mathbf{H}$ is the magnetic stray field from the sample. The system is operated in non-contact mode, detecting changes in the resonance frequency of the cantilever induced by the magnetic fields dependence on tip-to-sample separation (see Figure 5.59).

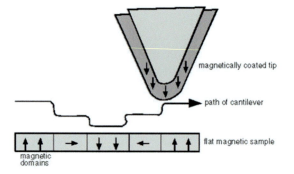

**FIGURE 5.59**   Schematic illustration of the magnetic force microscope tip and its interaction with a magnetic sample with domains.

MFM can be used to image magnetic domain structures. An image taken with a magnetic tip contains information about both topography and magnetic properties of a surface. Which effect dominates, depends upon the distance of the tip from the surface, because the interatomic magnetic force persists for greater tip-to-sample separations than the van der Waals force. If the tip is close to the surface, in the region where standard non-contact AFM is operated, the image will be predominantly topographic. As we increase the separation between tip and sample, magnetic effects will become apparent. One way to distinguish between these effects is to take a set of images at different tip heights.

Since the magnetic stray field from the sample will affect the magnetized state of the magnetic tip, and vice-versa, it can be difficult to obtain quantitative magnetic information from MFM measurements. To interpret the MFM information quantitatively, the tip configuration must be known. Such measurements can achieve spatial resolutions of 30 nm. In Figure 5.60(b), we show a MFM image of the magnetic bits on a hard disk. In Figure 5.61, we illustrate the variation of the magnetic domain structure in a thin epitaxial Fe (110) film under the application of an external magnetic field.

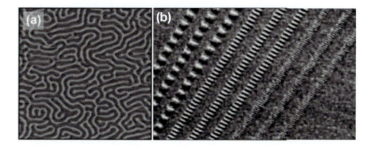

**FIGURE 5.60** (a) MFM images of magnetic "maze" domain patterns in GdFe multilayer films. (b) This magnetic force microscope image of a disk shows the individual data bits in tracks of different densities. The far right tracks can store 10 billion bits per square inch. Techniques using ultrafast lasers may allow recording on these tracks at extremely high speeds.

### 5.3.9 *Scanning Near-Field Optical Microscopy*

The diffraction-limited optics in conventional optical microscopy means, that the spatial resolution of an image is fundamentally set by the wavelength of the incident light as well as the numerical aperture of the condenser and objective lens systems. The development of near-field optical methods has allowed this limit to be swept aside. Scanning near-field optical microscopy (SNOM), also referred to as near-field scanning optical microscopy (NSOM), is yet another development, that owes its existence to the work of Binnig and Rohrer in the invention of the scanning probe techniques in the 1980s.

The first development of the SNOM techniques was published by D. Pohl et al. (1984), who announced, that they could achieve a spatial resolution, well beyond the diffraction limit, of $\lambda/20$ using near-field optics. For their work at $\lambda = 488$ nm, they obtained a resolution of around 25 nm using a quartz tip etched using hydrofluoric acid and coated with aluminum.

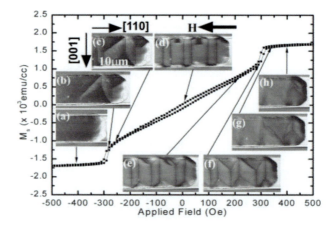

**FIGURE 5.61**   Magnetic domain structures of a patterned epitaxial (110) Fe film under a magnetic field. The film was magnetized at $-5,000$ Oe before the measurement. (a) At $-400$ Oe, the edge domain wall is formed. (b) At $-300$ Oe to $-270$ Oe, the domain wall starts to propagate from the edge into the sample. It shows a rapid change in magnetization. (c) At $-260$ Oe, the domain structure is formed in the film. (c)–(f) Between $-260$ Oe and 260 Oe, the domain walls only move within a short distance. This shows flux closure domains. (g) At 300 Oe, the domain walls start to disappear and exist only at the edge and at the surface. (The surface domain walls has a much lower signal, i.e., light color.) (h) At 400 Oe, it is close to the saturation magnetization, however, the surface domain wall is still visible. In contrast, there is no surface domain in the domain pattern (a) (that was magnetized at $-5,000$ Oe before the measurement). The surface domain disappears at an applied magnetic field of about 600 Oe.

The diffraction limit plays the role of a low-pass filter in terms of the spatial frequencies of a sample, since only plane waves can reach the detector, which is placed far from the object. However, information regarding the sub-wavelength optics remains confined to space immediately above the object's surface. This confined region of light is known as the near-field optical zone. It is in this region, that the evanescent electromagnetic field contributes significantly to the total electromagnetic field. We note, that the evanescent near-field light is a non-propagating field, that is permanently localized around the surface of the object. In fact, the first hint of such an approach was made much earlier with the works of Edward H. Synge in 1928. He suggested a new form of an optical microscope, that would overcome the diffraction limit, but required an aperture with dimensions much inferior to those of the wavelength used. Synge was also aware, that such an approach posed a number of complex difficulties such as the fabrication of such a small aperture and its positioning on a very precise scale. Indeed, these difficulties were not overcome until the 1970s, when E. A. Ash and G. Nicholls (1972) demonstrated the near-field resolution of a sub-wavelength aperture scanning microscope operating in the microwave region of the electromagnetic spectrum. Clearly, the longer wavelengths of the microwaves with respect to the visible spectrum meant, that the fabrication of the aperture could be in the range of millimeters and not the nanometer scale required for visible light.

There are many optical configurations possible to detect sub-wavelength information from the surface of an object. Firstly, it is possible to illuminate the surface with evanescent waves, which are diffracted by the surface, giving rise to plane waves containing near-field information, and which can be detecting in the traditional way. Alternatively, the surface can be illuminated with plane waves, which are again scattered from the surface in the form of evanescent waves, which again carry information from the near-field region. In either case, it is necessary to find a method of converting plane waves into evanescent waves or vice versa. The SNOM method uses a probe tip, which serves as a method of conversion. This tip often contains an aperture, though aperture less tips are also frequently used. The essential geometries of these approaches are illustrated in Figure 5.62. Furthermore, there are,

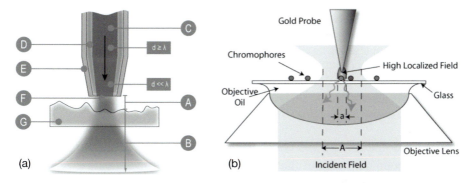

**FIGURE 5.62**    (a) Apertured SNOM tip: A – near-field area $(h \ll \lambda)$, B – far-field area $(h \geq \lambda)$, C – laser beam, D – optical fiber, E – metallic coating of optical fiber, F – aperture $(d \ll \lambda)$, G – sample. (b) Apertureless SNOM probe.

within the aperture approach, a number of alternative schemes for measuring the near-field radiation, as illustrated in Figure 5.63. As with other scanning probe microscopies, the probe-sample distance must be very small since the evanescent waves decay exponentially from the sample surface. The closer the tip to the sample, the stronger the signal. The image formed in the SNOM method relies, as with other scanning probe techniques, on the scanning of

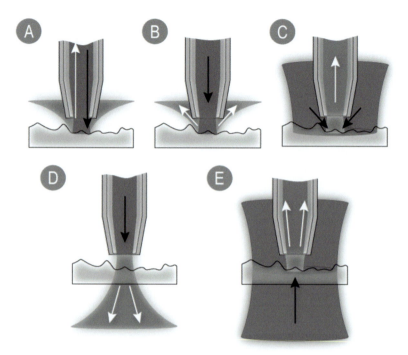

**FIGURE 5.63**   Basic techniques of illumination and/or signal detection: (A) Illuminate and collect reflected radiation using the same optical fiber; (B) Illuminate
using an optical fiber probe, collect the reflected light from the sample surface; (C)
Illuminate sample using an external source of laser radiation. To collect reflected radiation, use probe; (D) Illuminate the sample using a probe, collect transmitted light.
The technique is applicable only to transparent samples; (E) Illuminate sample using
an external source of laser radiation. To collect transmitted light, use the probe. The
technique is applicable only to transparent samples.

the tip over the surface and collecting the signal (light) emitted from the surface to obtain an image of the sample. This is performed using piezoelectric
driven $x - y - z$ scanners to control the tip position, which forms a raster
pattern over the sample. The collection and formation of the image itself
will depend on the collection scheme, as illustrated in Figure 5.63. The scan
system can be either coupled to the sample or to the tip. The signal data is
correlated with the relative tip position over the sample to form a computer-
generated 2D or 3D representation of the sample surface. Feature scales can
be as little as 1 nm, giving excellent image resolution. The tip of the SNOM
set-up has similar attributes to the AFM tip and can allow measurements with
constant height mode or constant intensity mode in an analogous fashion.
To improve the signal-to-noise ratio, the tip can be operated in oscillation,

where the frequency of oscillation corresponds to the resonance mode of the tip. This allows lock-in detection to be used, permitting the filtering of low-frequency noise. Furthermore, for good Q-factors, sensitivity can be greatly enhanced. As with the AFM, the tip-sample interaction will change the resonance frequency and Q-factor. The tip oscillation amplitude and frequency can be monitored, such as by the shear-force, which uses lateral motion to control the tip-sample separation. A tapping mode can also be utilized in an analogous manner to the AFM.

Other forms of scanning microscopies do not benefit from the broad range of contrast mechanisms that are available to optical microscopy. Most are limited to morphological studies. Furthermore, contrast enhancement methods are available such as staining, fluorescence, polarization, phase contrast, and differential interference contrast. Optical methods can also have spectroscopic and temporal resolution capabilities and will depend on the nature of the illumination used. In the latter, a pulsed laser is required.

A schematic illustration of the essential components of a SNOM system, showing the control and information flow of an inverted microscope is depicted in Figure 5.64. The system has a laser excitation source, which is

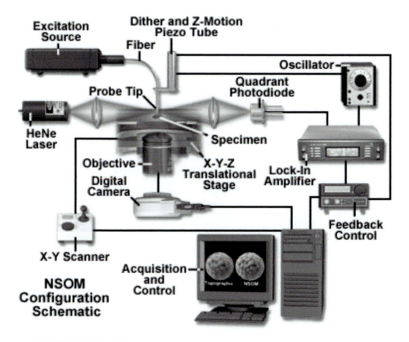

**FIGURE 5.64**   Illustration of a modern SNOM microscope.

coupled into an optical fiber for illumination of the sample, and the probe tip motion is monitored through a feedback loop coupled to a second laser (HeNe). As with other microscopy techniques, many researchers adapt their systems to perform specific tasks and image objects to highlight their properties. There are a great number of experimental configurations, though they all rely on the basic principles outlined above.

By placing a nanoparticle on the tip of an apertureless SNOM in proximity with a flat gold surface, it is possible to study the properties of a single nanocrystal. This is illustrated in Figure 5.65. This has been performed by Jazi et al., (2018), who studied a single CdSe/CdS core-shell nanoparticle. The measurements were made as a series of scans as a function of the distance between the particle and the gold surface. The results were compared with models of the expected response as a function of the nanocrystal orientation. Once the orientation has been determined, the tip-nanocrystal ensemble can be used as a probe for plasmonic nanostructures. In this way, it is possible to make a near-field probe of the size of the nanocrystal (15 nm in the case of this study).

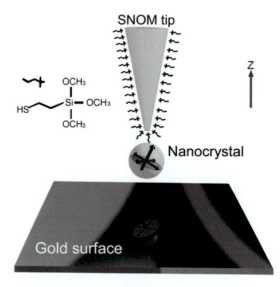

**FIGURE 5.65**    Scheme of the NC attached at the end of the SNOM tip covered with a silane/thiol polymer over a flat gold layer. The c-axis (black arrow) is perpendicular to the 2D dipole (red arrows). Reproduced from Jazi et al., 2018 with permission from the PCCP Owner Societies.

## 5.4 Summary

Surface analysis is crucial for an understanding of the physical properties of surfaces and thin films. We have classified the techniques for surface analysis into diffraction, spectroscopies, and microscopies. Each has a specific range of applicability and provides specific information about the surface. Diffraction techniques are used to determine the surface crystalline order, spectroscopies provide information on the surface chemical composition as well as surface excitations, while microscopies are used to give a direct image of the surface, allowing morphological features to be imaged.

Diffraction techniques have been used for many decades as an analytical tool for determining the crystalline structures of solids. The information they provide allows the determination of both the crystalline order of materials and of the lattice parameters, which define the atomic order. In the case of surface analysis, it is necessary to limit the penetration of the incident radiation that only information from the surface atomic layers is probed. This can be generally achieved by either reducing the energy of the radiation or by making it incident at a grazing angle. In this chapter, we have concentrated on the use of electrons to produce the diffraction images since they are the most commonly available for in-situ studies. Electrons are particularly well suited to such studies since it is rather easy to control their energies and wavelengths using the accelerating potential of the electron gun. A low energy of the incident electron beam will limit the penetration to the uppermost atomic layers, as will be noted from the inelastic mean free path lengths for energies in the range of a few hundred eV, see Figure 5.13. This fact is exploited in the low-energy electron diffraction method, where the incident beam is normal to the surface. Electrons that are elastically scattered from the surface are imaged on a fluorescent screen. Here the geometry of the set-up provides a top view of the reciprocal lattice from the elastically backscattered electrons. For the case of RHEED, the grazing incidence of the electron beam means that the surface sensitivity is maintained within the upper atomic layers of the surface. The incident beam geometry means that we are obliged to image in the forward direction, giving an end-on view of the reciprocal lattice structure. This also allows the study of morphological features of the surface, making RHEED very sensitive to the surface roughness. Another consequence of the grazing incidence is that the space above the sample surface is "free" and thin film deposition can be performed while imaging the RHEED pattern. This means that dynamic studies of film growth can be made. This is exploited commonly in MBE, where oscillations in the specular and diffracted beams are correlated with the deposition of a monolayer of material.

Electron spectroscopies cover a broad range of surface analytical techniques, which can provide a wealth of information on surface composition, surface electronic states, and excitations. While the diffraction methods rely on the collection of the elastically scattered electrons, spectroscopic techniques are principally concerned with the measurement of inelastically scattered electrons. Once again we need to ensure that the information gathered using electron spectroscopies arises from the surface of the sample, this can either be done by using primary beam energies which lie in the 20–200 eV range or by filtering the energies of the secondary electrons. The exchange of energy from the primary beam with the sample can produce a large number of phenomena and the experimental technique used will allow different processes to be studied.

One of the most commonly used of the electron spectroscopies is Auger electron spectroscopy (AES). This technique is based on the emission of characteristic electrons due to the excitation of core-level electron levels within atoms at the sample surface. The electron exchange between the levels within an atom depends on the specific energies of those levels. Since these are specific to the different elements, the Auger electrons emitted provide a chemical fingerprint allowing us to identify the composition at a solid surface. This is the principal application of AES and can also provide quantitative information on the chemical composition of the atomic species at the surface. Quantitative analysis can be complex, though used correctly it can allow the study of thin film growth modes as a function of deposition thickness. This technique can also be exploited to provide the chemical mapping. Electron energy loss spectroscopy measures the energy transferred to the sample from the primary electron beam. Loss peaks can be identified by adjusting the primary energy and observing which spectral features move in the same quantity as the primary energy. Auger peaks are energy-specific and will not shift with the primary beam energy. EELS can be adapted to measure core-level states in the conduction band as well as plasmons and surface localized plasmons. High-resolution EELS is also capable of detecting the vibrational modes of adsorbed species at the surface of substrates.

Photoelectron spectroscopies rely on the use of ultraviolet light or X-rays to excite secondary electrons. The techniques of ultra-violet and X-ray photoelectron spectroscopies (UPS and XPS, respectively) are based on the photoelectric effect, where the energy of the emitted electrons can be used to determine the valence states of solids or as a means of characterizing the chemical composition of a sample. In the latter case, the information obtained is similar to that of AES. Angular resolved measurements are often used in UPS to determine the band structure of solids.

Microscopies are a powerful means of making direct images of the real space surface of a solid. There are a host of techniques available and regularly used to study the nature of surfaces and nanostructured materials. One of the earliest methods used to obtain a detailed image of a surface is via the field emission of electrons from a pointed tip. The tip allows a large field gradient to be established which can draw out surface electrons via a quantum tunneling effect. While the images obtained from the field emission microscope appear somewhat blurred, they allow the surface planes of the tip to be observed. The related field ion microscopy technique works on a similar principle, the imaging is performed by the emission of ions adsorbed on the tip surface. Since the ions are in much closer contact with the surface atoms, the imaging process is of far greater resolution, allowing individual atoms to be observed. These techniques are limited to tip shaped samples which are typically metallic and can support large electric field gradients.

Electron microscopes are among the most used methods for the study of materials. Both scanning and transmission techniques are extensively used to characterize materials and their surfaces. TEM was invented in the 1930s and has developed into an extremely important tool in the study of materials. The high-resolution version allows imaging at the atomic scale. In this case, a beam of mono-energetic electrons is incident on a thinned sample, allowing the electrons to penetrate completely. These electrons can be used to obtain the diffraction pattern of the sample or can be focused on using electron lenses to obtain a real-space image of the sample. The scanning electron microscope is used to obtain a direct real-space image of a surface via a number of effects. The incident electron beam interacts with the sample, much as is the case with the electron beams used in electron spectroscopies. The imaging of the surface is made by measuring the intensity of the secondary electrons or X-rays emitted from the sample surface. By scanning the incident or primary beam across the surface, the imaging is performed by measuring the intensity of the secondary electron/X-ray beam as a function of position. This provides a 2D image of the surface.

Scanning probe microscopies have become an essential tool in the study of nanosystems. These were initiated in the early 1980s with the work of Binning, Rohrer, and co-workers on the scanning tunneling microscope (STM).

The STM works on the principle that a metallic tip placed in close proximity with a conducting surface will allow a small current of electrons, which tunnel from one to the other, to be measured. The tunnel current is measured as a function of the position of the probe tip over the sample surface. What is quite astounding about this method is that it is extremely sensitive to the separation of the tip and surface and can measure the difference between the position above and between atoms, providing truly atomic resolution microscopy. This has been used to image both metal and semiconducting materials, illustrating the crystalline reconstructions that occur at sample surfaces. The STM technique is sensitive to the density of electronic states of both the tip and the surface materials and can thus be used as a kind of spectroscopy, which measures their density of states. Soon to follow the STM was the atomic force microscope (AFM), also developed by Binnig and co-workers. The electrostatic forces between the tip and surface mean that even non-conducting surfaces can be imaged. The resolution is not as good as that of the STM, but allows the surface morphology to be imaged on a wider range of materials. AFM has become a standard measurement of surface morphologies for the full range of thin films deposited by a broad range of methods. It is both cheaper and easier to use, making it a much more popular method to the STM, which requires a vacuum environment, unlike the AFM which can work in atmospheric conditions. Further adaptations of the AFM method have allowed the tip to be adapted to other types of forces. For example, a magnetic tip can be used to measure magnetic domains, this technique is known as magnetic force microscopy (MFM). Another adaptation using a tip made from an optical fiber or a gold tip can be used to measure the near optical field. This technique, which is referred to as scanning near-field optical microscopy (SNOM), allows surface imaging on a scale well below the diffraction limit for light since it makes use of the near-field evanescent waves and can image objects on the nanoscale.

## 5.5  Problems

(1) Demonstrate the equivalence of the Bragg law: $n\lambda = 2d_{hkl} \sin\theta_{hkl}$, and the von Laue condition as expressed by $\mathbf{k} - \mathbf{k}_0 = \mathbf{G}_{hkl}$, where $\mathbf{k}_0$ is the wave vector of the incident beam and $\mathbf{k}$ that of the diffracted beam. These have amplitudes which can be expressed as: $|\mathbf{k}_0| = |\mathbf{k}| = 2\pi/\lambda$. $\mathbf{G}_{hkl}$ is the reciprocal lattice vector for the (hkl) planes with separation $d_{hkl}$.

(2) Show the form of the LEED pattern for the Si(111)-(7×7) surface.

(3) Compare and contrast the techniques of LEED and RHEED, giving advantages and disadvantages of each.

(4) Interpret the following images:

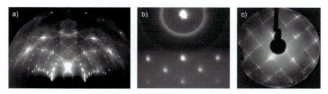

(5) Using image (b) in the previous problem, determine the lattice parameter of the structure, which for this purpose can be assumed to be bcc, with the image being taken along the [110] direction. To calculate this you will need the following experimental conditions: The electron beam is accelerated by a 10 kV potential; sample to screen distance is 75 cm; separation between nearest neighbor diffraction maxima on the screen is 2.5 cm.

(6) Describe how you would experimentally distinguish between Auger and loss peaks in the secondary electron spectrum. Explain.

(7) Aluminum is deposited onto a Si(100) substrate at low temperatures to produce a continuous and uniform Al film with no Al-Si intermixing. Estimate the decay of the Si LVV Auger signal (92 eV) after the deposition of the Al film with thicknesses of 0.25, 1, 5 and 10 ml. Take the attenuation length of the 92 eV electrons in Al to be $\lambda_{Al}(92\ eV) = 4.03$ Å and the thickness of 1 ml of Al to be $d_{Al}(1\ ml) = 1.13$ Å.

(8) Determine the escape depths for Auger electrons in Si and Cr and their oxides ($SiO_2$, $Cr_2O_3$). The following table gives some data that will be required:

| Material | Molecular Mass (g mol$^{-1}$) | Density (g cm$^{-3}$) |
|---|---|---|
| Si | 28.085 | 2.33 |
| Cr | 51.996 | 7.20 |
| $SiO_2$ | 60.08 | 2.26 |
| $Cr_2O_3$ | 151.99 | 5.21 |

(9) Determine the AES escape depth of Ta for its NNN peak at 179 eV as a function of the emission angle at $\theta = 10°$ and $90°$.

(10) Consider the dielectric response $\varepsilon(\omega)$ of a free electron gas and show that the plasma frequency takes the form:

$$\omega_p = \sqrt{\frac{ne^2}{e_0 m}}$$

(11) Check the validity of the plasma frequency for Si, which has four valence electrons per atom and has a plasmon loss peak at 16.9 eV. The lattice constant for Si is 5.43 Å. What is the corresponding surface plasmon loss of energy?

(12) Describe the physical processes involved in the UPS and XPS electron spectroscopies. Explain exactly what information can be obtained from these techniques.

(13) Consider the shape of a spectrum in the second derivative mode $(-d^2N(E)/dE^2)$ for the case of a small peak on an intense background in the $N(E)$ spectrum. Prove that the energy position of the main peak in the $-d^2N(E)/dE^2$ spectrum coincides with that in the $N(E)$ spectrum.

(14) Explain how the experimental set-up of the HREELS method attains the necessary electron energy resolution ($\sim$ meV).

(15) A FEM has a tungsten tip (with work function 4.5 eV) with a radius of curvature of 500 Å and is maintained at a potential of $-5$ kV relative to its surroundings. Estimate:

    (a) the electric field strength just outside the tip.
    (b) over what distance would an electron have to tunnel to escape the tip.

(16) A TEM objective lens has a focal length of 2 mm and forms an image at a distance of 10 cm from the center of the lens. Calculate the magnification assuming thin lens behavior.

(17) Describe in detail the TEM apparatus and how an image is formed which magnifies the object/sample under investigation. Explain how the standard TEM must be modified in order to obtain a high-resolution version of this apparatus (HRTEM).

(18) Describe how an image of magnetic domains is formed using the Lorentz microscopy technique. Explain which factors will be important in determining the spatial resolution of this technique.

(19) A very recent study has used STM to examine the desorption rate of H from a hydrogen-terminated Si(100) surface. After preparing a Si(100) surface with a 1 ml termination of hydrogen atoms, the surface was annealed at 611 K for a prolonged period of time. The number of sites from

which hydrogen desorbed has been counted as a function of annealing time and the following results obtained:

| Coverage of dangling bond sites* (ML) | Annealing time (s) |
|:---:|:---:|
| 0.038 | 2000 |
| 0.065 | 4000 |
| 0.165 | 11000 |
| 0.200 | 13000 |

* i.e., sites, from which hydrogen has been desorbed.

(a) Show that these results are consistent with the following simple rate equation:

$$\frac{d\theta}{dt} = -R_d\theta$$

where $\theta$ is the coverage of desorbing species and $R_d$ is the desorption rate. ($R_d = n_d e^{-E_d/k_B T}$, where $n_d$ is the desorption prefactor and $E_d$ is the desorption energy).

(b) From the data above and given the value of $R_d$ at 668 K, $8 \times 10^{-4}$ s$^{-1}$, estimate the desorption energy.

(20) The high spatial resolution of the STM technique is a consequence of the exponential dependence of the tunnel current with the tip â sample separation, $d$, as given by the relationship:

$$I \propto e^{-2kd}$$

The total measured current is a sum of all the tunneling interactions between the tip and the sample; however, the tip apex and the sample have much higher probabilities than those between other regions of the tip and sample surface. Explore this dependence by considering the following tip profiles:

(a) $y = 2 + x^2$
(b) $y = 7 - (25 - x)^{0.5}$
(c) $y = 2 + x^2/4$

Assume that the tunnel current is the sum of the elemental tunneling currents $dI$ between regions of the tip $dx$ and the sample. Plot the elemental tunneling currents for each tip, where you will need to assume that the current, at any point $x$ is proportional to $e^y$. Which tip will have the highest resolution?

(21) A silicon cantilever used in non-contact AFM has a spring constant of $k \sim 50$ N/m and has a mechanical resonant frequency of $v_0 = 175$ kHz. Estimate the mass of the cantilever.

(22) AFM tips are typically much broader than those used in STM and can be modeled as having a spherical shape. Since the radius of curvature can be substantial, the apparent size of an object can appear somewhat larger than its actual size. One model uses the following equation to relate the radius of tip curvature, $R_t$, the AFM image $R$ and the real radius, $r$, of an object:

$$r = \frac{R^2}{4R_t}$$

(a) Calculate the measured radius of a DNA molecule (real radius 2 nm), when imaged by tips of the radius of curvature 45 nm and 100 nm. Evaluate the broadening for each tip.

(b) For a tip of the radius of curvature 45 nm, calculate the measured radius for features with diameters 20 and 5 nm. Evaluate the broadening for each feature.

(23) From the previous problem, it should be clear that the effective contact area between tip and sample could be quite substantial on the atomic scale. For flat surfaces "atomic" resolution images have been reported, with arrays of points with atomic dimensions of similar appearance to those of STM images. The precise nature of these points has been the subject of debate (see, for example, Amerin, M., et al., (1988). *Science, 240*, 514). Even if a sufficiently sharp tip could be produced, there would be theoretical difficulties in attempting to image individual atoms. Suppose that a minimum load of 1 nN is applied to and AFM tip, with a contact area of radius 10 pm. Estimate the mean pressure applied. Compare this figure with the yield strength for silicon nitride, 550 MPa. What is the likelihood of single-atom contacts giving rise to the AFM image?

## References and Further Reading

Ash, E. A., & Nicholls, G., (1972). *Nature, 237*, 510.

Bauer, E., (2014). *Surface Microscopy with Low Energy Electrons*, Springer, Berlin–Heidelberg.

Binning, G., Rohrer, H., Gerber, Ch., & Weibel, E., (1982). *Phys. Rev. Lett., 49*, 57.

Briggs, D., & Seah, M., (1983). *Practical Surface Analysis by Auger and Photoelectron Spectroscopy*, John Wiley, Chichester.

Julian Chen, C. (2007). *Introduction to Scanning Tunneling Microscopy*, Second Edition, Oxford University Press, Oxford.

Courjon, D., & Bainier, C., (2001). *Le Champ Proche Optique: Théorie et Applications*, Springer-Verlag, France.

Egerton, R. F., (2009). *Rep. Prog. Phys., 72*, 016502.

Goodhew, P. J., Humphreys, J., & Beanland, R., (2000). *Electron Microscopy and Analysis*, 3rd Edition, Wiley.

Hall, P. M., & Morabito, J. M. (1979). *Surface Sci., 83*, 391.

Haugstad, G., (2012). *Atomic Force Microscopy: Understanding the Basic Modes and Advanced Applications*, Wiley.

Henzler, M., (1977). *Electron Spectroscopy for Surface Analysis*, H. Ibach (Ed.), Springer, Berlin.

Hofer, F., Schmidt, F. P., Grogger, W., & Kothleitner, G., (2016). *Mater. Sci. and Engineering, 109*, 012007.

Jazi, R., Ung, T. P. L., Maso, P., Colas des Francs, G., Nasilowski, M., Dubertret, B., Hermier, J.-P., Quélin, X., & Buil, S., (2018). *Phys. Chem. Chem. Phys., 20*, 16444.

Lagally, M. G., (1985). *Methods of Experimental Physics*, Vol. 22, Park, R. L., & M. G. Lagally (Eds.), Academic Press.

Larsen, P. K., Meyer-Ehmsen, G., Bölger, B., & Hoeven, A.-J., (1987). *J. Vac. Sci. Technol. A, 5*, 611.

Larsen, P. K., & Dobson, P. J., (1988). *Reflection High-Energy Electron Diffraction*, NATO-ASI 188, Plenum Press, New York.

Pendry, J. B., (1974). *Low-Energy Electron Diffraction*, Wiley-Interscience.

Pohl, D., Denk, W., & Lanz, M., (1984). *Appl. Phys. Lett., 4*, 651.

Prutton, M., (1998). *Introduction to Surface Physics*, Oxford University Press.

Prutton, M., & Gomati, M. M. El. (Eds.), (2006). *Scanning Auger Electron Microscopy*, John Wiley and Sons.

Samori, P. (Ed.), (2006). *Scanning Probe Microscopies Beyond Imaging: Manipulation of Molecules and Nanostructures*, Wiley-VCH.

Seah, M. P., & Dench, W. A., (1979). *Surf. Int. Analysis, 1*, 2.

Seah, M. P., (1984). *Vacuum, 34*, 463.

Smith, G. C., (1990). *Mat. Charact., 25*, 37.

Synge, E. M., (1928). *Phil. Mag., 6*, 356.

Tsong, T. T., (1990). *Atom-Probe Field Ion Microscopy: Field Ion Emission, and Surfaces and Interfaces at Atomic Resolution*, Cambridge University Press, Cambridge, U.K.

Venables, J. A., (2000). *Introduction to Surface and Thin Film Processes*, Cambridge University Press, Cambridge, U.K.

Watts, J. F., & Wolstenholme, J., (2003). *An Introduction to Surface Analysis by XPS and AES*, Wiley.

Williams, D. B., & Barry Carter, C. (1996). *Transmission Electron Microscopy: A Textbook for Materials Sciences*, 4 volumes, Springer, Berlin—Heidelberg.

Woodruff, D. P., & Delchar, T. A., (1994). *Modern Techniques of Surface Science*, Second Edition, Cambridge University Press.

# PART II

# NANOFABRICATION TECHNIQUES

# Chapter 6

# Lithographic Technologies

Since the early days of the semiconductor and microelectronics industry, miniaturization technologies have been developed and have become one of the cornerstones of the modern high-tech industries. Such developments were particularly important in the success of the integrated circuits (IC) that have reached such levels of miniaturization that chips with over a million transistors can be fabricated in ultra-large-scale integration (ULSI)—the so-called superchip. Technology has advanced so much in recent decades that 3D-ICs are now available, in which two or more active layers of components are integrated both vertically and horizontally in a single circuit.

Much of this technology stems from the use of lithographic techniques for the production of miniaturized components. Today, device fabrication uses various lithographic and nano-fabrication techniques. We will discuss some of the most important techniques, describing their basic principles and applications.

## 6.1 Optical Lithography

Optical lithography, or photolithography, was the first and the earliest microfabrication technology used in semiconductor IC manufacturing. Since the invention of planar technology in the early 1960s, optical lithography has been the main fabrication tool in very-large-scale integration (VLSI). Optical lithography enables the projection of a series of 2D patterns, layer-by-layer, onto a (silicon) substrate and precise alignment of each projection layer. A very complex IC structure is then possible by the patterning technique. Modern IC processing involves normally 20–30 layers of patterning on silicon by optical lithography.

The key to high-density integration of semiconductor devices is to make smaller and smaller images of circuit patterns on substrates, typically silicon, by optical lithography. The smallest feature size by optical lithography in the 1970s was around 4–6 m, which had reduced to $\sim$ 1 m in the early 1980s. It appeared at that time that feature sizes by optical methods were limited and would be replaced by other techniques, such as electron-beam lithography or X-ray lithography, in the coming years. However, optical lithographic techniques are still the dominant patterning technique for VLSI, which is due to improvements in the technique. Today, feature sizes of around 40 nm can be fabricated by optical lithography, though 90 nm technology for ICs is in mass production.

As well as the major fabrication tool for VLSI, optical lithography is also the main production technique for microsystems or MEMS (Micro-Electro-Mechanical Systems), where optical lithography of very thick photoresists are employed.

There are two ways to perform optical lithography: mask aligning and projection. Mask-aligning lithography includes contact and proximity lithography, while projection will allow 1:1 and reduction projection lithography.

Optical lithography is based on a similar principle to photography, where a photoresist coated substrate, usually a silicon wafer, is used in place of a photographic plate or film. A schematic representation of the stages in this planar microfabrication procedure is illustrated in Figure 6.1.

Complex 3D integrated circuits can be produced using a series of 2D planar structures. Each 2D circuit layout forms an optical mask pattern. The purpose of optical lithography is to illuminate through the optical mask, so that the mask layout is imaged onto the photoresist coated substrate. After exposure to light and development, the mask pattern is printed onto the photoresist. The next step is to transfer the photoresist pattern into the substrate. Techniques for pattern transfer will be discussed in further detail at a later stage. We will now discuss the process of mask alignment and projection lithographies in more detail. These processes are illustrated in Figure 6.2.

### 6.1.1 Mask Aligning Lithography

As can be seen in the above figure (Figure 6.2), mask aligning prints the mask image in a 1:1 manner (contact and proximity) and can truly reproduce the image on a substrate. In order to minimize image blurring, mask alignment lithography normally will require that the photomask is in complete contact with the substrate. Such contact printing can be subdivided into hard and soft contact. The former is to apply a pressure between mask and substrate to ensure complete contact, while the latter regards a reduced contact pressure.

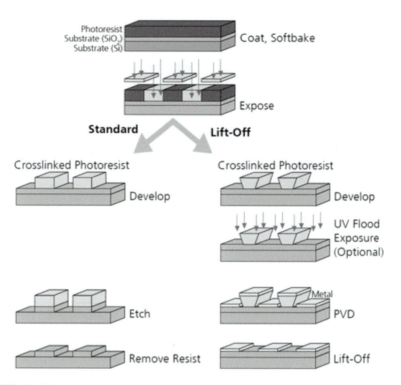

**FIGURE 6.1** The standard procedure for positive and negative mask photolithography. The left-hand side shows how the mask defines the positive features to be produced, while on the right the features of the lithographic structure form a negative with respect to the mask structure.

If a small gap is maintained between a mask and a substrate, the resulting lithography is termed proximity printing. The distances involved are typically between a few micrometers to a few tens of micrometers. It should be noted that the gap between mask and substrate in proximity printing will have a strong influence on the quality of optical imaging. The fidelity of printed image versus proximity gap can be expressed as:

$$W = k\sqrt{\lambda z} \tag{6.1}$$

where $W$ is the width of image blurring, or difference between the actual image size in the photoresist and the designed image size in the mask, $\lambda$ is the wavelength of the illuminating radiation used, $z$ is the gap between the mask and the surface of the photoresist and $k$ is a factor related to the resist process conditions. It will be seen from Eq. (6.1), that the image blurring can only be controlled by the wavelength and proximity gap. The effects of gap blur-

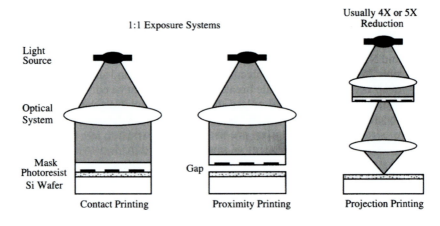

**FIGURE 6.2**    Three basic forms of optical photolithography: (a) Contact, (b) proximity and (c) projection.

ring are illustrated in the figure below (Figure 6.3), which shows simulated images for contact, proximity and projection printing. It may seem obvious that image quality should be superior for contact printing, however, this can cause damage to masks, especially in hard contact printing. Scratching can be common and this can prejudice the quality of images in a subsequent printing. The re-use of a photomask is an important consideration in the repeated production of identical images, where image integrity will always be an important consideration. As such, proximity printing can often be the favored method in mask aligning lithography and the photomask lifetime will be greatly increased.

The distance of gap in proximity printing can depend on several factors such as flatness of the silicon wafer, which can be bowed. This can limit distances to a few micrometers and bowing in the wafer can cause non-uniform exposure across the wafer.

Therefore, it will be seen that, while proximity printing can prolong the lifetime of photomasks, it will have a compromised resolution and uniformity.

The basic set-up of a mask aligner will include an illuminating light source (usually a mercury lamp), a mask holder and a wafer stage. A simple optical exposure can be done by using a lamp and a mask. However, modern mask aligners can be quite complex, including collimation of the light from the mercury lamp, which can improve the uniformity of illumination. The proximity gap can be finely controlled to enhance resolution.

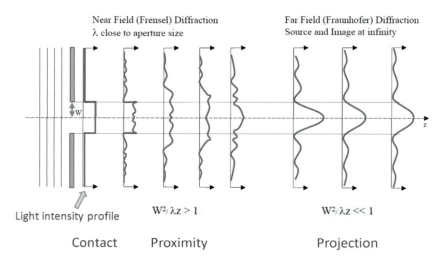

**FIGURE 6.3** Effects of image resolution for the different forms of optical photolithography. For proximity printing, the gap between the mask and substrate is an important parameter for determining the quality of the image. This is also true for projection lithography. Diffraction effects must be controlled to maintain pattern transfer integrity.

A vacuum contact option is frequently available in mask alignment and can allow good resolution for a feature size minimum of around 0.5 m. With hard contact lithography, the minimum printable feature is around 1 m. A mask aligner must be equipped with an alignment system to allow accurate overlap between different exposure layers. Alignment accuracy is normally within 1 m. Frequently, this will be performed using a double-side alignment system, which uses specific marks on the mask and wafer to ensure accurate alignment. Some mask aligners adopt an infrared system for backside alignment, utilizing the fact that silicon is transparent to infrared radiation, so that marks on the rear of the wafer can be seen at the same time and alignment can be made without marking the top side. While being cheaper and simpler than the double-sided alignment systems, this is only suitable for silicon substrates.

While mask aligners are no longer used in VLSI production, they are still widely used in microfabrication processes with less stringent resolution requirements, such as in the development of MEMS devices.

### 6.1.2 Projection Lithography

Mask aligners are not suitable for the high volume production of ICs because of the drawbacks detailed above. They were quickly replaced by projection lithography. This technique can print patterns on a substrate to the same size as the mask (1:1 projection) or smaller than the mask pattern (reduction projection). In a projection system, mask patterns are projected onto a substrate through an optical imaging system. Unlike a mask aligner, where image quality very much depends on the proximity gap between mask and substrate, the quality of the projected image depends only on the imaging optics, which avoids either mask damage (of contact printing) or the non-uniform exposure problem of proximity printing.

1:1 projection lithography was a good replacement for the mask aligner, though not a long term solution for future generations of ICs. With the continuous reduction of circuit feature size, it became increasingly difficult to produce 1:1 photomasks. The next revolution of optical lithography in IC manufacturing was the application of reduction projection lithography. This has become the principal lithography tool in VLSI production since the 1980s. Since the circuit pattern is smaller on the wafer, the same patterns can be repeated many times on the same wafer in a step-and-repeat manner, heralding the advent of the optical stepper.

The resolution $(R)$ of projection lithography depends on the illuminating radiation wavelength $(\lambda)$, the numerical aperture $(NA)$ of optical lenses and resist process conditions $(k_1)$:

$$R = k_1 \frac{\lambda}{NA} \tag{6.2}$$

From an optical imaging point of view, the increase in resolution can be achieved by reducing the wavelength and increasing the numerical aperture. This is illustrated in Figure 6.4, which shows the influence of wavelength and numerical aperture on the intensity distribution; so shorter wavelengths and higher numerical apertures are to be favored.

In the 1980s, the g-line of mercury ($\lambda = 436$ nm) was frequently used for exposure. With a NA of 0.3, and later 0.54, resolution of projection optical lithography was able to achieve 0.65 m.

Resolution is only one aspect of lithography quality. Another important aspect is the depth of focus (DOF). DOF also depends on the wavelength of illumination and the NA of the optical system, as seen by the following

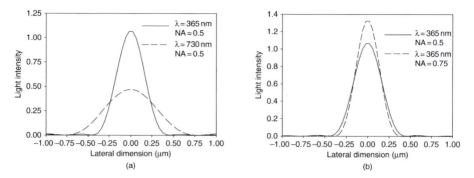

**FIGURE 6.4** Effects of illumination light wavelength (a) and numerical aperture (b) on light intensities at the photoresist.

definition:

$$DOF = k_2 \frac{\lambda}{(NA)^2} \tag{6.3}$$

where $k_2$ is related to a specific lithography system and photoresist process. The DOF is inversely proportional to the square of the numerical aperture. Therefore, pursuing resolution by increasing NA or reducing $\lambda$ will sacrifice the DOF. DOF is a more important parameter in the lithography of VLSI. As is known, ICs are fabricated on large silicon wafers, usually from 6 up to 12 inches in diameter. Such size wafers are rarely flat and we must also account for the topography of previously formed circuit structures in previous process steps. One can imagine that if the DOF is very small, the optical imaging can only be in focus in a very small range of height variation. There would be many places in a wafer, where defocus would be inevitable due to the topography or wafer bowing.

Every layer of a 2D circuit layout has a critical minimum feature, in which size or dimension is called a critical dimension (CD). This CD has to be controlled within a certain range during fabrication to ensure that the final circuit works properly. In the IC industry, the CD variation should be no more than 10%. For example, for a 100 nm critical feature dimension, any errors during fabrication should increase or reduce the feature dimension by no more than 10 nm. The DOF of a lithography system has to be large enough to ensure that the CD variation is within ±10% everywhere on the wafer.

With developments in the computer industry, VLSI was pushed to new heights in the 1990s, when g-line and i-line illumination from mercury lamps were no longer sufficient to maintain the size reduction necessary in IC production. Excimer lasers, which emit deep ultraviolet radiation, were developed into the illumination sources for such applications and the next generation optical lithography was called deep UV lithography. Excimer lasers produce radiation from excited gases; KrF gas produces radiation of $\lambda = 248$ nm and ArF gas has a wavelength of $\lambda = 193$ nm. By the late 1990s, deep UV lithography had achieved a resolution of about 0.25 m.

### 6.1.3   *Process of Optical Lithography*

Optical lithography is a complex process, which converts a design layout on a photomask into the functional structures on a substrate. It undergoes about 10 steps in the process, including photoresist coating, exposure, development, etching and a number of other processes depending on the materials and system requirements.

(1) *Pre-treatment of silicon wafer:* Apart from chemical cleaning of the wafer surface to remove all particles and stains, the wafer must be thoroughly dry so that the photoresist can adhere well to the surface. For this, a silicon wafer will be baked at $150 - 200°C$ for 15–30 minutes before use. Also, a coat of HMDS hexamethyldisilazane or bis(trimethylsilyl)amine may be used to promote good adhesion of the photoresist. Without good adhesion, photoresist patterns may peel off at later stages.

(2) *Photoresist coating:* Photoresist is coated onto the silicon wafer by the spin coating process. In this process, a drop of photoresist is placed at the center of the wafer and high-speed spinning will produce a uniform coat to spread over the wafer surface. The thickness of the photoresist can be controlled by the speed of spinning.

(3) *Pre-exposure bake:* The purpose of the pre-exposure bake (pre-bake) is to drive out all of the solvents in the photoresist layer so that the photoresist layer is completely dry and solid. The pre-bake is a soft bake, with time and temperature depending on the type of photoresist.

(4) *Exposure:* The photoresist silicon wafer is placed on an optical stepper for exposure. Typically alignment is necessary, especially if this is not the first layer. The alignment will ensure accurate overlay with the previous patterning on the wafer. Exposure is chosen by the type of photoresist used.

(5) *Post-exposure bake:* Multiple reflections in the exposure stage can give rise to standing wave interference, which can affect the exposed photoresist. Post-baking can partially eliminate this effect. Some Si pro-

cessing can use an anti-reflective coating before or after the photoresist to prevent the standing wave formation.

(6) *Development:* There are three methods of development: immersion, spray, and puddle. Immersion development is the simplest and does not require any special equipment. The exposed wafer is immersed in a developer filled container for a preset time, then taken out and washed to remove residual developer. Spray development requires spraying the developer onto the spinning wafer. Washing and drying follow. Puddle development combines both immersion and spray methods. The wafer is first immersed in the developer for a certain time, then undergoes spinning for further spray development, cleaning and drying.

(7) *Descum:* Development and washing do not remove all the photoresist. A thin layer of residual photoresist (scum) remains. This is more apparent for deep and narrow pattern structures. Though only a few nanometers, scum can prevent subsequent pattern transfer. Descumming usually takes the form of plasma etching for a short period of time in oxygen. This process must be performed with care, since it partially reduces the photoresist thickness and can have detrimental effects on the pattern structure.

(8) *Hard bake:* Hard bake enhances adhesion of the photoresist pattern structure, though the actual temperature used should not be too high, since it can make later photoresist removal difficult.

(9) *Pattern transfer:* The purpose of optical lithography is to transfer the mask design onto the silicon wafer. The photoresist only serves as a masking layer; i.e., to protect the substrate, so that only those areas of photoresist that have been developed and the substrate surface, which is now exposed, can be further processed, either by etching or deposition.

(10) *Removal of photoresist:* After pattern transfer, the photoresist has fulfilled its purpose and needs to be removed. This can be either a wet or dry process. The wet procedure involves using various acidic or alkaline or solvents to dissolve the remaining photoresist. Acetone is a common solvent, which can dissolve most photoresists. The dry removal of photoresist involves oxygen plasma etching.

These principal stages of optical photolithography are illustrated in Figure 6.5.

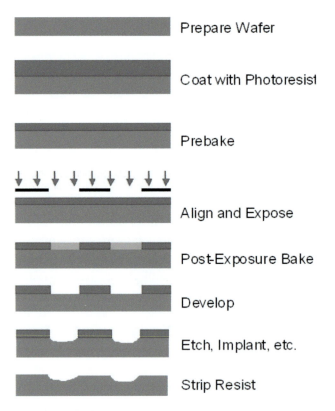

**FIGURE 6.5**     Main stages of optical lithography.

### 6.1.4   Resolution Enhancement

In the late 1980s, optical lithography reached the 0.65 m minimum feature size, which was at the limit of i-line optics at that time.  As indicated in Eq. (6.2), at a fixed wavelength, the resolution can be increased by an increase of the numerical aperture of the lens.  However, an increase in NA will sacrifice the depth of focus (DOF), see Eq. (6.3). There is then only one parameter in Eq. (6.2) that can be changed, that is the factor $k_1$. Normally, $k_1$ is about 0.8 in normal process conditions in volume IC production. If $k_1$ can be lowered, higher resolution can be achieved at i-line wavelengths. $k_1$ is a factor determined by many process parameters, which cannot be formulated by optical imaging theory. Since the 1990s, a series of resolution enhancement techniques (RET) have been developed, called wavefront engineering. These techniques can be grouped into those which (i) improve the optical system, including off-axis illumination and spatial filtering, and (ii) techniques, which improve the photomasks, including phase-shifting masks and optical proximity correction. These techniques aim to manipulate the optical wave outside the imaging lenses, as there is little that can be done with optical lenses apart from a change in numerical aperture and reduction of aberrations. For a lens system, the fundamental limit in resolution is already set by the optical wavelength. Modern optical lithography using wavefront engineering can significantly break the optical resolution limit of conventional optics.

### 6.1.4.1  Off-Axis Illumination

The light source, such as a mercury lamp in a conventional optical stepper, normally illuminates through a central (on-axis) circular aperture to a condenser lens system. Off-axis illumination (OAI) deliberately blocks the light on the central region (axis) and only allows off-axis light to enter the condenser lens. The main types of OAI aperture are annular illumination and quadrupole illumination. OAI can improve imaging resolution by 20% and the depth of focus by 40%. Quadrupole illumination is more effective than annular illumination, but only for exposure of periodic patterns. OAI has been widely adopted in modern optical steppers, which are usually equipped with interchangeable apertures.

### 6.1.4.2  Spatial Filtering

Spatial filtering modulates the amplitude and phase of a light wave in its frequency domain, which can improve image resolution and depth of focus of the imaging system. From the theory of Fourier optics, optical imaging can be mathematically expressed as a Fourier transform process; i.e., a mask pattern is transformed by a lens from object space to image space at the lens pupil plane, where the light intensity distribution through the patterned mask becomes amplitude and phase distribution. If an image is sharp and there is little loss of detail, it would be rich in high-frequency components in its frequency spectrum at the pupil plane. The concept of spatial filtering is to partially deter the low-frequency components, while enhance those at high frequency by using a spatial filter at the lens pupil plane, which can improve the sharpness of the image. By varying the transparency of the filter, phase and amplitude modulation can be achieved.

While spatial filtering can improve image quality in optical lithography, it is fairly difficult to implement, because the filter has to be inserted in the pupil plane of an optical lens system. A modification of the lens column is necessary.

### 6.1.4.3   Phase Shift Masks

The most significant advance in optical lithography in recent times is the application of phase-shift masks (PSM). This can be considered as an extension of spatial filtering, through which the light wave intensity and phase are modulated to improve the image contrast and depth of focus. The concept of phase-shifting was introduced in the early 1980s. However, due to the complexity and additional costs of making PSM, it did not draw much initial attention. Despite this, the limits in the IC industry by the 1990s meant that the use of PSM became a viable and necessary next step. In Figure 6.6, we show five different PSMs compared to a conventional mask.

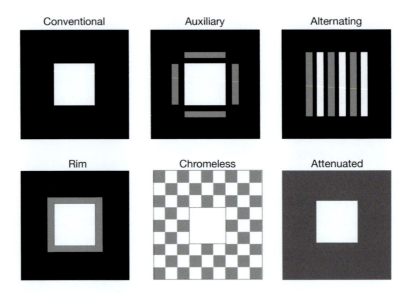

**FIGURE 6.6**   Comparison of the different types of phase shift masks with the conventional photomask.

Black regions are opaque, white are transparent and gray regions are phase shifting. The imaging process of these different PSMs are illustrated in Figure 6.7, where the phase shift, amplitude of electromagnetic radiation and light intensity distribution for alternating, rim and attenuating PSMs are compared to a conventional photomask.

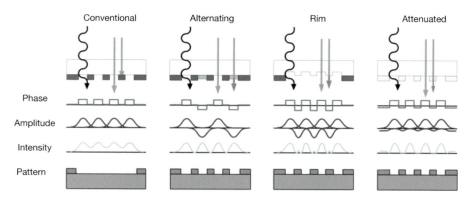

**FIGURE 6.7** Imaging process for the alternating, rim and attenuated PSMs in comparison with the conventional photomask.

Whatever the phase-shifting method, the purpose of a PSM is to introduce a light wave with 180° phase difference, which can partially cancel the incoming light wave (0° phase) at the edge of the pattern, therefore enhancing the image contrast as well as the additional benefits of increased image resolution and depth of focus. The advantage of PSM over conventional (binary image) masks (BIM) is shown in Figure 6.8, where the light intensity distribution of lines for the two types of mask are compared.

We summarize the main features of the different PSMs in Table 6.1.

There are two main difficulties in the practical implementation of PSM technology: design and fabrication. The design and production of PSMs are complex and usually need special design software. In the early days of PSM technology, only rim and attenuating PSMs were actually considered in IC manufacturing, since only these have the same design layout as the conventional mask. However, these two PSMs are weak phase-shifting and only bring improvements in contrast and depth of focus of about 10–20%. Strong phase-shifting types, such as alternating and chromeless PSMs, can give improvements up to 50%.

### 6.1.4.4 Optical Proximity Correction

The theoretical basis of projection optical lithography is diffractive optics or diffraction-limited optics. In Figure 6.3, the difference between ideal and diffracted images has been shown. Diffraction imaging will cause a loss of high-frequency components during the optical wave transmission and transformation in the lens system. The consequence of high-frequency loss is a blurring of the resulting image. This loss of sharpness is manifested in corner

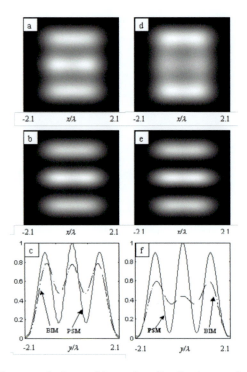

**FIGURE 6.8** Computed plots of intensity distribution at the wafer for a mask placed in an optical system (NA = 0.6, M = 1/5). (a) Image of the BIM obtained with s = 0.7. (b) Image of the PSM obtained with s = 0.7. (c) Cross-sections of the intensity patterns for the BIM (dashed) and the PSM (solid). (d)-(f) Same as the patterns in the left-hand column, but for s = 0.5.

rounding or line end shortening. This is the so-called optical proximity effect and is illustrated in Figure 6.9. This effect is usually of little importance if the pattern size is large. However, in recent times, it has become important, since pattern sizes are now so small and transistor density has increased dramatically.

In order to maintain the fidelity of IC designs, optical proximity effects must be corrected. The basic method of optical proximity correction (OPC) is to deliberately distort the original pattern to balance the image intensity. Over-exposure regions can be reduced, while under-exposure regions can be enhanced by taking away or adding extra features, see Figure 6.9. In Figure 6.10, we can see the actual effects of OPC.

An image of a real OPC mask is also illustrated in Figure 6.10. The smallest features (serifs) are only 100 nm. Proximity effects are also apparent in electron beam lithography, which is the pattering tool for mask making. It

**TABLE 6.1**   Summary of the Five Types of Phase Shift Mask (PSM)

| Type of PSM | Features |
|---|---|
| Alternating | Strong phase shifting, significant contrast improvement and depth of focus<br>Complex mask fabrication<br>Requires transitional phase patterns<br>Requires double exposure to remove transitional phase<br>Complicated process design and complex implementation |
| Chromeless | Strong phase shifting, significant contrast improvement and depth of focus<br>Requires large number of sub-resolution phase patterns ( 0° and 180° phase)<br>Complex mask fabrication<br>Complicated process design and complex implementation |
| Auxiliary | Weak phase shifting<br>Only useful for isolated and sparse patterns<br>Requires auxiliary 180° phase patterns<br>Complicated process design and complex implementation |
| Rim | Weak phase shifting<br>Easy to design<br>Easy to fabricate<br>Has side lobe effect |
| Attenuated | Weak phase shifting, but better performance than rim PSM<br>Design as with conventional mask<br>Easy to fabricate, but requires special mask blank material<br>Can have side lobe effect, depending on the degree of attenuation |

is very difficult to make the small serifs with high fidelity and this will affect the correction results.

### 6.1.4.5   *The Limit of Optical Lithography*

The downfall of optical lithography has long been predicted. However, it is still the main tool of lithography in today's VLSI manufacturing, owing to the continuous advances in technology. In order to process ever-smaller circuit structures, optical exposure wavelength has been driven down, from early g-line (436 nm) and i-line (365 nm) to current deep UV (248 nm, 193

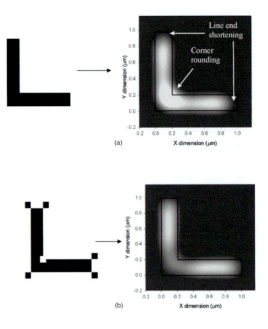

**FIGURE 6.9**    (a) Optical proximity effect results in corner rounding and line-end shortening. (b) Optical proximity correction modifies the mask design to restore the pattern fidelity.

nm). The next wavelength down the line is called vacuum ultra-violet (VUV) with a wavelength of 157 nm generated by Freon lasers.

Resolution enhancement techniques have been widely applied to IC mass production. New chemically amplified resists have provided high resolution, high contrast, and high sensitivity. The combination of chemically amplified resists and RET have reduced the $k_1$ factor from Eq. (6.2) to below 0.3. If the generation of IC technology is marked by the minimum circuit feature size, the 130 nm generation VLSI started to be mass-produced in 2001, in which deep UV (248 nm) steppers are used in conjunction with RET. The next 90 nm generation VLSI began mass production in 2004 using 193 nm deep UV steppers. The following generation of Freon-based optical lithography at $\lambda =$ 157 nm for 70 nm VLSI is now in production, where it is foreseen that with PSM techniques circuit feature sizes down to 50 nm can be produced.

In fact, 50 nm IC features have already been produced in the laboratory with deep UV (248 nm). In 2002, IBM produced transistors with a minimum gate length of 25 nm using deep UV optical lithography in combination

**Without OPC**                                    **With OPC**

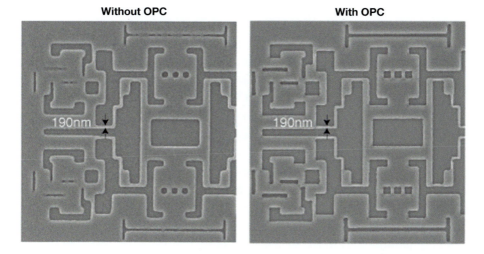

**FIGURE 6.10** Comparison of the pattern transfer with and without optical proximity correction: Scanning electron micrographs of pattern showing improvements from OPC. (Left) The exposure dose was uniform within the pattern. (Right) An optimum exposure dose was used. Patterns were exposed using zone plates at $\lambda =$ 400 nm, NA = 0.7, and an AOM (Reprinted with permission R. Menon et al., 2005. © Elsevier.).

with chromeless PSM. However, the industry usually lags behind the latest laboratory results by several years. The ultimate limit of optical lithography for industrial manufacture should be around 50–70 nm. The last candidate for a light source in optical lithography being Freon based lasers. For the post-optical era, next-generation lithography (NGL), there are a number of candidates: electron projection lithography, ion projection lithography, X-ray lithography, and extreme UV (EUV) lithography. So far only EUV has gained wide acceptance in the IC industry. EUV has a wavelength of 13 nm, which is no longer strictly UV, but soft X-rays.

   Optical lithography will eventually give way to other more advanced technologies. However, this does not mean that optical lithography will be completely abandoned. It will be a mature technology in VLSI fabrication for a long time to come. This is because only a few levels of lithography in VLSI fabrication demand the highest resolution patterning out of the total 20–30 levels. It is possible that mix-and-match lithography, i.e., a mix of advanced, medium and low-resolution level optical lithography, will be the norm in VLSI lithographic technology.

### 6.1.5   Alternative Methods of Optical Lithography

Optical lithographies have developed enormously over the decades and in the last three decades or more, there has been a sustained concerted effort to improve the quality and resolution using a number of methods. Indeed, optical lithography is one of the most important industrial processes for large scale production of ICs and represents a vital market for modern electronic goods on a global scale. The investment in improved technologies is therefore understandable in this climate.

Optical projection lithography, as we have noted, is a far-field technique, which has a fundamental resolution limit of $\lambda/2$, set by the wavelength of the light used. Contact lithography, on the other hand, is a near-field imaging process, since there is no space for the light wave to travel between the mask and the photoresist. In the near-field regime, the usual propagation of light is no longer via a sinusoidal wave, but as an evanescent wave, which decays exponentially with distance. The lithography based on the use of this information is referred to as *evanescent near-field optical lithography* or ENFOL. This method stems from an adaptation of contact printing using conformably masks, which are thinner and can *conform* to the substrate morphology, providing better contact and resolution. Within this subgroup of lithography, we can consider embedded amplitude masks, which have the mask structure embedded within the transparent mask, whose surface is flush between the patterned and transparent regions of the mask. This type of mask structure is referred to as a conformable embedded-amplitude mask or EAM. Alternatively, *light-coupling masks* (LCM) can be used, which are made of elastomer materials, allowing conformal contact. This type of mask is not a conventional amplitude mask, with absorbing and transparent regions, but has a surface relief pattern. There are two forms of masks for LCM; one which has an absorber and one that does not. The contrast achieved for the non-absorbing mask is provided by the phase difference near the odd multiples of $\pi$, as with a binary phase mask. For large features $(d \geq \lambda)$, optical modes are excited in the protruding region of the LCM that are propagated into the photoresist. In the case of higher resolution features $(d < \lambda)$, a wave-guiding effect fuses the light into the higher refractive index material, amplifying the intensity below the protruding mask regions and creating contrast within the photoresist, without the need for an absorber. Features of less than 200 nm have been reported using this technique.

ENFOL is also a conformal contact lithographic technique. The principal difference between this method and the above techniques concerns the diffracted optical region that the image of the mask. While both of these methods function in the near field, ENFOL is able to operate, where the contrast of the intensity patterns results from electromagnetic waves that are evanescent, i.e., they decay exponentially in the transmission direction. Traditionally, contact lithography has been dependent on propagating electromagnetic waves to obtain exposure contrast. The short-range nature of the evanescent fields (this typically lies in the order of tens of nanometers when illuminated with UV wavelengths) makes it essential that the photoresist coated substrate is sufficiently close to the mask to be able to take advantage of the image contrast in this region. Membrane masks ($\sim$2 m thick) have been developed that are flexible and able to conform over topography. ENFOL also requires that the photoresist layers are thin enough, such that the image capture occurs in the evanescent near field. Normally, this will be less than 100 nm. The style of the mask is similar to the standard contact masks with the absorber patterned on top of the mask substrate. During exposure, the mask is held in contact with the substrate using vacuum pressure. Figure 6.11 shows a schematic of the ENFOL technique.

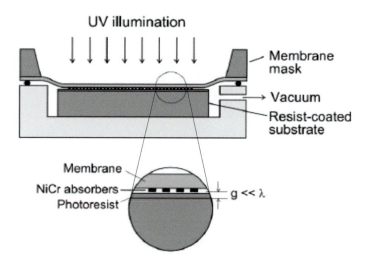

**FIGURE 6.11**   The evanescent near-field optical lithography (ENFOL) process. A patterned mask is held in intimate contact with an ultra-thin photoresist layer. UV illumination generates high-resolution, near-field (evanescent) light that is captured by the resist.

One of the principal factors, which makes ENFOL a sub-wavelength method, is related to the surface plasmon polariton (SPP) phenomenon. This effect in metallic objects is explained in more detail in Section 12.4.1. The

SPP is essentially a charge density wave localized at the surface of conducting material, silver, and gold being the most popular materials used in the study of SPPs. The incident radiation can excite a SPP, which propagates parallel to the interface of the conducting material and a dielectric medium (the photoresist). The SPP excitation has a wavelength, which is correlated with that of the incident radiation and which depends on the dielectric constant of the photoresist. Since the SPP is a form of electromagnetic radiation, it can form interference patterns, which are periodic and result in standing waves, that can then effectively expose the photoresist in regions of the maxima, while minima remain unexposed. In this way, the SPP can create feature sizes, which are controlled by the permittivities of the metal-photoresist ensemble and the wavelengths of the SPP can be considerably shorter than those of the incident electromagnetic radiation. The relation between the SPP and incident wavelength can be expressed as:

$$\lambda_{SPP} = \lambda_0 \sqrt{\frac{\varepsilon_m + \varepsilon_d}{\varepsilon_m \varepsilon_d}} \tag{6.4}$$

where $\varepsilon_m$ and $\varepsilon_d$ are the permittivities of the metal and the dielectric, respectively. The pattern revealed in the photoresist can, therefore, have a feature size, which is considerably smaller than the metallic mask pattern. This method has been shown to be capable of producing features sizes of around 100 nm and simulations indicate that gratings with periods of $\lambda/20$ are also feasible (McNab and Blaikie, 2000).

More recently, several researchers have shown that scanning near field optical lithography is also a feasible method of generating features with dimensions of well below 100 nm. For example, Credgington et al. (2010) have reported that reproducible structures of less than 60 nm (or $\lambda/5$) in areas of up to 20 m × 20 m are possible. McLeod and Ozcan (2012) have used the near field SNOM method to produce features of less than 10 nm.

Another effective method of producing sub-100 nm patterned features is via optical interferometric lithography (IL). Optical interference of two beams of light of the same wavelength can cause the distribution of light intensity with a periodic variation that depends on the wavelength of the light used and the angle of incidence. For interferometric lithography, the source of light is typically a coherence laser source, which is split into two equivalent beams that are then brought into convergence on the surface of a photoresist coated substrate. The variation of the light intensity (in 2D for thin resist layers or in 3D for thick photoresist layers), will be recorded by a photosensitive polymer. This method is also referred to as holographic lithography. The intensity variation across the surface can be expressed as:

$$I(x) = 2|E|^2[1 + A_{pol}\cos(2kx\sin\theta)] \tag{6.5}$$

where $k = 2n\pi/\lambda$ is the effective wave vector of the light used, $x$ denotes the position on the surface, $\theta$ is the angle of incidence of the light beam, $n$ is the refractive index (which can be varied depending on whether the interference is made in the air, where $n = 1$, or through a prism, where $n$ will depend on the refractive index of the prism glass) and $A_{pol} = 1$ for TE polarization and $A_{pol} = \cos(2\theta)$ for TM polarization. The periodicity of the interference pattern along the $x$-direction is given by:

$$\Lambda = \frac{\lambda}{2n\sin\theta} \tag{6.6}$$

For a particular wavelength of radiation, the periodic lines with different pitches can be obtained and can be set by the angle of incidence. Changing wavelength can also be used to control the periodicity. The minimum pitch available, for the maximum oblique angle incidence, will be $\lambda/2$. In Figure 6.12, we show a schematic illustration of the basic interferometer as well as interference regime for a two-beam interference pattern at the surface of the photoresist layer.

The basic apparatus for the IL and can be set-up quite cheaply in a laboratory. Feature dimensions can be in the sub-100 nm regime without too much difficulty in the air set-up, Figure 6.12(a). One of the main advantages of the IL method is that no mask is required for the lithographic process. This greatly implies the pattern definition in the photoresist. On the downside, this method can only really be used for periodic line and dot structures. If further beams are added, some control over the periodic arrangement can be performed. Improvement of the resolution in this method can be achieved using the solid immersion scheme, Figure 6.12(b). This set-up allows an improvement of the effective numerical aperture (NA) for the set-up, the so-called high-NA regime. By increasing the NA, we can improve the resolution, as given by Eq. (6.2). In this case, we require a prism to be placed above the photoresist layer, which is a disadvantage, adding a small complication to the set-up. Further improvement of the NA can be achieved using the set-up in Figure 6.12(c), where the fields in the photoresist are now evanescent and image depth becomes a severe physical constraint. The highest reported resolutions that have been obtained are features of 26 nm using 193 nm deep-UV

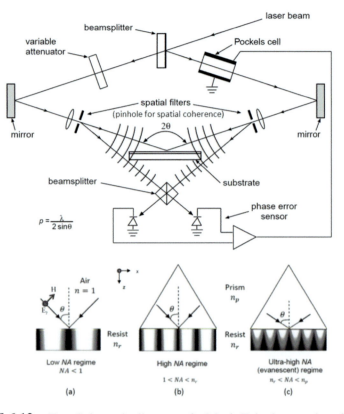

**FIGURE 6.12**   Top: Schematic diagram of a Mach-Zehnder type interferometer, which allows the interference of two light beams from the same source with a variable angle of incidence. Bottom: Schematic representations of (a) basic interference lithography carried out with air as the ambient medium, (b) inside a prism at a high numerical aperture, and (c) in a prism at an ultra-high numerical aperture (NA), with $\theta$ beyond the critical angle at the prism-resist interface. These illustrations are for transverse electric (TE) polarizations, with the electric and magnetic field directions defined in part (a). Electric and magnetic field orientations are exchanged in the transverse magnetic (TM) configuration (Reprinted with permission from Mehrotra et al., 2013. © The Optical Society.).

radiation, (Smith et al., 2006). Some examples of interference lithography defined structures are illustrated in Figure 6.13.

Maskless lithography (ML2) is an important development for optical methods of pattern transfer. In terms of industrial processing, the lack of a need for a mask is a big plus, since the costs associated with mask production can be significant and require a large amount of preparation. Indeed, the interferometric lithography method is an example of an ML2 technique.

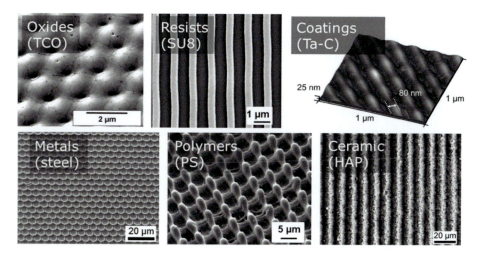

**FIGURE 6.13** Examples of periodic surface structures on a transparent conducting oxide (TCO), a photoresist (SU8 epoxy), tetrahedral amorphous carbon (Ta-C), i.e., a diamond-like coating, metal (stainless steel), polymer (polystyrene), and hydroxyapatite ceramic (HAP) [A. F. Lasagni et al., SPIE Newsroom 2016 DOI: 10.1117/2.1201602.006325].

Direct laser writing offers another form of massless lithography. In this case, a laser beam is scanned across the surface of the photoresist to expose it. While this is a feasible method of exposure for the photoresist, it is not considered to be fast enough to be adaptable as a commercially viable method of massless lithography for production purposes. It was, however, considered to be a suitable means for mask production. In the late 1980s, this method was more appropriate than electron beam methods of mask preparation, since it had higher throughput. Later, multi (laser) beam systems were developed with raster scanning systems, making this a commercially viable system for the 0.7 m generation of ICs. This was later to be replaced by a system with 32 individually addressable beams. As such, patterns could be *painted* onto the photoresist in the form of pixels on a rectangular grid, formed of 4096 by 32 pixels, using a 24 facet polygonal mirror. The pixel size corresponded to 96 nm in one axis and 160 nm on the other. This technology was adequate for the 90 nm generation of IC manufacturing.

MEMS technology was also adapted to the problem, producing micro-mirror arrays, which can be individually tilted by applying an electrostatic field. Large arrays of such mirrors are referred to as digital micro-mirror devices or DMDs. Each mirror in the array effectively acts as a pixel. Since each mirror can vary the light intensity of each pixel individually from the full light intensity (white) to no light (black) and all shades between the appropriate tilting, this system is often referred to as a spatial light modulator (SLM). A SLM based laser lithography tool can work like an optical sweeper. A SLM chip, with an array of $2048 \times 512$ pixels formed by micro-mirrors can be controlled by a pattern generator, where the pattern is defined by electrical signals to each of the mirrors and can be programmed. The SLM is illuminated with light from a pulsed excimer laser, which is reflected onto the substrate via a demagnifying lens system. The substrate is housed on a moving stage, which allows the full pattern to be stitched with the exposing field from frame-to-frame. With current technologies (at the time of writing), it is possible to write pixels of 30 nm, which can be used for the massless lithography of ICs at the sub-100 nm resolution.

Two-photon processes can be used to manufacture three-dimensional objects in a polymer. Such a process is referred to as multi-photon lithography (MPL). The basic idea is based on stereo-lithography, which essentially exposes the photoresist at the focal point of a laser beam, which has sufficient energy to photo-polymerize the resist material. In the case of MPL, two (or more) laser beams are used to focus on the same point inside a pool of the liquid polymer. When the two beams strike the same point in space, the focusing point will have the highest energy and, if sufficient, will solidify the polymer at this position. The three-dimensional object is then formed by moving the focal point systematically within the liquid polymer to define the 3D object. In Figure 6.14(a), we show a schematic of the apparatus, which employs a femtosecond laser as one source and a LED as a second source. The latter forms a constant illumination, while the femtosecond laser is focused using a SLM to form the 3D image. In this way, the two photons are ensured to arrive at the correct spatial positions to form the desired image. Structures with extraordinary complexity can be produced in this manner, some examples are illustrated in Figure 6.14(b). The key to this technique is to ensure that the coincidence of the two light beams is focused correctly in the smallest spot possible to obtain the best resolution possible. The sensitivity of the polymerizable liquid resin is also of great importance. The best-reported resolutions obtained with this method were in the region of 60–70 nm diameter woodpile structures (Haske et al., 2007).

## 6.2   Electron Beam Lithography

Electron beam (e-beam) lithography is based on the principle that certain macromolecular polymers are sensitive to electrons and can be patterned by

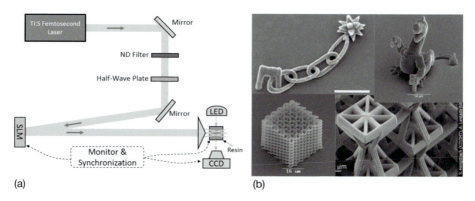

**FIGURE 6.14** Examples of periodic surface structures on a transparent conducting oxide (TCO), a photoresist (SU8 epoxy), tetrahedral amorphous carbon (Ta-C), i.e., a diamond-like coating, metal (stainless steel), polymer (polystyrene), and hydroxyapatite ceramic (HAP) [A. F. Lasagni et al., SPIE Newsroom 2016 DOI: 10.1117/2.1201602.006325].

electron exposure, very much like optical lithography. The basic structure of the SEM has no fundamental difference from an e-beam lithography system.

The application of e-beam microlithography began with the discovery of suitable electron sensitive photoresists. These are similar to those of optical lithography and are based on the same principles of exposure. Since electrons of a specific energy can be readily produced by an electron gun—(the wavelength depends only on the accelerating potential used):

$$\lambda = \frac{1.226}{\sqrt{V}} \tag{6.7}$$

in units of nm. $V$ being the potential (or energy in eV) of the electrons— there is no real limit as far as wavelength considerations are concerned in the lithography process. For example, at 100 eV, the corresponding electron wavelength is 1.23 Å. In fact, the electron wavelength is not the resolution limiting factor in e-beam lithography. This depends on factors such as the various electron aberrations and electron scattering in resists.

Electron beam lithography began in the 1960s and the polymer PMMA poly(methyl methacrylate) was introduced in 1968 as an electron resist. By the early 1970s, structures of 60 nm could be patterned. With the doom predicted for optical lithography in the 1980s, electron beam lithography was proposed as an alternative for VLSI manufacturing. However, this never materialized due to the low throughput compared to optical lithography. This does not prevent the e-beam technology from becoming an important and versatile tool for micro-patterning. While patterning by optical methods can only be done by masking, electron beams can be directed with magnetic lenses onto the resist enabling a direct write process for patterning. It will be noted that e-beam lithography is already used for mask fabrication in VLSI manufacturing. In the current nanotechnology boom, e-beam lithography systems combined with special resist technology can fabricate structures of less than 10 nm. This ultra-high-resolution capability and the direct write flexibility makes e-beam lithography an important technology in research and development of all micro- and nanosciences.

### 6.2.1   Principles of Electron Optics

The core of an electron beam lithography system is the electron optical system which focuses and deflects the electron beam. To understand the operation of e-beam lithography, an understanding of the principles of electron optics is important. Electron optics describes the movement of electrons in an electric or magnetic field. There are some similarities between an electron and light optics. For example, electrons can change speed and direction in different electromagnetic fields, as an analogy to refraction in light optics. The main differences between electron and light optics can be summarized as follows:

(1) The refractive index in optical media is a discontinuous function of position (space); i.e., light entering (from the air) into an optical lens undergoes discontinuous refraction. The refractive index of electron optics is directly proportional to the square root of electron energy, which is a continuous function of spatial coordinates, because the electric potential is always continuously distributed in space.

(2) The refractive index of optical media is between 1–2.5. The refractive index of electron optics depends on the spatial potential, which can be very large.

(3) The geometrical aberrations in light optics can be eliminated by changing arbitrarily the curvature of the refractive surface (shape of lens). However, electrical potential in space cannot be arbitrarily changed. Correction of aberration in electron optics is completely different from that in light optics.

(4) Electrons repel each other due to their electrical charge. For a high current density electron beam, there is a space charge effect, which can deteriorate the focusing of an electron beam.

Once the relation between electron refraction and space distribution of electrical potential is understood, the principles of electron optics can also be understood. In Figure 6.15, we show the space potential distribution inside two cylindrical electrodes at different potentials.

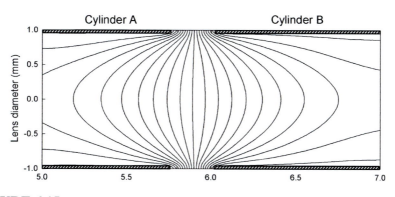

**FIGURE 6.15** Spatial distribution of the electric potential inside two cylindrical electrodes with different voltages; cylinder A: 0 V, and cylinder B: 100 V.

The curved equipotentials act like an optical lens, the difference, however, is that an optical lens has a fixed shape of curvature, while the curved equipotentials inside the cylinder are distributed continuously. An electron coming from one end of the cylindrical electrodes will experience an axial force around the gap region, which can force the electrons from different incident angles and positions to converge towards the cylinder axis. Therefore, the electric field at the gap region acts like an electron lens. This electron type of lens system is called an immersion lens. Another type of lens, the Einzel lens, is shown in Figure 6.16. This assembly has three cylindrical electrodes, which are usually used by holding the outer electrodes at the earth, while the central electrode is biased with a negative potential for electron focusing. The focal length of the system is controlled with the applied potential.

To analyze an electron optical system, it is necessary to solve the electromagnetic field equation and then to calculate the electron trajectory. The equation below is the equation of motion in a 3D electromagnetic field:

$$\frac{\partial^2 \mathbf{r}(x,y,z)}{\partial t^2} = \frac{e}{m}\mathbf{E}(x,y,z) \tag{6.8}$$

This equation can be solved numerically to obtain potential distributions and electron trajectories, as shown in Figure 6.15, for example.

Focusing by electric fields is called an electrostatic lens, while using a magnetic field is a magnetic lens. Electrostatic lenses are normally of the

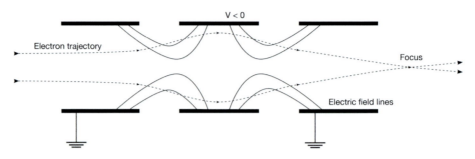

**FIGURE 6.16** Diagram of an Einzel lens used to focus a beam of electrons. The focal length of the lens system is adjusted with the voltage applied to the central cylindrical electrode.

type described above. The structure of a magnetic lens is shown in Figure 6.17. The magnetic field is typically generated by an electric current flowing through coils, which wind over a soft iron core, called pole pieces. As the electrons pass through the region of the non-uniform magnetic field, they are deviated from their path via the Lorentz force. The distribution of the magnetic field in the bore region is such that the electron beam is focused after the magnetic coil, where the focal distance is controlled by the strength of the field created by the coils.

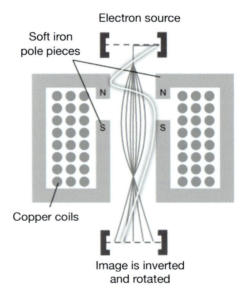

**FIGURE 6.17** A magnetic lens for electron beams. When we pass a current through the coil, a magnetic field is created in the bore. The strength of the field in a magnetic lens controls the ray paths, bringing off-axis rays back to focus.

Electrostatic or magnetic lenses will normally have axial or rotational symmetry giving a similar symmetry to the field distribution. In an axially symmetric field, the electron movement, given by Eq. (6.8), can be simplified to the following:

$$\frac{d^2r}{dz^2} = \frac{(1+r'^2)}{2V(r,z)}\left[\frac{\partial}{\partial r}V(r,z) - r'\frac{\partial}{\partial z}V(r,z)\right] \tag{6.9}$$

where $r' = \frac{\partial r}{\partial z}$. From Eq. (6.8), the focal length of an electron lens can be calculated as:

$$\frac{1}{f_i} = \frac{1}{4\sqrt{V_i}}\int\frac{V''(z)}{\sqrt{V(z)}}dz \tag{6.10}$$

where $V''(z) = \frac{d^2V(z)}{dz^2}$. This equation shows that the focusing power of an electrostatic lens is directly proportional to the second derivative of the axial potential distribution; i.e., proportional to the first derivative of the axial electric field. From this, it is clear that a uniform electric field cannot function as an electron lens. The focusing power will thus increase with greater variations of electric field. For a magnetic lens, the larger the electric current in the coils, and the smaller the gap between pole pieces, the greater the variation of magnetic field and hence the stronger the magnetic lens. For practical reasons, it is simpler to produce magnetic lenses, which are the main type of lenses used in electron optics.

Besides the axially symmetric lenses discussed, there are also axially asymmetric lenses, such as quadrupole or multipole lenses, though these will not focus the electron beam and can be used as a stigmator. The treatment of such lenses is more complex than that of axially symmetric lens systems.

In addition to electron lenses, electrons can be easily deflected using a deflector. The simplest form of deflector would be a pair of parallel plates. In any case, as long as the force, generated by either an electrostatic or magnetic dipole, is perpendicular to the electrons trajectory, they will be deflected. Further to this, an electron optical system requires an electron gun. This generally consists of two parts: a cathode for emitting electrons and a lens to focus them into a beam. There are two types of cathode: the thermionic cathode and the field emission cathode, also called the cold cathode. The former, usually tungsten or $LaB_6$, works at high temperature (2700 K for W), where thermal kinetic energy is sufficient to overcome the work function of the cathode material and escape the surface. They are then extracted by the electric field of a nearby electrode. The emission area of a thermionic cathode is around 10–100 m$^2$. Cold cathodes are very sharp needles, which typically have an apex of about 0.5 m. The sharper the needle, the higher the electric field at the needle tip, even at low voltage. The typical field strength is of the order of $10^8$ V m$^{-1}$. Such a high field is sufficient to pull electrons from the cathode surface; this process is called field emission. Since the tip is so fine, no lens system is usually required to focus the emitted electrons. This allows for very high resolution and as such, field emission sources are common in electron microscopes, though not in e-beam lithography. This is because generally high vacuum is required and this is generally incompatible with e-beam lithography. As such, thermionic emission guns have been adopted for e-beam lithography techniques.

### 6.2.2 Electron Beam Lithography Systems

The three basic components in an electron optical system are: electron gun, electron lens and electron deflector. A practical e-beam lithography system will be somewhat more complex as shown in Figs. 6.18 and 6.19.

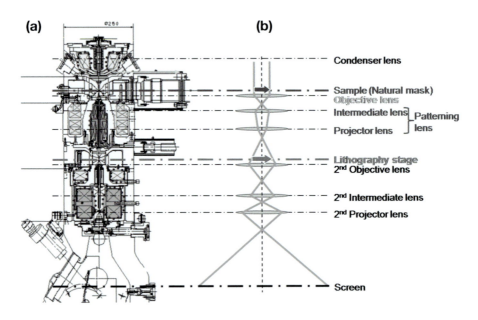

**FIGURE 6.18** Schematic diagram of a projection electron-beam lithography system. (a) Design diagram. (b) Ray diagram equivalent.

The electron column consists of the following sub-systems:

**Electron gun**: Usually of energy in the range 10–100 keV.

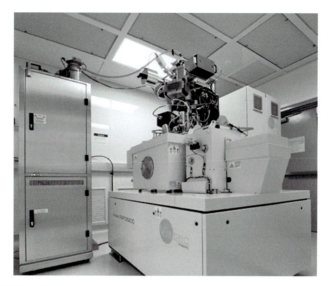

**FIGURE 6.19**   Exterior image of an electron beam lithography system.

**Electron gun alignment system**: This is usually required for small corrections in misalignment of the cathode tip. A deflection system at the electron gun section is normally installed to collimate the electron beam if an alignment is required.

**Condenser lens**:  As in an optical system, this is to converge the emitted electrons from the cathode.

**Beam blanking**: This is to switch the electron beam on and off, so that it can only pass through the column to reach the substrate when exposure is required. This is an electrostatic deflector to take the beam off-axis when it is not required.

**Zoom lens**: This is for adjusting the focal plane, including dynamical focusing of the electron beam.

**Stigmator**: Astigmatism occurs, when focusing on a beam is different in the $x$ and $y$ directions, resulting in an elliptical image of the beam instead of circular. It is usually caused by errors in the mechanical machining of the lens components. Stigmators are generally composed of multiple lenses, which can focus a beam laterally from different directions.

**Apertures**: Several apertures are used in an electron column. These are generally used to limit the lateral spread of an electron beam. The main aberration of an electron lens is the spherical aberration, which is directly proportional to the cubic power of the beam spread angle. By limiting this angle, we can effectively improve the electron optical resolution. However, small apertures can reduce beam current. In practice, aperture size can be adjusted. The position of the aperture can be finely tuned to ensure that the center of the aperture coincides exactly with the axis of the electron column.

**Projection lens**: This is the final lens of an electron column, which focuses the electron beam coming out of the aperture to form the final beam spot at the exposure surface.

**Deflectors**: Electron beam deflection and scanning are realized by deflectors. These can be placed before, after or inside the projection lens and can be electrostatic or magnetic. Usually, the latter are preferred as they produce less aberration and distortion, whereas electrostatic deflectors are high speed. An electron beam lithography system is usually equipped with two sets of deflectors.

In addition to the above, other accessories are common, such as a beam current detector and a backscattering electron detection system to inspect the exposure surface. This can be used in finding alignment masks on the surface.

    There are a number of ways to classify an e-beam lithography system:

(1) exposure methods—vector scan and raster scan systems;
(2) electron beam shape—Gaussian (round) and shaped beam systems;
(3) imaging methods—direct write and mask projection systems.

    Some of the more important aspects of an e-beam lithography system are as follows:

(1) **Minimum spot size**, which is an indication of system resolution.
(2) **Acceleration voltage.** Higher energy equates to higher resolution due to smaller electron proximity effect and wavelength considerations. Higher energy also means thicker resist layers can be effectively exposed.
(3) **Beam current.** Larger current (intensity) means higher speed lithography. However, exposure speed is limited by scanning speed and response

of deflectors. High currents can usually mean large spot size, so for high resolution, a lower beam current must be used, while a large beam current will be used for large pattern exposure, which gives high throughput.

(4) **Scanning speed.** High scanning speeds result in shorter exposure times. This is usually expressed in frequency. The maximum frequency of current vector scans is 50 MHz while for raster scans it is up to 500 MHz.

(5) **Scanning field.** For a large scanning field most patterns can be exposed within the field which can avoid problems like stitching

(6) **Other considerations** include accuracy of work stage, the accuracy of overlay between different exposure levels and accuracy of field stitching.

An illustration of a vector scan and a raster scan is shown in Figure 6.20. Essentially, the vector scan will only scan the electron beam over the exposure pattern, whereas the raster scan will scan the electron beam over the entire field and the beam blanker switches on for a definition of the exposure pattern. A work stage can be used in synchronization with a raster scan to increase scanning speeds.

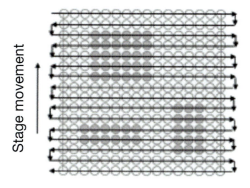
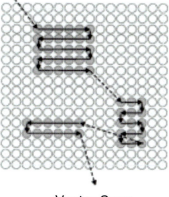

Raster Scan

Beam is constantly rastered over sample with the beam being switched on and off for the relevant exposure.

Vector Scan

Beam is only scanned in the locations where exposure is desired and switched off between these regions.

**FIGURE 6.20**   Comparison of the raster scan and the vector scan methods.

Although e-beam lithography has a very high resolution, its throughput is far inferior to that of optical lithography and this is the principal reason, why e-beam lithography has not been adopted in VLSI mass production. As one

**TABLE 6.2**    Summary of Types of Photoresist Used in Electron Beam Lithography (*measured at 20 keV and measured in $C\ cm^{-2}$)

| Type of resist | Resist tone | Resolution (nm) | Sensitivity [1]* |
|---|---|---|---|
| PMMA | + | 10 | 100 |
| ZEP – 520 | + | 10 | 30 |
| ma – N 2400 | – | 80 | 60 |
| EBR – 9 | + | 200 | 10 |
| PBS | + | 250 | 1 |
| COP | – | 1000 | 0.3 |

candidate for the next-generation lithography, e-beam projection lithography (EPL) was being developed. This works on the same basic principles as its optical counterpart. EPL can achieve a resolution of better than 100 nm due to the short wavelength of electrons and the strong focusing capability of electron optical systems.

### 6.2.3    Electron Beam Resists and Processes

E-beam resists are a group of polymers which are sensitive to electrons. Chain secession and cross-linking can occur in photoresists when molecules absorb photons. The same is true in electron beam resists, which can have both positive and negative tone and can be conventionally or chemically amplified. Much of what we discussed for optical photoresists holds for e-beam resists. The following are some properties unique to the e-beam resist.

There are two main types of e-beam resists: those for optical mask patterning and others for high-resolution direct writing.

For high-resolution e-beam write, the resolution is of primary concern as it aims to produce patterning of microstructures, which cannot be produced by optical methods. Some of the main conventional e-beam resists are listed in Table 6.2, indicating their sensitivity and resolution capability.

#### 6.2.3.1    Chemically Amplified Resists

Chemically amplified resists (CAR) were originally developed for deep UV optical lithography. While originally used for optical use, they are also sensitive to e-beam exposure. The concept of chemical amplification was proposed in the early 1980s. An acid generator in the resist polymer can release acid molecules when irradiated. At an appropriate baking temperature after exposure (post-exposure bake, PEB) and before development, the acid molecules can act as catalysts to induce more acid molecules, causing a chain reaction in the polymer and resulting in a change of resist solubility in developers. Although the initial exposure is low, the baking process has effectively amplified the chemical reaction, leading to an effective amplification of resist sensitivity. Compared to PMMA, chemically amplified resists can have a few tens of times greater sensitivity. Chemically amplified resists are also of high contrast. These properties are illustrated in Figure 6.21.

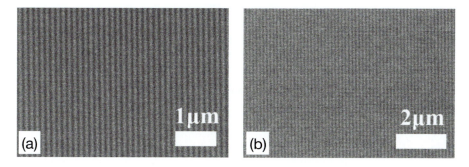

**FIGURE 6.21** SEM images of 70 nm half-pitch gratings exposed at 3 keV, 7.5 m aperture, no development, on (a) 72 nm 950k PMMA on Si, 600 C cm$^{-2}$ (b) 64 nm ZEP 520A on Si, 100 C cm$^{-2}$. Reprinted with permission from D. A. Z. Zheng et al., (2011). *J. Vac. Sci. Technol. B, 29*, 06F303, ©(2011), American Vacuum Society.

Although the resists have distinct advantages, they also have some shortcomings, such as being very sensitive to post-exposure baking conditions, restriction on delay time between exposure and post-exposure baking. The effects of post-exposure baking (PEB) are illustrated in Figure 6.22, for two different temperatures, showing the great sensitivity that can be observed. In general, exposure sensitivity and PEB latitude are conflicting properties of chemically amplified resists. Large PEB latitude is desirable for large scale production.

Another problem in CAR is the post-exposure delay (PED) effect. This is the resist sensitivity shift if the exposed resist patterns do not undergo PEB as soon as exposure is finished. This effect can be important in e-beam lithography, where the processing of a large pattern area can take a few hours up to 10 or 20 hours.

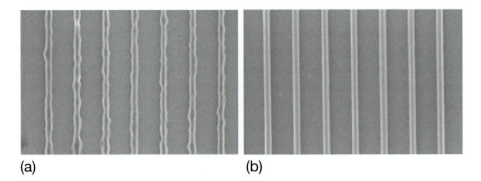

(a)                                (b)

**FIGURE 6.22**    Comparison of edge roughness of ZEP-520 resist lines (40-nm wide) developed at (a) room temperature and (b) at $-4°C$ temperature.

### 6.2.3.2    *Multilayer Resist Process*

In general, e-beam lithography requires only a single layer of resist. However, multilayer resist is needed in the following:

(1) To obtain a suitable resist pattern profile for easy lift-off;
(2) To planarize substrate surface topography;
(3) To achieve high-resolution lithography.

Lift-off is the main technique to fabricate metallic microstructures. The lift-off process is illustrated in Figure 6.23.

An ideal resists pattern profile to facilitate easy lift-off of a metal thin film is one of a narrow opening at the top with a broader one at the bottom, the so-called undercut profile. Metallic materials are thermally evaporated and deposited onto the resist surface as well as onto the substrate in exposed regions of the resist pattern. The metal film covering the resist surface has to disconnect from the metal film landing at the substrate surface, so that the metal pattern on the substrate will stay, when the resist layer is dissolved.

A bilayer resists stack can easily form an undercut profile. The most commonly used bilayer resist stack consists of PMMA of two different molecular weights. High molecular weight PMMA forms the top layer and low molecular weight PMMA forms the bottom layer. The lower molecular weight corresponds to a greater sensitivity of photoresist, so for the same exposure, the lower molecular weight resist will develop faster. This will result in a broadened profile in the lower section of the resist pattern, as shown in Figure 6.23. One of the features of a bilayer stack is that different developers are used for the two resists, so there is no interference of one developer on the resist of the

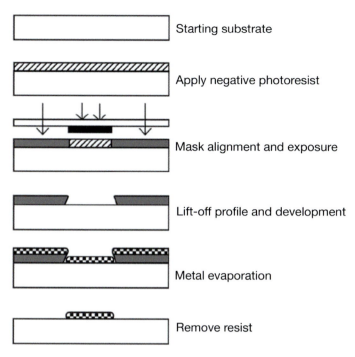

Starting substrate

Apply negative photoresist

Mask alignment and exposure

Lift-off profile and development

Metal evaporation

Remove resist

**FIGURE 6.23** Schematic illustration of the lift-off process for metal deposition.

other, making undercut profiles easier to control. Care must be taken that no intermixing occurs between the two resists. This can be achieved by baking the first layer after coating, or introducing a stopper layer between the two resists. The latter can be simply a layer of titanium or aluminum of 10–20 nm.

### 6.2.4 Ultimate Resolution of E-Beam Lithography

E-beam lithography is by far the most flexible and the highest resolution microlithography technique. However, the development of nanotechnology presents ever more demanding tasks on the patterning resolution of nanostructures. Although commercial e-beam systems can easily pattern structures of less than 20 nm, there are a number of factors, which may influence the ultimate resolution of an e-beam lithography system. These will be summarized below:

**E-beam lithography system**: Resolution of e-beam lithography is directly linked with spot size. The beam spot size is limited by aberrations in the electron optic system, which are mainly spherical, chromatic and astigmatic aberrations. The spherical aberration is proportional to the cubic power of the beam spread angle. The most effective way to reduce this angle is to use a smaller aperture. However, a small aperture will severely affect the beam current, resulting in a need to increase exposure time. In addition, a small beam current will make the detection of beam focusing marks and alignment marks difficult because of the reduced intensity of backscattered electrons. Employing a field emission source can greatly improve beam current density, thus facilitating the use of a small aperture. A field emission source also has a low electron energy spread, which will be advantageous for reducing chromatic aberrations. With the contrast improvement in system performance, modern advanced e-beam lithography systems can achieve spot sizes of the order of 10 nm. The industry is searching for further improvements to obtain a spot size of 1 nm.

**Secondary electron scattering effect**: There are two types of electron-scattering: the forward scattering and the backscattering. Whether forward or backward, they are all for primary electrons, i.e., incident electrons from the source e-beam, and there is the proximity effect caused by the primary electron scattering. An experiment using a modified STEM (scanning transmission electron microscope) was conducted with an accelerating voltage of 300 keV. The electron beam diameter from the STEM was only 2–3 nm and the thickness of resist was only 10 nm. At such conditions, it was expected that forward and backscattering effects would be negligible, and the exposed feature would be roughly the size of the beam spot. However, the minimum achievable feature size was still around 10 nm. So it was concluded, that the ultimate resolution of e-beam lithography could not be less than 10 nm. There is a theoretical basis for this conclusion. Primary electrons are of high energy and do not interact directly with resist molecules. Their function is to excite low energy secondary electrons, which are much more efficient in their interactions with resist molecules, causing chain secession and cross-linking. Secondary electrons have energies below 400 eV and 80% below 200 eV. Though at low energy, these secondary electrons can still travel some distance. For example, 400 eV electrons can travel about 12 nm, while 200 eV electrons can travel about 5 nm in the resist. The important characteristics of these low energy secondary electrons are that they always travel, or spread, in a path vertical to the direction of primary electrons. Such lateral diffusion is, on average, within a range of 10 nm, causing the smallest exposure feature to be no less than 10 nm, even though the forward and backscattering of primary electrons are excluded.

**Resist process**: The 10 nm resolution limit discussed above is derived from the lateral spread of low energy secondary electrons. E-beam lithography of a resist is not only related to the electron energy, but also to the resist process conditions. It has been shown that 10 nm is not the resolution limit of e-beam lithography. Linewidths of around 4 nm have been achieved, being by far the smallest feature obtained by conventional e-beam lithography to date. The key to achieving such small exposure features is the use of ultrasonically assisted development and a low developer strength. There is still no satisfactory explanation as to why the ultrasonic method can enhance the resolution of resist development. One theory suggests ultrasonic vibrations can speed up the interaction between resist and developer, giving a shorter development time and enhanced development contrast. Enhancement of contrast can effectively improve the resolution. Although low energy secondary electrons have a finite spread range, a resist can only be exposed above a certain threshold exposure dose. If the exposure dose is low, a small number of low energy secondary electrons with small lateral spread may induce resist molecule chain reactions with the help of external ultrasonic energy. In this way, the exposed resist feature size is limited to the lateral spread of those lower energy secondary electrons, hence an increase in the resist image resolution. Low strength developers can reduce the swelling of resist molecule chains; i.e., reduce the gyration radius of polymer molecular chains, which can also effectively reduce the exposure feature size.

## 6.3   Focused Ion Beam Lithography

A focused ion beam has much in common with a focused electron beam. They are charged particles focused by an electromagnetic field into a fine beam. The principal difference between the two relates to their relative masses. The lightest ion being hydrogen, which is 1840 times heavier than the electron. While ion beams can be used for lithography, the heavy mass of the ions can also directly sputter atoms off the solid material, which has made the focused ion beam a widely used and more versatile direct micromachining tool.

The application of the ion beam was developed in the early 20th century, when Thomson developed the gas discharge ion source (1910). One of the major breakthroughs in focused ion beam (FIB) technology was attributed to the advent of liquid metal ion sources. The liquid metal ion source, which is superior to other ion sources in brightness and small source size, has become the ion source for all current FIB systems. Liquid metal ion sources combined with advanced ion-optical systems have enabled FIB with a minimum diameter of only 5 nm. On the other hand, a FIB can machine a material by sputtering away the material. A FIB combined with chemically active gases can deposit materials onto a substrate. The FIB can also implant different atomic species into a substrate by using ion sources consisting of different liquid materials. All of the above processes are based on the high-resolution capability of a FIB system, which enables the processed structures to be on the micrometer and nanometer scale. FIB systems are a true microfabrication tool with many applications.

### 6.3.1   *Liquid Metal Ion Sources*

The source of ions is the main component in a FIB system. Early studies used ion sources for the development of mass spectrometry. Later applications in ion implantation in semiconductors led to the development of ion sources. Sources can be characterized as:

**Electron bombardment**: Electrons thermionically emitted are accelerated through a gas chamber, bombarding the atoms which can become ionized and produce a stream of ions. These typically have low ion currents and a small ion energy spread.

**Gas discharge ion sources**: Gas discharges (glow, arc and spark) can generate ions. These can produce high currents and are widely used in nuclear physics.

**Field ionization sources**: These are based on the principle that gas molecules can be directly ionized in an extremely high electric field, for example at the tip of a sharp needle at high potential. Gas molecules adsorbed at the surface become ionized and then can be emitted by the tip. This is the basic principle of the FIB.

FIB is only possible with a liquid metal ion source (LMIS), which is a field emission source from a liquid metal under a strong electric field. LMIS was developed in the 1970s and its basic structure is shown in Figure 6.24.

A tungsten wire of about 0.5 mm diameter is electrochemically etched into a needle with a tip radius of around 5–10 m. This is whetted with a liquid metal, which adheres to the needle. When subject to a high voltage, an electrostatic force, the molten metal is forced onto the tip apex, where the field can reach $10^{10}$ V m$^{-1}$. This field is large enough to ionize the metal and

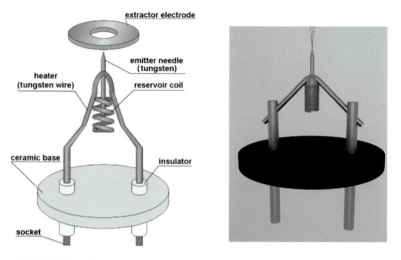

**FIGURE 6.24** Schematic diagram of a liquid metal ion source.

will then be repelled by the tip, resulting in ion emission. Total emission currents will be of the order of a few A, which corresponds to current densities of $\sim 10^6$ A cm$^{-2}$. The angular ion emission is up to $\sim 20$ A sr$^{-1}$.

The liquid metal tip has a continuous supply of material in order to compensate the loss of liquid due to emission. The analysis of field ion emission is based on three conditions to maintain stable ion emission in LMIS:

(1) Emission surface must have a certain shape to establish the surface electric field.
(2) The surface electric field must be high enough to maintain a certain level of ion emission current (plus a certain rate of liquid metal supply).
(3) Liquid flow rate sufficient to compensate the rate of loss of material. This requires a good whetting of the tungsten tip.

LMIS is characterized by:

(1) Threshold voltage for ion emission: $\sim 2$ kV. Above this, the emission current increases rapidly.
(2) Large emission angle: $\sim 30°$. This increases with emission current and the usable current will decrease with the increase of emission angle.
(3) The angular distribution of the emission current is relatively uniform.
(4) Large energy distribution of ions: $\Delta E \sim 15$ eV. This also increases with the emission current.

(5) For low currents ($<$ 10 A), most ions are singly charged, almost 100%, as measured by mass spectrometry.

LMIS started with Ga metal, which has a very low melting point (29.8°C) and is the predominant source in FIB systems. Many other metals have been used, such as those used for semiconductor doping (Al, In, As, etc.).

## 6.3.2   *Focused Ion Beam Systems*

Since the focused ion beam (FIB) system deals with the focusing of charged particles, the optical system is similar to that used in e-beam lithography systems. The principal differences regard the sign of the charge and the mass of the particles concerned. The forces at play are the magnetic and electric forces due to the applied fields (magnetic and electric):

$$\mathbf{F}_e = q\mathbf{E} \tag{6.11}$$

$$\mathbf{F}_m = q(\mathbf{v} \times \mathbf{B}) \tag{6.12}$$

The spread of energy of emitted ions from LMIS is relatively important and is due to space charge effects. The lower the speed, the larger the repulsion force and the larger the spread. For Ga ions, the space charge effect is 350 times greater than for electrons. Such effects will broaden the beamwidth. For a FIB system, chromatic aberration will be more serious than a spherical aberration. This can be reduced by using lower emission currents. Commercial FIB systems will typically run at $<$ 2 A.

The basic set-up of the FIB system, which is a two-lens system, is shown in Figure 6.25. The focusing lenses are electrostatic, as described in the previous section. The ion beam deflector is electrostatic and octopole. The typical acceleration voltage used in FIB is between 20–50 keV, which is also the energy the ion beam carries to sputter at a target surface. The typical working current is $\sim$ 1 pA – 30 nA. There is a minimum working current for which the LMIS works stably. For Ga, this is above 0.5 A. Reduction of ion beam current is then achieved by using apertures. At the minimum working current, a FIB system can achieve a resolution of about 5 nm. If a FIB system is used for direct implantation, an alloy LMIS will be used. There should be a mass filter between the two lenses in order to select different ion species. The mass filter is normally an $\mathbf{E} \times \mathbf{B}$ field mass selector.

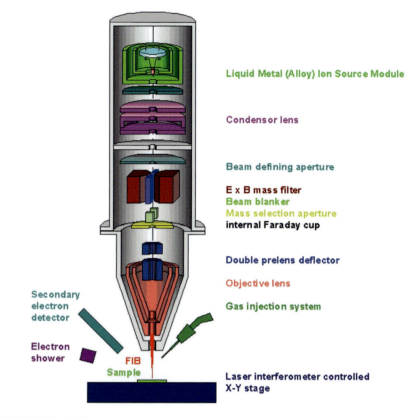

Liquid Metal (Alloy) Ion Source Module

Condensor lens

Beam defining aperture

E x B mass filter
Beam blanker
Mass selection aperture
internal Faraday cup

Double prelens deflector

Objective lens

Secondary
electron
detector

Gas injection system

Electron
shower

FIB
Sample

Laser interferometer controlled
X-Y stage

**FIGURE 6.25**    Schematic diagram of the focused ion beam (FIB) apparatus.

## 6.3.3    *Principle of Focused Ion Beam Processing*

Sputtering is the main application and function of the FIB tool. This is a process whereby incident ions transfer energy to atoms in a target, which, upon acquiring enough (kinetic) energy, can escape from the solid target surface. Sputtering is not a one-to-one process: an ion beam bombarding a target material can generate a large number of recoiling atoms. These recoiling atoms can further interact with surrounding atoms. When such atoms gain enough energy, they can overcome the binding energy and move off into free space and are then termed sputtered atoms. The sputtering yield, Y, is the term coined to specify the number of sputtered atoms. This yield can be evaluated using Sigmund's linear collision cascade model giving the yield as:

$$Y(E,\theta) = \frac{0.042}{U_s}\alpha S_n(\varepsilon)\cos^{-f}(\theta) \qquad (6.13)$$

**TABLE 6.3**  Ion Sputter Yields for Various Target Materials ($Y^*$, sputter yield measured in sputtering atoms per incident; $R^{**}$, sputter yield measured as sputtering material in volume ($m^3$) divided by ion dose (nC))

| Target | $Ga^+$ | | $Al^+$ | | $In^+$ | |
|--------|-------|------|------|------|------|------|
| | $Y^*$ | $R^{**}$ | $Y$ | $R$ | $Y$ | $R$ |
| Si | 2.1 | 0.26 | 0.9 | 0.11 | 2.9 | 0.36 |
| Cu | 1.8 | 0.13 | 0.9 | 0.07 | 5.2 | 0.39 |
| Zr | 2.5 | 0.36 | 1.2 | 0.18 | 1.9 | 0.28 |
| Mo | 1.1 | 0.10 | 0.8 | 0.07 | 2.0 | 0.19 |
| Ta | 3.5 | 0.40 | 1.3 | 0.15 | 2.0 | 0.22 |
| W | 1.1 | 0.11 | 0.8 | 0.08 | 2.1 | 0.21 |

The model predicts that the ion sputtering yield depends not only on the incident ion energy, but also on the angle of incidence, atomic density and mass of target material. Although ion sputtering yield can be calculated theoretically, most of the sputtering yield data are taken experimentally at specific sputter conditions (see Table 6.3).

Sputtering yields will be given for specific cases and conditions. The sputter yield has the following features:

(1) Ion sputtering yield increases with beam incident angle, with a maximum of around 80°. Further increase towards 90° (parallel to the surface) results in a rapid decrease of the sputter yield. This is illustrated in Figure 6.26.

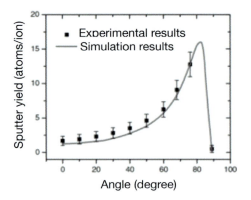

**FIGURE 6.26**    Sputter yield as a function of the incident angle of the ion beam.

(2) High energy ions do not necessarily generate a high sputter yield. For $Ga^+$, the sputter yield becomes saturated above 30 keV of ion energy. This is one of the reasons that commercial systems work in the range of 10–30 keV.

(3) Redeposition of sputter material can reduce the net sputtering yield. This can be significant when making deep holes with FIB. As the hole increases in depth, more and more sputter atoms redeposit on the sidewalls of the hole. This can be reduced by decreasing the dwelling time of the ion beam; i.e., to sputter with a fast, multiple scans of the ion beam.

(4) Ion bombardment combined with chemically active gases can greatly enhance the sputtering yield. The chemical gas-assisted sputtering process is a combination of ion sputtering and chemical gas-phase etching, because ionized chemically active gases, such as chlorine and fluorine, can react violently with Si, GaAs, Al, etc. to produce volatile compounds. The process works as follows: a small amount of chemically active gas is introduced into the chamber; gas molecules then absorb onto the target surface; ion bombardments ionize these absorbed atoms; these gas ions react with the target surface material and convert this into volatile compounds, which are then evacuated from the FIB system. Experiments show, that this process can be tens of times more efficient than the normal ion sputtering and can also reduce redeposition.

When sputtering multilayer materials, endpoint detection is required to stop ion sputtering once a layer of material has been removed. This can be achieved by observing the contrast of the secondary electron (ion) image. Different materials will have different brightnesses. Another method is to use mass spectrometry.

### 6.3.4 Ion Beam Assisted Deposition

If the gas introduced into the chamber is not chemically active, the gas ions, produced by ion beam bombardment of adsorbed gas atoms, reacting with the target material will form non-volatile compounds, which are not removed by evacuation, but remain on the target surface. This is ion beam assisted deposition. While the material is deposited by ion sputtering of adsorbed gas atoms, ion sputtering also removes the deposited atoms. The deposition process and ion sputtering coexist and compete with each other. In this case, only a fine-tuning of the ion energy, ion dose, gas flow rate and chamber pressure can adjust the deposition rate to make it greater than the sputtering

**TABLE 6.4**   Commonly Used Organic Metal Compound Gas Precursors for Film Deposition

| Organic metal compound gas | Sputtering ion | Ion energy (keV) | Deposits | Composition |
|---|---|---|---|---|
| $W(CO)_6$ | $Ga^+$ | 24 | W : C : Ga : O | 75 : 10 : 10 : 5 |
| $C_7H_7F_6O_2Au$ | $Ga^+$ | 40 | Au : C : Ga | 80 : 10 : 10 |
| $(CH_3)_3Al$ | $Ga^+$ | 20 | Al : O : Ga : C | 37 : 27 : 26 : 10 |
| $C_7H_{17}Pt$ | $Ga^+$ | 35 | Pt : C : Ga : O | 45 : 24 : 28 : 10 |

rate, giving a net deposition of material. Since material deposition will only take place where there is ion bombardment, controlling the ion beam scan can be used to fabricate 3D structures. This is shown in Figure 6.27, where an example of deposition and machining (sputtering) are shown in the same sample.

**FIGURE 6.27**   Microstructures made on the same surface by ion beam sputtering (lower) and ion beam assisted deposition (upper).

The purpose of ion beam assisted deposition is to make functional microstructures such as metal interconnections in IC chips. To deposit metal ions, organic metal compounds are used, such as shown in Table 6.4).

As can be seen, the deposited film is not pure metal. FIB assisted deposition can also deposit non-metals, such as $SiO_2$, which can be formed by the introduction of water vapor in the chamber.

### 6.3.5  Applications of FIB Technology

FIB technology has advanced greatly making it a versatile fabrication tool, used in both industrial applications and as a research method. Among other applications of FIB technology are the following:

**Inspecting and editing ICs**: High-density ICs contain from a few to a hundred million transistors with interconnections. With such complex fabrication techniques, errors in production are inevitable. FIB sputtering and deposition allow corrections to be made to fabricated ICs. So-called debugging can be carried out directly on the IC. In the example shown (Figure 6.28), part of an IC chip has been modified: one metal track is cut and a new connection has been made by a deposited bypass. FIB technology can also assist in the diagnosis of problems occurring in the manufacturing process. For example, cutting through IC chips to look at the cross-section of different process layers, to help find out, where process steps go wrong. Step-wise profiling allows a cross-section to be observed using SEM.

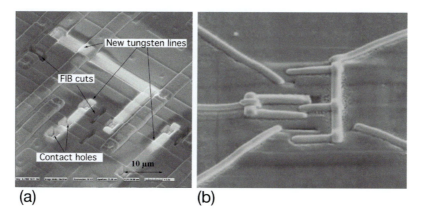

(a)                                              (b)

**FIGURE 6.28**   (a) Circuit debugging using FIB sputtering and deposition. (b) Circuit fabrication using FIB deposition.

**Repairing defects in optical masks**: Another major application of FIB is the repair of defects in optical masks. Defects can frequently occur in optical masks used in the photolithographic process causing failures to IC chips. This can be very important especially for correcting small defects, which other techniques cannot perform. One potential problem with FIB repair is Ga staining. Ga ions can become implanted in the FIB processing and, since it is a heavy metal, Ga ions can alter the light transmission of glass. The implantation depth is usually only a few tens of nanometers and Ga staining can be removed by reactive ion etching (RIE) after FIB repair.

**Preparation of TEM samples**: TEM or STEM requires very thin samples in order to allow the electrons in the e-beam to penetrate the sample to form electron diffraction patterns. Conventional methods to prepare TEM samples to consist of mechanical cutting and grinding. Such a technique can only be used to prepare large-sized samples. With FIB technology, a sample can be sliced locally for observation and analysis. The preparation of TEM samples by FIB is similar to the cross-sectioning process. Ion beam sputtering is carried out from the front and back directions, eventually leaving a thin slice in the middle, as the TEM sample. Such a section is shown in the following image (Figure 6.29). The sample has a thickness of only 100 nm. Since the direction of the ion sputtering is perpendicular to the surface of a TEM sample, there will be no implantation of gallium ions. However, other damages to the sample are possible. For example, if the ion beam diameter and scanning step size are large, often the case for high beam currents required for removing a large quantity of material, microscale ripples can be formed along the sputtering direction. This can degrade the image contrast in TEM and the periodic nature of the ripples can interfere with the diffraction image of the crystalline structure. To reduce such problems, firstly, a low-resolution ion beam will be used to sputter away the bulk of the material and then a fine ion beam will be used to scan small steps and carry out a final polishing of the sample.

**Microfabrication**: A FIB is like a scalpel with a sharp end of a few tens of nanometers and operates under high-resolution SEM in the FIB system. FIB can be used as a micro-milling station working under a microscope. Below (Figure 6.30) are some examples of micro-milling using FIB technology, which no other tools can perform.

The micro-machining capability of FIB has been applied to large scale manufacturing. For example, read-write heads for hard disks in modern computers are machined partly by FIB. This usually in the final stages of production, to improve the structure precision of the optical lithographically produced heads. FIB trimming can take only a few seconds to reduce the heads to about 100 nm.

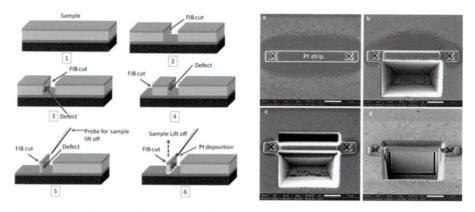

**FIGURE 6.29** (Left) Schematic illustration of TEM sample preparation. (Right) SEM images of the basic FIB milling steps. (a) The deposition of platinum onto the area of the TEM foil will be cut to protect the foil from sputtering. Milling of crosses for pattern recognition during the automated milling process. (b) Trench milled with high current (2700 pA) in front of the foil. (c) Rough milling of trenches in front and in the back of the foil accomplished. Bright contrast is due to redeposited material. (d) The specimen is tilted at 45° with respect to the ion beam and foil thickness is about 500 nm. At that stage of the automated milling process, the foil is cut free at its base and at both sides leaving a small strip of material to keep the foil in its position (Wirth, 2004)].

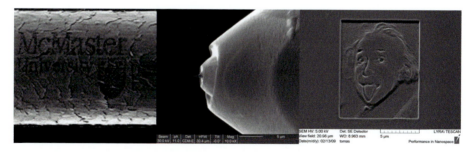

**FIGURE 6.30** Examples of FIB micro-machining: (left) Inscription on a human hair, (center) FIB machined Diamond super-tip of 20 nm radius, (right) Image of Einstein created on a surface using FIB milling.

FIB is used in high precision machining of MEMS devices such as resonators, whose natural frequency will depend on the resonator structure and mass. Once made, the structure can be fine-tuned by adding or removing material. FIB processing is reported to be capable of changing the resonance frequency of a micro-resonator by up to 12%.

**Imaging**: The ion beam in the FIB system interacts with the surface in a similar way to the electron beam in an electron microscope. This allows the FIB system to also image the surface, upon which structures have been fabricated. One of the effects that are particular to the FIB imaging is the strong channeling, that can be observed and allows for enhanced contrast that can permit the visualization of grains in metallic structures. Examples of these channeling effects can be seen in Figure 6.31.

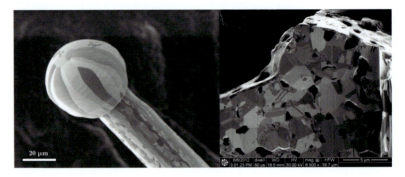

**FIGURE 6.31**    Examples of FIB imaging: (left) Ion channeling contrast is seen in a gold wire. (right) Focused ion beam channeling contrast image. The black spots are pores and the geometric shapes of varying shades of gray are the grains.

### 6.3.6  *Focused Ion Beam Lithography*

Like electron beam lithography, a focused ion beam can also be used for lithography. Ion beam lithography (IBL) is of very high sensitivity, because the energy transfer between ions and the solid's atoms is much more efficient than that using electrons. Commonly used electron beam resists are 100 times more sensitive to ion beams than to electron beams. The same resist for 20 keV EBL requires 20 C cm$^{-2}$, but needs only exposure of 0.1 C cm$^{-2}$ with Ga ion beams at 100 keV. This high sensitivity can be both an advantage and a disadvantage. A high sensitivity will reduce exposure. If it is too sensitive, the number of ions required to expose a resist is less and can lead to static noise. Since the ion scattering is random, a smaller number of ions can cause a fluctuation of the statistical distribution of the exposure dose. This is manifested as edge roughness of exposed patterns. In addition to high sensitivity, IBL has almost no proximity effect. Because ions are much heavier than electrons, the scattering range of ions in the resist is much lower than that of electrons. There is little spread of ions due to forward scattering and almost no backscattered ions. Due to these factors, feature dimensions with IBL can be of the order of 10–15 nm.

One of the main drawbacks of IBL is the very small penetration depth in the resist. For example, a Ga beam of ions of energy 100 keV will have a penetration depth of about 0.1 m. Penetration depth can only be increased by increasing the beam energy or using lighter ions. Doubly ionized ions will have twice the acquired kinetic energy for the same accelerating potential of singly ionized species.

Focused ion beam lithography with LMIS suffered from its limited exposure depth in resists. However, in R&D, FIB can still be found as a lithographic tool. In the 1990s, ion beam projection lithography (IBPL) was considered as next-generation lithography (NGL) for VLSI technology. This technique has much in common with the e-beam counterpart. The main differences can be summarized as follows:

**Ion source**: The ion source used in IBPL is a gas plasma source such as hydrogen or helium gas plasma source, instead of the liquid metal ion source used in most FIB systems. These ions are evidently lighter and can penetrate deeper into the resist and, as such, are more suitable for lithography. Ion projection needs a broad ion beam instead of a fine focused one and a gas plasma source also better lends itself to this task.

**Mask**: The difference between an e-beam projection lithography mask and one for ion beam projection lithography is that mask patterns must be unblocked from the ion beam; i.e., they must be see-through patterns. The format is the same as for e-beam projection lithography. They are made of a silicon membrane (2–3 m). Patterns are etched through the membrane. The mask is divided into many cells, with each being 2 mm$^2$. Cells are separated by reinforced ribs. The whole mask can be 4–6 inches in diameter, so that it can meet the requirements of VLSI.

**IBPL has a large DOF**: Ion beams produced by gas plasma have good parallelism and the NA of the projection lens is only 0.001, resulting in a depth of focus of up to 100 m (c.f. optical lithography where DOF ~1-2 m. The figure below (Figure 34) shows line features exposed by IBL, with lines crossing a step on a silicon substrate and have no change in linewidth. Another advantage of ion beam projection lithography is the adjustable exposure depth. This is possible by controlling the ion energy.

Ion beam projection lithography is principally a research and development tool, though it was previously considered as an industrial manufacturing technique for VLSI.

### 6.3.7    *Focused Ion Beam Implantation*

Ion implantation is an important process in semiconductor manufacturing. Transistors are formed essentially by ion beam implantation. Ion implantation in semiconductor processing is large area implantation. The implantation areas are selected by a mask, very much like the optical lithography process. Focused ion beam implantation, on the other hand, is maskless implantation. The doping elements required by the formation of transistors, such as B and As for silicon transistors, or Si and Be for GaAs devices, can be made into a liquid alloy ion source. With the help of a mass filter, the FIB system can select different ion species and separate them, so implantation can be localized by focusing and scanning the ion beam in the area of interest. The implantation depth and distribution can be controlled by ion energy and dose. In industrial ion implantation, ion species can only be implanted one at a time, while FIB with an alloy source can implant several different ion species at the same time. Flexibility is the major advantage of FIB implantation, whereas low productivity is its Achilles' heel. As such, FIB implantation is really only used as a laboratory technique.

## 6.4    X-ray Lithography

The X-ray lithography (XRL) technique was first developed by Henry Smith in the early 1970s. We noted earlier, that optical lithography is limited by the illumination wavelength due to diffraction effects. However, X-rays have a much shorter wavelength, which means that smaller feature sizes are accessible and make it a natural successor to optical methods in the preparation of lithographically defined objects. By the early 1990s, X-ray lithographic techniques were capable of producing 85 nm MOSFET device structures in the laboratory and by 1995, IBM was able to manufacture VLSI (very large scale integration) chips with minimum feature sizes of 0.25 m using XRL. Despite this, X-ray lithography did not become a mainstream manufacturing tool for VLSI, though it can be found in other smaller production applications.

The preparation of microstructures with the smallest feature size is just one of the possible applications of XRL. Another important application is making structures with high aspect ratios, which requires deep lithography. This is a process called LIGA, which we shall discuss in a later section. Such processes are important in the fabrication of MEMS type structures.

### 6.4.1 Fundamentals of X-Ray Lithography

X-rays are electromagnetic radiation with wavelengths ranging from 0.1–100 Å. From the electromagnetic spectrum, see Figure 6.32, we see that X-rays extend from the soft regime ($\lambda \sim 10$ Å, $E \sim 1$ keV) to the hard regime ($\lambda < 1$ Å, $E > 10$ keV). Besides the difference in wavelength, the fundamental difference between X-rays and optical (visible) wavelength radiation is that X-rays can penetrate the majority of materials. Only those elements with a high atomic number can absorb X-ray radiation. Furthermore, X-rays cannot be focused because the refractive index of all materials for X-rays is approximately the same, ($n \simeq 1$). As such, there can be no refraction of X-rays and and therefore, XRL can only be a 1:1 proximity type lithography.

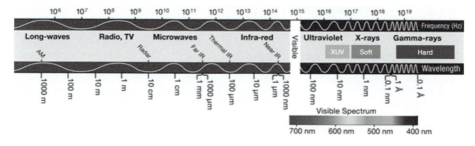

**FIGURE 6.32**   Electromagnetic radiation spectrum.

The basic set-up for XRL is illustrated in Figure 6.33. The principal component consists of the X-ray mask, which is made up of a membrane of a low atomic number material, typically Si or SiC, with a patterned high atomic number material, that acts as the X-ray absorber, typically Au, W or Ta. For a X-ray beam of wavelength 1 nm, the Si membrane thickness is of 1–2 m and the absorber thickness is about 300–500 nm. The exposure depth in the resist will be of about 1 m and the proximity gap, $G$, between the X-ray mask and the substrate typically lies in the range of 5–50 m. The proximity gap is necessary due to the poor mechanical strength of the X-ray mask. The large thin membrane is very fragile and cannot be in mechanical contact with the exposure surface, ruling out the possibility of contact type lithographies.

The resolution of proximity lithography is dependent on the gap distance. Eq. (6.1) is also valid for X-ray proximity lithography. In terms of the gap distance, we can write this as:

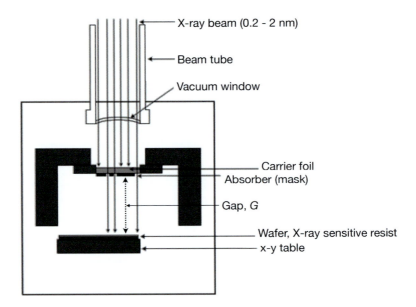

**FIGURE 6.33**   Schematic illustration of the X-ray lithography apparatus.

$$G = \frac{\alpha W^2}{\lambda} \tag{6.14}$$

where $W$ is the exposure width, $\lambda$ is the X-ray wavelength and $\alpha$ represents a constant of proportionality, which depends on the process conditions and is equivalent to $1/k$ from Eq. (6.1). High $\alpha$ means strict requirements on the resist process conditions, which are difficult to implement in a production environment. The usually accepted value of $\alpha$ is unity. From this relation, it is possible to determine the required gap necessary to achieve specific line features; for example, line features of 100 nm require a gap distance of 10 m.

Similar to optical proximity printing, there are diffraction effects such as Fresnel diffraction, when X-rays pass through the mask pattern. This means that the XRL image will be related to the spectrum width spatial coherence of the X-ray radiation source. In Figure 6.34, the image intensity distribution of four sources are shown. The images show that a monochromatic source has a more serious diffraction effect and that an extended source is preferable to a point source.

The diffraction effect can blur the boundary of transparent and non-transparent regions of the X-ray mask, as is shown in Figure 6.35. This shows the intensity distribution of a 0.1 m opaque line pattern. The effect

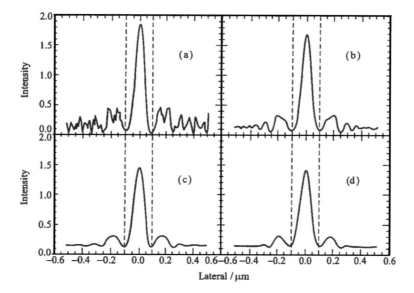

**FIGURE 6.34** Intensity distributions from four different sources: (a) monochromatic point source, (b) polychromatic point source, (c) monochromatic extended source, (d) polychromatic extended source.

of diffraction is to cause the dark region to appear wider than the desired 0.1 m feature. In order to make the exposed feature meet with the design dimension, the mask pattern must be adjusted to become larger or smaller than the desired feature size and thus compensate for the diffraction effect.

The diffraction induced distortion has been taken advantage of to make exposure patterns smaller than the mask pattern. This allows, to a certain extent, a de facto demagnification of the features in XRL. The demagnification mechanism can be explained from the Fresnel diffraction image in Figure 6.35. Due to the Fresnel diffraction on both sides of the boundary, between the opaque and transparent regions, the intensity is neither 0 nor 1 (normalized). In fact, the position, where the normalized intensity becomes unity (at around 6 in Figure 6.35), is displaced from the boundary by an amount, which depends on the gap and the wavelength of the radiation. Evidently, for different intensity values, the feature dimension will also be modified. The higher the intensity, the narrower the intensity distribution. Therefore, by selecting a low exposure dose and the sensitivity of the resist, it is possible to manipulate the feature size. In Figure 6.36, we show the comparison of the feature sizes on the X-ray mask and the printed pattern on the resist. The transparent lines on the mask are 152 nm, while those on the resist are

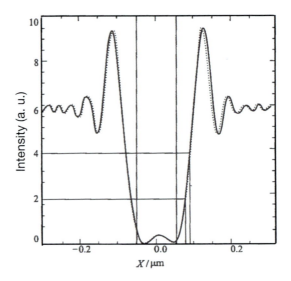

**FIGURE 6.35**    Mask pattern dimension compared with diffracted X-ray intensity distribution.

reduced to 61 nm. Such demagnification effects are capable of producing feature sizes of down to 25 nm.

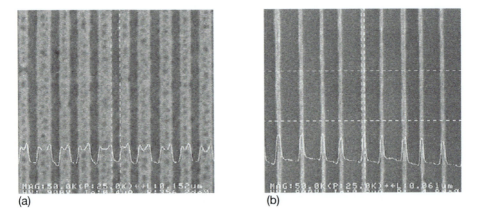

(a)                                                                (b)

**FIGURE 6.36**    Comparison of the mask and resist feature size. (a) Features on the X-ray mask, where the transparent lines are measured as 152 nm, and (b) SAL605 negative tone resist line features with a line width measurement of 61 nm.

For a point X-ray source, the X-ray radiation will not be parallel. This will cause a shadowing effect (or penumbra) on the substrate. This will cause

**TABLE 6.5**  Common X-Ray Sources and Their Corresponding Irradiation Power

| Source type | Power (mJ cm$^{-2}$ s$^{-1}$) |
|---|---|
| Electron bombardment | 0.01–0.1 |
| Laser plasma | 0.01–1 |
| Synchrotron radiation | 10–100 |

both blurring and lateral shifts in the features. Generally, this effect will be minimized for synchrotron radiation, which has good parallelism.

## 6.4.2  X-Ray Lithography Apparatus

The basic elements of an X-ray lithography system include the X-ray source, a mask aligner, an X-ray mask, and the X-ray photoresist.

**1) X-ray source:**

X-rays can be generated by two methods. One way is to make the electrons surrounding an atomic nucleus transit from a high energy level to a lower one. The energy released between these levels can correspond to sufficient energy differences, that give rise to photons in the X-ray regime. In this case, the energy requirement is that a core level be involved, which will give rise to short wavelengths and X-ray radiation. Another method for producing X-ray radiation is via the deceleration of a charged particle (such as the electron). This can be performed for example using a synchrotron source.

In an X-ray point source, a metal, typically copper, is used as a target, where a high-energy electron beam excites the electrons in the metal to high energy levels. Many X-ray sources, such as is medical applications, rely on this method. Another form of the point source is a laser-plasma X-ray source. In this case, a high energy density laser beam is directed at the target to produce the excitation of core electrons. Typically, a Nd:YAG solid-state laser is used ($\lambda = 1064$ nm). Frequency doubling is often used to halve the wavelength to 532 nm. The laser pulse width is of the order of ns. X-rays produced by a laser-plasma source typically have wavelengths in the region of 130–140 Å, which belong to the soft X-ray and XUV region of the electromagnetic spectrum. Since the laser is pulsed, so too will be the X-ray radiation. So while the pulsed X-rays can have an elevated power, the continuous power is relatively low. In Table 6.5, we compare the power for different X-ray sources.

Synchrotron radiation sources (SRS) are of practical value as a source of X-ray radiation. The principal of synchrotron radiation is that high-energy electrons traveling close to the speed of light in circular orbits (storage ring) can radiate a brand spectrum of electromagnetic radiation along the tangential direction to the orbit. The radiation spectrum covered ranges from IR to hard X-rays. SRS radiation has the following characteristics: (i) High brightness ($10^4$ greater than a conventional X-ray tube), (ii) Broad radiation spectrum—these can be selected from the output by an alternating magnetic field (wiggler), (iii) Highly polarized radiation, and (iv) Ultra-short pulsed radiation ($\sim$ ns). Modern synchrotron sources can generate electron beams with energies up to 8 GeV.

For XRL applications, good matching is required between the X-ray radiation energy spectrum and the resist sensitivity spectrum. Any mismatch could lead to heating and damage to the delicate X-ray mask. The energy to cause effective resist exposure (sensitivity spectrum) is the product of the mask transmission spectrum and the resist absorption spectrum, which is described by the mask—resist filter function (MRFF). This is shown in Figure 6.37 and indicates that the energy spectrum for the resist exposure is in the region of 1–2 keV. The matching issue concerns the matching of the SRS output energy spectrum and the MRFF. By careful selection of the beryllium window and the angle of the reflection mirror, the energy spectrum can be adjusted to match as much as possible the MRFF.

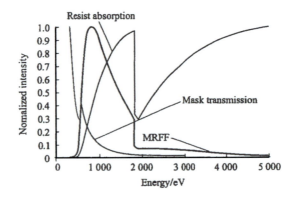

**FIGURE 6.37**     Mask-resist filter function for soft X-rays.

**2) X-ray mask aligner and stepper:**
This is essentially the same as for the optical mask alignment, except that it is structurally simpler, since no complex imaging optics is required. There

are two components to a X-ray mask aligner: (i) a work stage system and (ii) an alignment system. The system is generally mounted vertically for SRS illumination. The work stage requires accurate precision for a gap region of around 10–20 m. Since the SRS beam size is roughly 20 mm$^2$, the stage must be able to move precisely to perform step and repeat exposures; this is referred to as an (X-ray) stepper. The mask must be free from mechanical stresses, as it can easily distort and affect exposure accuracy.

**3) X-ray mask:**
This is one of the most critical components in any XRL system. XRL is a 1:1 lithography and as we stated earlier, the mask depends on the density and thickness of the absorber as well as the supporting membrane. The typical X-ray mask is made of a 25 mm × 25 mm membrane of thickness 1–2 m, with a patterned heavy metal layer.

The X-ray mask fabrication process is important and can be summarized as follows:

(1) Deposit supporting membrane (1–2 m) on to a Si substrate.
(2) Fix coated Si substrate on to the frame.
(3) Remove Si substrate in the central region by KOH chemical etching.
(4) Deposit heavy metal on to the supporting membrane. Patterning can be performed by wet chemical etching (low resolution). Electroplating can be used with e-beam lithography. Reactive ion etching can also be employed (for W or Ta). The layer thickness of the absorber can be controlled to the required thickness to obtain a satisfactory contrast (N.B. steps (iii) and (iv) can be reversed).
(5) Inspection of the mask to test for defects. These can be repaired using FIB if necessary.

Pattern accuracy is a crucial factor for pattern transfer since XRL is a 1:1 process. Internal stresses can have a great influence on the pattern accuracy. Since the mask is a large area membrane, residual or thermal stresses can deform the membrane and distort the mask pattern. This is also true of the heavy metal absorber. The pattern distortion caused by absorber stress can be modeled using the following relation:

$$\delta \geq \frac{\sigma_a t_a}{2 E_s t_s} W \qquad (6.15)$$

where the distortion, $\delta$, is directly proportional to the mask size, $W$, the stress in the metal absorber, $\sigma_a$ and its thickness, $t_a$, while being inversely proportional to the membrane thickness, $t_s$, and its Young's modulus, $E_s$. The distortion is also related to the pattern density and geometry. Large area stress can be relieved by improvements to the film deposition process and reduce any internal stresses that may arise during the deposition of the absorber layer.

In addition to the internal stresses, external stresses can also have a detrimental effect causing distortion o the mask structure. To avoid this, care must be taken not to introduce thermal variations and forces in mounting the mask on the frame. Heating can be caused by exposure to the X-rays, so the frame should be a good thermal conductor. In SRS, there will be He gas between the output Be window and the mask. The He gas is a good thermal conductor and can help in dissipating heat away from the mask.

**4) X-ray resists:**

The process of exposing a resist to X-rays is almost the same as in e-beam lithography. The difference is that electrons directly interact with the resist molecules in exposure, whereas X-rays generate photoelectrons and Auger electrons, which interact with the resist molecules. Any electron sensitive resist can also serve as a X-ray resist. The most commonly used resists being PMMA, SAL601, AZPN114, AZPF514, and ZEP520. The X-ray resists process, in general, is simpler than for optical lithography and EBL, because X-rays have a stronger penetration capability and it is thus easier to achieve a vertical sidewall in the resist pattern.

### *6.4.3   High-Resolution X-Ray Lithography*

The greatest advantage of X-ray lithography is its short wavelength. The goal of XRL has mainly been geared towards the production of ultra-fine integrated circuits and can have resolution superior to optical lithography. One of the most effective processes in XRL is the interaction of the X-rays in the resist, creating Auger electrons, which are more efficient than secondary photoelectrons. The scattering range of Auger electrons is somewhat reduced and less than 5 nm. This augurs well for the lithographic process and essentially means that this should allow very high resolution. By limiting the X-ray energy, the production of photoelectrons can be reduced. The optimum wavelength for X-rays in XRL being between 0.7 and 1.2 nm. Another advantage of XRL over other processes is the relatively large exposure and process latitudes. XRL has good tolerance to exposure dose and resists process parameters.

In theory, XRL can reach minimum feature sizes of less than 20 nm. However, in practice, the challenge of high-resolution XRL is in the production of masks. This is of particular importance, since the XRL process is a 1:1 pattern transfer. Such small object sizes can be achieved using electron beam lithography. Another challenge in XRL is the accurate control of the mask-to-wafer gap. As we saw earlier, high-resolution XRL requires the gap to be controlled to within 10 m. A variation in the gap can cause a degradation of the image resolution. This also requires that the Si wafer to be flat within $\pm 0.25$ m, while the mask should be flat to within $\pm 0.5$ m. These are rigorous restrictions. The gap control accuracy can be performed using laser interferometry, however, this track is made difficult due to the fragility of the mask membrane.

Although X-rays have short wavelengths, there can still be diffraction effects, which become more problematic for coherent sources such as synchrotron radiation, which has a divergence of $< 1$ mrad. Diffraction can cause proximity effects in the exposed pattern in much the same way as in optical lithography. The exposure pattern resolution and fidelity are dependent on the pattern density and area. Commonly used corrections of the proximity effect in XRL are the biasing of the pattern design by enlarging or reducing the original design dimensions. For example, to expose a 0.1 m grating, a bias of 20 nm to each absorber line is necessary. To achieve high resolution in XRL, an optimization of various parameters is required: substrate gap, absorber material, and thickness, mask feature bias, etc.

Despite being an excellent candidate for VLSI manufacturing, processing costs are far in excess of those of more standard optical photolithography. The advances in optical lithography, phase shift masks, and optical proximity correction have meant that it has been able to rise to the challenge of most VLSI processing and XRL is not (yet) an economically viable alternative. This is not to say that XRL has been abandoned as a technique. XRL can be important in specific tasks such as in GaAs ICs. Another important area for XRL applications is the fabrication of ultra-deep microstructures. Indeed, the additional costs for XRL with respect to the optical methods reside in the mask preparation and X-ray source production costs.

### 6.4.4 High Aspect Ratio XRL–LIGA Technology

LIGA is a German acronym that has been adopted comes from *Lithographie* (LI) *Galvanoformung* (G) *Abformung* (A), which means lithography + electroforming + molding. LIGA technology was developed in Germany in the early 1980s. The principal motivation was to develop this technology to fabricate micro-nozzles for the refinement of uranium isotopes. These micro-nozzles have minimum features in the micrometer range, but the overall structure is on the millimeter scale. Using the great penetration capacity of hard X-rays, structures can be made of 1 mm depth and still have lateral dimensions on the micrometer scale. LIGA technology can fabricate metallic or plastic structures, which are beyond the capability of conventional precision machining. Since its development, LIGA technology has rapidly become an important microfabrication tool for the manufacture of MEMS devices.

The LIGA process is schematically illustrated in Figure 6.38. X-rays from a synchrotron source irradiate a thick resist layer through a mask. The deep resist structure, after development, is filled with a metal by electroforming. The metal microstructure is formed after the removal of the resist mold. The

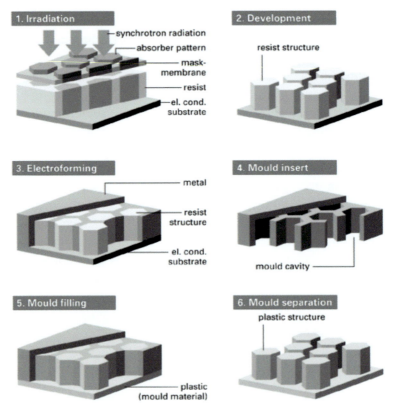

**FIGURE 6.38** Process steps for LIGA technology.

metal structure can then be used as a mold for molding or embossing plastic. Finally, a plastic replication of the original resist structure is obtained. Once the mold has been made, it can be used to produce hundreds or thousands of replication structures. In Figure 6.39, we show some example structures that have been made using the LIGA process. The distinctive feature of LIGA

**FIGURE 6.39** Microstructures fabricated using LIGA technology. (a) Active part of a 2x2 optical switch, only the electrostatic actuator. (b) A 517 m tall copper coplanar waveguide. (c) A nozzle for uranium enrichment.

technology is its ability to make high aspect ratio structures. The aspect ratio (structure height or depth to lateral dimension) can be as high as 100:1. The key components for LIGA technology are the X-ray source, the LIGA mask, and the X-ray sensitive resist.

Although LIGA technology is able to make very high aspect ratio objects with good vertical sidewalls, there are some factors, which limit its progress in pattern accuracy, minimum feature size, and sidewall verticality. These factors include X-ray diffraction effects, photoelectron scattering, penumbra effect, non-vertical sidewall in absorber pattern, mask distortion and secondary electrons induced by X-rays in the substrate material. Despite this, LIGA is still a very useful technique and is principally used in the fabrication of microstructures used in MEMS devices. Since these have dimensions of a few tens to a few hundred microns, inaccuracies at the sub-micron level are generally negligible on the dimensions of the microstructures.

## 6.5 Etching Techniques

We have already discussed the various techniques used for the realization of pattern transfer. In the cases of lithography, this simply means the reproduction of the pattern in the form of a resist pattern. Frequently, this is the start of a new step; pattern transfer into the fundamental material. In many fabrication processes, this will be performed by using an etching technique, where the resist forms an etching mask. Functional materials will be those of interest for the final device, such as silicon, silicon dioxide, silicon nitride for ICs and MEMS devices, GaAs for high-frequency electronics, III–V semiconductors for optoelectronic devices and quartz or glass for optical masks or microfluidic devices, to name some examples.

Etching can be a physical or a chemical process, or even a combination of the two, which removes part of the functional material unprotected by the polymer resist pattern. Etching techniques include chemical wet etching, plasma dry etching as well as other forms of etching. At a certain etch rate, the etch depth in a functional material is a function of the etch rate and this can depend on many factors.

### 6.5.1   *Wet Chemical Etching*

Wet chemical etching was the earliest pattern transfer technique used in the semiconductor industry. The *wet* implies that the etching is based on a liquid chemical erosion. The most distinctive feature of wet etching is its isotropic nature, i.e., the etching rate is equal in depth and in lateral directions. Some exceptions to this can occur, most notable for certain crystals such as silicon, which is indeed quite anisotropic, as we shall discuss in the following section. Isotropic etching means that the resolution of the etched patterns is larger than those of the actual patterns defined by the lithographic mask. It is not surprising that the anisotropic dry etch quickly replaced wet etching to become the preferred method for pattern transfer in VLSI manufacturing. However, that is not to say that wet chemical etching does not have its uses and applications. In fact, wet etching is still used regularly in IC processing, mostly in the silicon cleaning process. Wet chemical etching has also found applications in the production of MEMS and microsystem fabrication. Since structures in MEMS are significantly larger than those in the IC industry, the low resolution is not a critical limitation. Also, the relatively low costs mean that it is rather advantageous for industrial processing. Si and $SiO_2$ are the most widely used materials in MEMS and microfluidic systems as well as the basic materials for the semiconductor industry. In fact, wet chemical etching is mainly limited to these two materials. We will focus on these materials in the following.

## 6.5.2 Anisotropic Wet Chemical Etching in Silicon

When we refer to the anisotropic wet etching of crystalline silicon, we mean that the etching rates in the various crystalline directions ($\langle 100 \rangle, \langle 110 \rangle, \langle 111 \rangle$, etc.) are different. This is due to the differences in atomic densities along with the different directions in the crystalline lattice. For example, the $\langle 111 \rangle$ direction has the highest atomic density.

To certain chemical etching solutions, the rate of chemical erosion is highly dependent on the atomic density of the crystalline plane. The higher the density, the lower the etch rate in general. One of the consequences of the difference in etch rate will lead to anisotropy in the etched features. For the case of silicon, the etch rate is fastest in the $\langle 100 \rangle$ directions. In Figure 6.40(a), we show an etched Si wafer with "$V$" grooves. Instead of vertical sidewalls, the sloped edge has an angle of 54.74°. This corresponds to the [111] planes, where the angle between the $\langle 100 \rangle$ and $\langle 111 \rangle$ direction is 54.74°. This is schematically illustrated in Figure 6.40(b). Since the etch

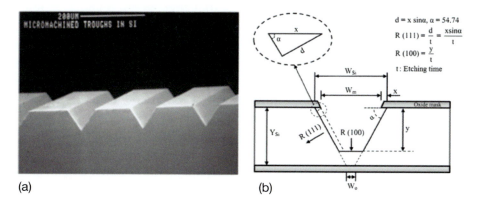

(a)   (b)

**FIGURE 6.40** Anisotropic etching in (100) Si. (a) Etching in the $\langle 100 \rangle$ direction perpendicular to the surface leads to "$V$" grooves on the Si substrate. (b) Schematic illustration of the etching process on the Si (100) surface.

rate in the $\langle 111 \rangle$ direction is much lower than that in the $\langle 100 \rangle$ direction, a groove profile is formed. As the etching follows the crystalline axes, the etched profile can be evaluated from the angle, from which we can write:

$$W_o = W_{Si} - \sqrt{2} Y_{Si} \qquad (6.16)$$

where we note that $2\cot(54.74°) \simeq \sqrt{2}$. Symbols are defined in Figure 6.40(b). Chemical etching in the $\langle 110 \rangle$ direction is also anisotropic. This is more complex since the $\langle 100 \rangle$ and $\langle 111 \rangle$ directions intersect at different angles. Specific etching profiles for the various directions will depend on the relative etching rates for different directions.

Once this difference is well understood, we can use these facts to create specific etched profiles and structures. For example, from the (110) oriented wafer, it is possible to produce deep trenches with high aspect ratios and vertical sidewalls. However, (100) oriented silicon wafers have been widely used to make the supporting frame for membrane structures such as X-ray masks.

The chemicals used in the anisotropic wet etching of silicon are an alkaline group of liquid solutions such as KOH/IPA (Potassium hydroxide/Isopropyl alcohol), EDP (ethylenediamine pyrocatechol) and TMAH (tetramethylammonium hydroxide). KOH is the most commonly used of the solutions. The etch rate in KOH depends on the concentration and temperature, and can be calculated from the formula:

$$R(100) = 2.6 \times 10^6 W^{2.5} e^{-(W/300+0.48)/(k_B(T+273))} \tag{6.17}$$

This gives the etching rate for the $\langle 100 \rangle$ direction. In this relation, $W$ refers to the width of the opening in the mask. This is an empirical formula, from which the relation between etch rate, the concentration of the etching solution and the temperature of the solution are calculated. These are shown in Figure 6.41. It can be seen from the curves in Figure 6.41 that the etch rate reaches a maximum at around 20% KOH concentration. The higher the KOH solution temperature, the greater the etch rate. The etch rate for the $\langle 110 \rangle$ direction in Si can be obtained by multiplying the rate in Eq. (6.17) by a factor of two.

Wet chemical etching is a common technique for removing the bulk of Si substrates, a process referred to as *bulk silicon micro-machining*, which differentiates it from surface Si micro-machining. The former is used in the fabrication of membranes. Si wafers typically have a thickness of around 400–600 m. So with typical etching rates of 1 m/min, the etching process can take many hours. Most photoresists are soluble in alkaline solution and therefore cannot be used as etch masks. SiO$_2$ is a good masking material for dry etching, but has a high etching rate for KOH. Pinholes in mask materials such as SiO$_2$ can be a major problem in wet chemical etching. The best masking material is silicon nitride deposited by low-pressure CVD. Other techniques do not produce a dense enough layer, which is necessary to resist KOH erosion. The etch rate in silicon nitride is about $10^4$ times slower than in silicon. Therefore, a 40 nm SiN layer is sufficient for an etch mask for the etching of 380 m in a silicon wafer.

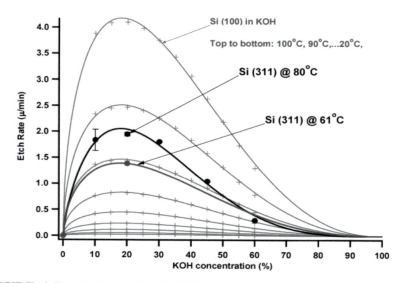

**FIGURE 6.41**  Etch rates for Si (100) (crosses) and Si (311) (dots) in solutions of various concentrations and temperatures. The values for Si (311) are determined experimentally.

The anisotropic nature of chemical etchants in Si makes the control of etching profiles somewhat problematic, since the final etched structure may differ from the initial design pattern. To overcome such problems, design compensation is necessary. The dependence of etching on the crystalline orientation highlights another critical issue in the wet chemical etching of silicon, namely, the alignment of etch masks with the crystalline axes. Mis-alignment of the mask pattern edge with the crystalline axes can result in the etched edge following the crystalline axis and not the mask pattern edge. In order to avoid this problem, commercial silicon wafers have *flats* marked on the edges, see Figure 6.42. This allows both the identification of the type of wafer surface and the alignment along the edges, though it is preferable to use more accurate alignment techniques. This can be done by etching trenches along with different directions.

### 6.5.3  *Isotropic Etching in Silicon*

Wet etching is generally an inherently isotropic process, with anisotropic sil-icon etching by alkaline etchants being the main exception. Etching silicon with acidic solutions is isotropic. The most widely used acidic etchant is a mixture of HF, $HNO_3$ and acetic acid ($CH_3COOH$)—known as the HNA

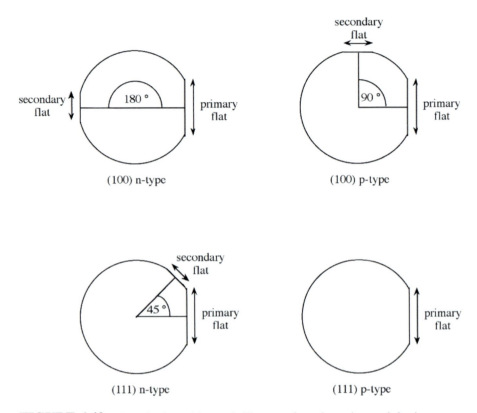

**FIGURE 6.42**    Standard markings of silicon wafer orientation and doping type.

etchant. The mechanism of etching silicon using HNA is that first the $HNO_3$ oxidizes the Si, then HF dissolves the $SiO_2$. The acetic acid in this solution acts as a buffer solution and can be replaced by water. The etch rate depends on the concentration ratio of each of the acids and can be represented on the three-phase diagram, as illustrated in Figure 6.43. For example, the diagram indicates that an etchant composed of two parts HF and one part $HNO_3$ with a small part of acetic acid can produce an etch rate of 470 m/min. HNA has been used as the standard etchant in the semiconductor industry since the 1960s. High etch rates are not necessary for most production processes.

HF is a strong etchant of $SiO_2$. Though the HF is diluted in HNA, the $SiO_2$ etch rate is still around 300–800 Å/min. In general, $SiO_2$ is not suitable as an etch mask unless used in very low concentrations ($\sim 1\%$). Photoresists are also not suitable for HNA to etch masking, since $HNO_3$ is a strong oxidant to photoresists. The best masking material is SiN, with an etch rate of less

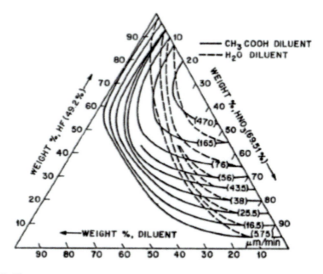

**FIGURE 6.43**   Three phase diagram of HNA concentration ratio and etch rates.

than 100 Å/min in HNA. Chrome or gold films can also be used as an HNA etch mask.

One important issue in wet chemical etching is that the etch rate is strongly influenced by the dynamics of the fluid the etchant liquid. Etching with and without agitation can result in quite different etch profiles.

### 6.5.4   *Isotropic Etching in Silicon Dioxide*

Silicon dioxide is the second most used material in the semiconductor industry, after Si. In the fabrication of ICs, silicon dioxide, $SiO_2$, is widely used for insulation and passivation. $SiO_2$ is also commonly used as a sacrificial layer in MEMS technologies. In silicon surface micro-machining, sacrificial layers are the key to the realization of movable micromechanical structures. In Figure 6.44, we illustrate the steps in the micro-machining of a Si surface using $SiO_2$ as a sacrificial layer. A polysilicon layer is deposited onto the $SiO_2$ film, part of which has been opened by lithography and etching. As such, the polysilicon is deposited directly on the Si substrate surface, forming an anchor for the structure. After patterning the polysilicon into the desired structure, the oxide layer is removed by chemical etching. The polysilicon structures are now suspended and become movable when subject to an applied force.

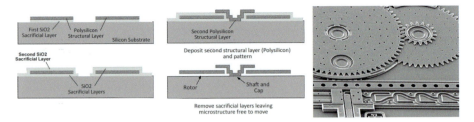

**FIGURE 6.44**   Left: Process steps of surface micromachining, where silicon dioxide is used as a sacrificial layer. Right: SEM image of a MEMS structure that can be fabricated in this manner.

Removing the oxide sacrificial layer requires isotropic etching, because the etchant needs to reach the oxide, which is sandwiched between the polysilicon layers and the silicon substrate. The etchant used is HF. Without dilution (49 % weight), HF has an etch rate of 1.8 m/min. In practice, buffered HF is used to dilute the solution and control the etch rate.

The lateral etch rate is very important in MEMS device fabrication. This can be improved by two methods: (i) using a mixture of HF and HCl acids or (ii) the more effective technique to remove the sacrificial layer is to deposit $P^+$ or $B^+$ doped oxide as the sacrificial layer. These are called phosphosilicate glass (PSG) and boron-phosphosilicate glass (BPSG). The etch rate of PSG in HF:HCL is 1.133 m/min, while for BPSG it can be as high as 4.17 m/min.

Glass is also a form of $SiO_2$ or amorphous oxide. Glasses have found many applications in microfluidic systems and devices. Glass is one of the base materials for making a variety of *lab-on-a-chip* devices, such as microfluidic channels in the glass. The procedure for making micro-channels is similar to that of making microstructures in Si.

### 6.5.5   *Dry Etching: Reactive Ion Etching*

Dry etching is a very broad concept. All etching processes, which do not use wet chemicals, can be termed *dry etching*. Etching implies that the *machining* commences from the material surface and removes the material layer-by-layer. A narrower definition of dry etching would be a surface machining by physical or chemical processes taking place in a plasma discharge. The broader definition includes not only plasma methods, but also other physical and chemical micro-machining processes such as laser micromachining, electro-discharge machining, chemical vapor etching, and powder blasting. Among the various techniques termed dry etching, the most widely used is the reactive ion etching method, which is commonly used in micro- and nanotechnologies.

Reactive ion etching (RIE) takes place in a plasma. This fundamental process of RIE is that plasma is established through gas discharge and contains a number of chemically active gaseous ions. These ions interact with the surface atoms of the sample under consideration, producing new compounds, which are volatile and will ultimately be pumped away using the vacuum system, in which the gas discharge is established. The cycle of ion reaction to gaseous by-products to be pumped away repeats, as the material is removed from the surface, until a certain depth of etch has been achieved. Apart from the chemical reaction, energetic ions bombarding the surface of the material can also sputter the surface atoms, resulting in a combination of chemical and physical etching. A good RIE process can be measured by the following two factors:

(1) *Etching selectivity:* the ratio of substrate etching rate to the mask etching rate. High selectivity means that there is little relative loss of the mask materials and this can endure long etching processes in the plasma.

(2) *Etching anisotropy:* anisotropy in RIE is represented by the ratio of etching depth to the lateral etching. Highly anisotropic RIE signifies that there is little or negligible lateral etching. The etching is a pattern transfer process and requires a mask pattern to be faithfully reproduced on the substrate surface. Different anisotropies in RIE can result in different etching profiles compared to the original mask pattern, as is illustrated in Figure 6.45.

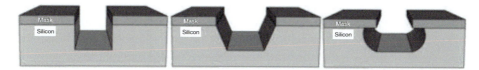

**FIGURE 6.45** Anisotropy in RIE: Fully anisotropic etching (left), partially anisotropic (center) and isotropic etching (right).

The majority of applications demand anisotropic etching. Other indicators include the etch rate and etch uniformity. The performance of RIE is dependent on many factors. The principal factors are listed below:

(1) *Flow of reactive gas:* The reaction rate at the material surface is directly dependent on the supply rate of the reactive gas species. How-

ever, a fast flow rate of the reactive gas at constant pressure can lead to short dwelling times of the gas molecules in the reaction chamber. If the plasma discharge power is not increased accordingly, the number of gaseous ions generated in the chamber will decrease, thus reducing the etch rate. If, on the other hand, the gas flow rate is too low, the consumed reactive gases cannot be replenished quickly enough, which will again lead to a reduction in the etch rate.

(2) *Discharge power:* The simplest RIE system is a planar diode electrode system, where the cathode is connected to a RF power supply by capacitance coupling. The high-frequency electric field between the cathode and the anode will accelerate and resonate with a small number of free electrons, which in turn will acquire sufficient kinetic energy to ionize the gas molecules in the chamber. Recombination of gas ions releases photons, which is seen as a glow discharge in the plasma chamber. The sample to be etched is placed on the cathode. The etching rate increases with the RF power. Etching anisotropy results from the directionality of the ions sputtering the surface, with almost no lateral etching.

(3) *Chamber pressure:* RIE is normally conducted in a vacuum chamber at low pressures, typically of the order of $10^{-3}$ mbar. This can be adjusted to optimize the etching process by controlling the electron mean free path. Higher acceleration energies lead to increased ionization probability. The low chamber pressure means high pumping speeds, which help to rapidly remove volatile products. At low chamber pressure, RIE is dominated by ion sputter etching, while at higher pressures, RIE is dominated by chemical reactions. The latter leads to a more isotropic etch profile.

(4) *Substrate temperature:* In general, high temperatures promote chemical reactions and help to remove reaction products. In RIE, ion bombardment will cause the substrate temperature to rise and can reach temperatures of $100 - 200°C$. This can cause a softening of the masking photoresist.

(5) *Electrode material:* This concerns both the cathode and the anode. Since the substrate to be etched is placed on the cathode (anode chamber walls are maintained at ground potential), the cathode material, which is generally partially exposed, needs to be of an inert material. Both cathode and anode suffer from ion bombardment during RIE. This is generally negligible compared to the etching of the substrate.

(6) *Loading effect:* This refers to the non-uniformity of etching caused by a variation of the etching area. When an etching area is increased while the gas flow is maintained constant, more gaseous ions are consumed and the etch rate decreases. There is also a micro-loading effect, which occurs when a variation of etch rate is set up due to the variation of the local pattern dimensions. This is illustrated in Figure 6.46. The trenches reduce in depth as their width is decreased despite being etched under the same RIE conditions. This is because the reaction rate is limited by the narrow etch region, resulting in a less efficient exchange of reaction products.

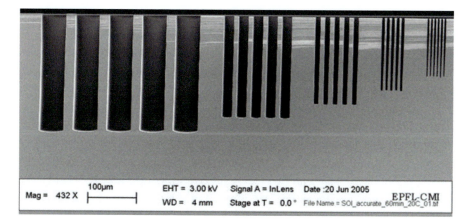

**FIGURE 6.46** Anisotropy in RIE: Fully anisotropic etching (left), partially anisotropic (center) and isotropic etching (right).

The fact, that RIE is not a completely anisotropic process, means that we can exploit this property for the fabrication of specific structures. One good example is the formation of Si tips, which can be used in the AFM set-up.

### 6.5.6 Dry Etching: Deep Reactive Ion Etching

With progress in VLSI technology, deeper trenches are required. In MEMS structures, devices are of a few hundred microns. For both of these applications, good directionality is imperative. Pure chemical etching is isotropic, while physical etching has good directionality, but poor selectivity. In the late 1990s, two techniques were invented, which made high aspect ratio silicon deep etching possible: induction coupling plasma (ICP) source and the *Bosch* etching process.

The first requirement for deep RIE is a high etching rate. The RIE rate is dependent on the plasma density. This can be increased by increasing the RF power. However, increasing the RF power will also increase the self-biasing voltage on the cathode, which in turn will increase the ion bombardment energy. This will deteriorate the etch selectivity. The ICP source solves this problem: the RF power in the ICP source is fed into the reaction chamber via an induction coil coupling from the outside. The plasma is separated from the etching chamber. The sample is connected to a second RF source an auxiliary RF supply to enhance plasma production. This has the effect of increasing the ionization probability. The second RF source the sample stage means that self-biasing can be controlled independently. Thus, the ICP source can produce very high plasma densities $5 \times 10^{11}$ cm$^{-2}$ and maintain a low ion bombardment energy. ICP sources can produce etching rates of over 20 mm/min. Another method for the production of high-density plasmas is the electron cyclotron resonance source, in which electrons are accelerated in a microwave cavity and plasma densities can be comparable to those of the ICP sources. However, cyclotron sources are technically complex and the ICP is the preferred method for applications.

In the Bosch process, the verticality of the sidewalls is maintained by periodically coating the sidewalls with an etch-resistant film. In this process, the deep silicon etching can produce walls of almost perfect verticality. The principle of this process is illustrated in Figure 6.47. Further improvements to the technique have focused on increasing the rate of etching while maintaining the high anisotropy and high selectivity. Etch rates of over 20 m/min in silicon have been reported. One drawback of the Bosch process is the distinctive sidewall scalloping or rippling, which is caused by the periodic switching between isotropic etching and passivation.

### 6.5.7    Dry Etching: Ion Sputter Etching

This is a purely physical etching technique, which typically uses argon as a gaseous source for ion production. The sputtering process has been outlined in Section 4.8.5, where the only major difference here is that the sample acts as the sputter target. Etching rates are strongly dependent on the atomic weight. The principal considerations that should be taken into account are the following:

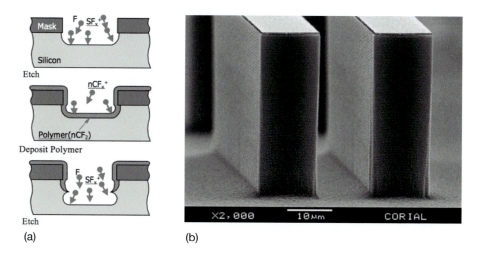

**FIGURE 6.47** (a) Illustration of the Bosch process for the deep etching of Si. In the cycle, the structure is isotropically etched, then passivated. The number of cycles used will define the depth of the etch. (b) Example of a monolithic structure prepared using the Bosch process. Good sidewall verticality is clearly observed. Note the sidewall rippling, which is caused by the periodic etching/passivation cycle.

(1) Ion sputtering is non-selective and will etch both material and mask. Therefore, we generally require thicker masks for deep structures.

(2) The ion sputter yield depends on the bombardment angle. To achieve uniform sputtering, the sample holder should rotate around a central axis to compensate for the effects of sample tilting.

(3) Ion sputtering is a physical etching process and does not form volatile compounds, so material sputtered off the sample surface may redeposit anywhere, including on the chamber walls and on other regions of the sample itself. Rotation, while tilting the sample during sputtering, can help clean up the redeposition of sputtered debris on the structure sidewalls.

### 6.5.8 *Dry Etching: Reactive Gas Etching*

The dry etching techniques discussed above all use ions generated from a plasma. Dry etching based on a reactive gas does not require any form of plasma. It is based on the principle that $XeF_2$ in a gaseous state can directly react with silicon to turn it into gaseous $SiF_4$ products. This etching technique is highly isotropic. The chemical reaction is given by:

$$2XeF_2 + Si \rightarrow 2Xe(gas) + SiF_4(gas) \tag{6.18}$$

The etching system only requires a vacuum chamber and pumping system. Compared to other techniques, $XeF_2$ gaseous etching has a number of advantages, such as:

(1) $XeF_2$ only reacts with Si. This allows an excellent level of selectivity.
(2) $XeF_2$ etching rate for Si is typically in the range of 1–3 m/min. This is independent of the doping level of the Si and its crystalline orientation. Therefore, the etch depth will only be a function of the etching time. It is not possible to have etch stops, as is the case in wet etching.
(3) $XeF_2$ Si etching is completely isotropic. The lateral etching to create undercut is very effective. This is a distinct advantage when making suspended structures.
(4) The etching surface can be very rough (on the order of a few m). By the addition of other halogen gases such as $BrF_3$ and $ClF_3$, smoother etching by $XeF_2$ can be achieved.

The implementation of this technique can be performed either by a direct and constant supply of $XeF_2$ gas into the etching chamber or by using a pulsed gas source.


### 6.5.9  *Dry Etching: Laser Micro-Machining*

This process is virtually identical to the laser ablation process used in pulsed laser deposition that was discussed in Section 4.8.4. The lasers employed in this method of dry etching include Nd:YAG solid-state lasers, Ti:sapphire, $CO_2$ and excimer lasers. A broad range of materials can be machined using the laser ablation process, again with the sample used in the target position. Conventional laser machining is based on thermal processing; i.e., a high energy density laser beam generates highly localized heating, causing the material to melt and vaporize. Patterns can be produced by scanning the laser head across the surface of the sample. Thermal processing inevitably causes scorching with rugged edges. To avoid these problems, cold micro-machining by laser ablation can remove material without scorching. Lasers can be used either for direct patterning by scanning the focused beam or by projection through a mask. Ablation by projection can produce large areas of patterned microstructures simultaneously. As shown in Figure 6.48, laser micro-machining can produce a variety of microstructures with good structure definition on a range of materials as well as on non-flat surfaces.

**FIGURE 6.48** Various microstructures produced by laser micro-machining.

Significant development of laser micro-machining has been made using femtosecond lasers. Using an ultrashort pulse width, the laser energy has no time to convert into heat via diffusion to adjacent areas and no melting occurs. This gives rise to much smoother edges in the machining process. An example is illustrated in Figure 6.49, where we show two holes, one made using an ultrafast laser with a pulse width of 200 fs, Figure 6.49(a) and in (b) a hole produced by a laser with a pulse width of 3.3 ns.

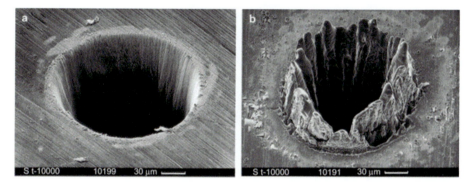

**FIGURE 6.49** Holes drilled in 100 m thick steel foils by ablation using laser pulses with the following parameters: (a) pulse width: 200 fs, pulse energy: 120 J, fluence: $0.5 \, \text{J cm}^{-2}$, wavelength: 780 nm; and (b) pulse width: 3.3 ns, pulse energy: 1 mJ, fluence: $4.2 \, \text{J cm}^{-2}$, wavelength: 780 nm. The scale bars represent 30 m. Reprinted by permission from B. N. Chichkov et al., *Applied Physics A: Materials Science & Processing, 63,* 109. ©(1996). Springer, Nature.

Some of the principal characteristics of the more commonly used laser in micro-machining are summarized in Table 6.6.

**TABLE 6.6** Specifications for Some Commonly Used Lasers in Laser Micro-Machining

| Laser | Wavelength (nm) | Pulse repetition rate | Pulse width | Pulse energy mJ | Minimum spot size m at $f = 100$ mm |
|---|---|---|---|---|---|
| Ti:sapphire | 750–850 Frequency doubling, tripling | 1 kHz, 5 kHz | ~ 120 fs –30 ps | 2 (1 kHz) 0.5 (5 kHz) 4.1 (260 nm) | 12.4 (780 nm) 6.2 (390 nm) |
| Nd:YAG | 1064 Frequency doubling, tripling quadrupling | 1 Hz– 2 kHz | 10 ns | 8 (1064 nm) 5 (532 nm) 3 (355 nm) 1 (266 nm) | 45.6 (1064 nm) 22.6 (532 nm) 15.1 (355 nm) 11.3 (266 nm) |
| Excimer | 157, 193, 248, 308, 351 | ~ 1–250 Hz | 25 ns | 25 (157 nm) 400 (193 nm) 600 (248 nm) 400 (308 nm) 320 (351 nm) | 2.5 (157 nm) 3.0 (193 nm) 4.0 (248 nm) 4.9 (308 nm) 5.6 (351 nm) |

Larger microstructures can be produced by other techniques such as electro-discharge machining (EDM), which uses a spark discharge between two electrodes a high voltage to generate local heating. Melted debris is removed by a cooling liquid. Powder blasting is another conventional machining technique. Typically, a fine powder of $Al_2O_3$ is used, where grains of the order of 3–30 m are used. Clearly, these techniques are only employed for fabricating larger structures of the order of hundreds of microns in size.

## 6.6 Summary

Lithographic techniques are extremely versatile in the fabrication of micro- and nanostructures. Their importance can not be overestimated. They form the principal methodologies for a large majority of the structures produced for commercial applications, including the microelectronics and data storage media industries. The markets are huge and global. Lithographies can take many forms and can be distinguished by the nature of the radiation used in the exposure of the photoresist.

By far the most common and cheapest of the lithographies is optical lithography. Much of the technology used today stem from this method and it is still used in much the same way as it has been for the last half-century or so. Its principal use was based on the development of ICs and has greatly improved over the decades in terms of its resolution and procedures. This is partly due to the use of shorter and shorter wavelengths to reduce the ultimate feature size and partly in the development of the photoresists and etching procedures. The standard photolithographic process is based on the pattern transfer, from a mask to the photoresist using an exposure and development procedure. Structures can then be finally transferred to the substrate by etching the pattern or by deposition and lift-off procedures. These define the positive and negative mask alternatives to photolithography. Within the optical lithography methodology, pattern transfer can be made using contact, proximity and projection printing. The pros and cons must be weighed up against the desired structures and their feature size. Projection lithography allows the mask pattern to be reduced using imaging optics and has been one of the principal methods used in the IC industry since the 1980s.

A number of resolution enhancement procedures have been developed over the years with the aim of improving the quality and resolution of the pattern transfer. This includes spatial filtering, phase shift masks, optical proximity correction, and chemically enhanced resists. Near-field optical lithography has also been developed using a proximity methodology, as well as interferometric methods to perform maskless procedures.

To obtain higher spatial resolutions, alternative methods using electrons and X-rays have also found many adherents. However, costs are high and these techniques are mainly used as research tools. The electron beam lithography technique uses much the same technology as the SEM, where an electron beam can be scanned across the surface of a photoresist coated substrate. The beam size and its interaction with the photoresist define the ultimate resolution of the method. Minimum feature sizes can be in the 10s of nm. X-ray lithography can only be used as a proximity type method since no optics are available for projection lithographies. However, due to the short wavelengths available, this method can produce an excellent spatial resolution. X-ray lithography is particularly well suited to the fabrication of high-aspect-ratio features in a procedure known as LIGA. This has been extensively used for the manufacture of MEMS devices.

The focused ion beam (FIB) apparatus is rather special. It typically uses metallic ions (Ga) to bombard a surface. The ions are controlled in much the same way as an electron beam, i.e., using electrostatic lenses, etc. The application of this technique is usually based on research applications. Nanostructures can be defined by focusing on and moving the ion beam over the surface of a substrate. The ions can be made sufficiently energetic to sputter the surface atoms to etch away the substrate material. This is referred to as nano-machining. Alternatively, by introducing a precursor with a metal, the ion beam can be used to perform deposition in the regions where the ions are incident on the substrate. Thus the FIB can be used to remove and deposit materials on the surface of a substrate. A further advantage of this method is that the ions interact with the surface atoms in a similar manner to the electron beam, thus producing secondary electrons which can be used for imaging the surface. The FIB can also be used as a source of radiation for ion beam lithography, with features sizes as small as 10–15 nm being possible. Implantation provides yet another application for FIB technologies, allowing the ions to be used as a dopant, for example.

## 6.7   Problems

(1) A symmetric convex lens made from glass with a refractive index of 1.52 has a radius of curvature of 10 cm and a diameter of 7 cm. This is used as a condenser lens in the projection lithography apparatus with $k_1 = 0.75$.

Calculate the ultimate resolution of this system using g-line illumination ($\lambda = 436$ nm).

(2) Two photoresists (A and B) are compared for suitability in a VLSI manufacturing process. The linewidth of variation (contrast) with light exposure is shown in the figure below. Discuss, in terms of the exposure latitude, which is the more appropriate of the photoresists.

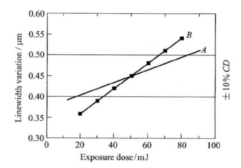

(3) Show that the optical resolution of a 0.54-NA ArF lens is the same as that of a 0.694-NA KrF lens. Which has a greater depth of focus?

(4) Explain the physical limitations of optical lithography.

(5) How are clean rooms classified? What is a class 100 cleanroom?

(6) An exposure is performed with coherent light using a step-and-repeat projection printing system. The light source has a wavelength of 365 nm (I-line of a Hg arc lamp). The pattern is a grating with a line-to-line spacing of 1 m.

   (a) Calculate the minimum value of the numerical aperture (NA) which will provide contrast at the image plane (the plane of the resist).

   (b) What is the maximum value of the numerical aperture, above which there will be no improvement in image quality?

   (c) Calculate the depth of field of the image for cases (a) and (b).

(7) Present a comparison of negative and positive photoresists. Also, explain what a permanent resist is. Describe what happens chemically to both positive and negative resists when exposed to UV radiation.

(8) A polyimide photoresist requires 100 mJ/cm$^2$ per m of thickness to be developed properly. Your lamp provides 1000 W/m$^2$. How long do you need to expose a 20 m thick film?

(9) Why is the resolution with incoherent light larger than that for coherent illumination? How is the depth of focus (DOF) of an imaging system

influenced by the numerical aperture of the imaging lens, the resolution of the system and the wavelength of the exposing light?

(10) Demonstrate with some simple sketches of how you would pattern a small Pt electrode for an electrochemical sensor using wet etching. The substrate is an oxidized Si wafer. (Remember that Pt is very difficult to wet etch).

(11) A 0.6 m film of silicon dioxide is to be etched with a buffered oxide etchant (BOE) with an etch rate of 750 Å/min. Process data show that the thickness may vary up to 10% and the etch rate may vary up to 15%.

(a) Specify a time for the etch process. Compare the etch time for an errorless case.

(b) Predict how much undercut will occur at the top of the film.

(12) Using the figure below as a model, assume that the silicon wafer has a bias of 150 V. A reactive ion etch is used to etch 10 m wide grooves. After 1 hour, the shape of the grooves is measured and found to have the following topology:

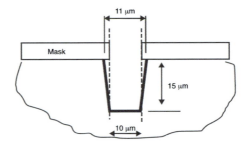

(a) If the bias were increased to 200 V and the depth of the groove was maintained at 15 m, would the width at the top of the groove be less than or greater than 11 m. Explain.

(b) If the same etch procedure was used to etch $SiO_2$, qualitatively describe the shape of the grooves in the $SiO_2$ if a 150 V bias is applied.

(13) If a 50 mm diameter GaAs wafer is left under a class 1000 laminar flow hood for 1 minute, calculate the number of dust particles that will land on the wafer for an airflow of 0.5 m s$^{-1}$. If the sticking coefficient of these particles is 0.05, and they are uniformly distributed, calculate the likely yield from devices with a surface area of 1 mm$^2$ if one adhering particle is enough to cause device failure.

(14) In electron beam lithography the term Gaussian beam diameter ($d_G$) describes the diameter of an electron beam in the absence of aberrations.

The current density of the beam is given by $J = J_p e^{-(r/s)^2}$, where $J_p$ is the peak current density, $r$ is the radius from the center of the beam and $s$ is the standard deviation of the electron distribution in the beam. Defining $d_G = 2s$, derive a relationship between $d_G$, the peak current density, $J_p$ and the total current in the beam, $I$.

(15) Consider a single electron in an electric field between two parallel plates located 10 cm apart. Assume the potential varies sinusoidally between $-1000$ V and 1000 V at a frequency of 13.5 MHz. Calculate the maximum kinetic energy of the electron.

(16) Why can only proximity masking be used in the case of X-ray lithography? What about projection printing with X-rays? Sketch the process for fabricating an X-ray mask. What are some of the positive attributes of X-ray lithography? What are the negative attributes?

(17) Compare UV, X-ray, ion-beam, and electron-beam lithography. Summarize in a comparison table. Which techniques are used mostly in the IC industry today? How are the photons or charged particles created in each case?

(18) Describe why a synchrotron is the best tool for producing the X-rays required to make the most accurate LIGA parts and make a table comparing X-ray lithography with UV lithography.

(19) Calculate the minimum grating feature size that can be lithographically patterned by using a light source with a wavelength of 0.4 nm and employing proximity printing with a resist thickness of 100 m. The gap between the resist and the absorber equals 100 m.

(20) Describe in detail how the time of flight (TOF) spectrometer is used to measure the size distribution of nanoparticles in a gas aggregation source. Use equations to support your arguments.

(21) Describe the three main operating modes of the FIB (focused ion beam) technique. Give three advantages and three disadvantages of this technique over other NGL technologies.

(22) The LMIS of a FIB source uses $Ga^+$ ions accelerated by a potential of 15 kV and then passes through a region (5 cm) of the uniform magnetic field of 100 Oe. Determine the deflection of the ion beam, from the undeflected position, at the sample stage a distance of 75 cm away.

(23) Explain how you would obtain a high aspect ratio structure with good vertical sidewalls.

(24) What is meant by the terms "isotropic" and "anisotropic" when applied to wet chemical etching? Sketch the cross-sectional profiles expected for these two cases under a non-erodable mask.

## References and Further Reading

Chichkov, B. N., et al., (1996). *Appl. Phys. A, 63,* 109.

Credgington, D., Fenwick, O., Charas, A., Morgado, J., Suhling, K., & Cacialli, F., (2010). *Adv. Funct. Mater., 20,* 2842.

Cui, Z., (2008). *Nanofabrication: Principles, Capabilities, and Limits,* Springer Science–Business Media, LLC, New York.

Chichkov, B. N., Momma, C., Nolte, S., von Alvensleben, F., & Tinnermann, A., (1996). *Apps. Phys. A, 63,* 109.

Haske, W. et al., (2007).*Opt. Express, 15,* 3426.

Lasagni, A. F. et al., (2016). SPIE Newsroom, DOI: 10.1117/2.1201602.006325.

McLeod, E., & Ozcan, A., (2012). *J. Lab. Autom., 17,* 248.

McNab, S. J., Blaikie, R., (2000). *J., Appl. Opt., 39,* 20.

Mehrotra, P., Mack, C. A., & Blaikie, R. (2013). *J. Opt. Exp., 21,* 13710.

Menon, R., Patel, A., Gil, D., & Smith, H. I., (2005). *Mater. Today, 8,* 26.

Madou, M., (2002). *Fundamentals of Microfabrication: The Science of Miniaturization,* CRC Press, Boca Raton.

Smith, B. W., Fan, Y., Zhou, J., Lafferty, N., & Estroff, A., (2006). *Proc. SPIE, 6154,* U200.

Werkmeister, J., Gonsalvez, M. A., Willoughby, P., Slocum, A. H., & Sato, K., (2006). *J. Microelectromechanical Systems, 15,*1671.

Wirth, R., (2004). *J. Eur. Mineralogy, 16,* 863.

Yao, N. (Ed.), (2007). *Focused Ion Beam Systems: Basics and Applications* 1st Edition, Cambridge University Press, Cambridge.

Zheng, D. A. Z., Mohammad, M. A., Dew, S. K., & Stepanova, M., (2011). *J. Vac. Sci. Technol. B, 29,* 06F303.

# Chapter 7

# Replication Techniques

In the 1990s, a new form of microlithography was developed based on the direct physical imprinting of structures. Imprinting as a method of reproduction and replication has long existed as a conventional industrial technique, where surface relief patterns can be made onto a plastic surface for example. The manufacture of compact discs is an example of this technique. The general idea can be considered to be a form of printing, where a large number of prints can be produced from a mold. Using a heating process can aid the manufacture of such structures in a process called embossing. This technique for printing structures can be readily adapted to an industrial process, with feature sizes of the order of micrometers and above. For structures of say below 100 nm, the procedure is referred to as *nano-imprinting*. In the course of its development, nano-imprinting has had several derivatives which are similar in their manner of pattern replication, but with different working principles. These include *soft lithography* and *step and flash lithography*. In addition to the hot embossing process and nano-imprinting, other replication techniques have also been experimented for the preparation of microstructures such as microinjection molding, micro-stereolithography, and nanoink printing.

In this chapter, we shall outline some of the principal technologies used in the preparation of micro and nano-sized objects. While most of the methods described have evolved over the years into true nanotechnologies, some are limited in terms of resolution and are not necessarily adapted to the fabrication of nano-objects. The decision has been to leave them in the discussion since they also form part of the evolution of technologies applied to nanotechnology as a whole.

As we shall see in Volume 2 of this book, periodic nanostructures offer a wide range of interesting and useful properties, such as photonic and magnonic materials. The methods outlined in this chapter offer some simple and effective tools for the fabrication of large-area periodic structures with excellent control over the size and separation of the nanostructure features.

## 7.1  Nano-Imprint Lithography

Nano-imprint lithography (NIL) was first developed in the mid-1990s by Steven Y. Chou at Princeton University and is now a method of nano- and microlithography. It can have an ultra-high-resolution, high throughput and is low-cost. The high-resolution comes from the fact, that there is neither light diffraction nor electron scattering in the processing. High throughput is possible since hundreds and thousands of structures can be made at the same time. A nano-imprinting system requires no complex optical imaging system, nor does it need sophisticated electromagnetic focusing, which makes this a low-cost technique for micro-imprinting in general. NIL has all the necessary attributes for an industrial processing tool and many applications have been reported for this technique.

The NIL method has a number of variants, which can use heating or UV light exposure at the pressing stage. The principal steps of the NIL technique are illustrated in Figure 7.1. The substrate should be clean before a photoresist is deposited, which can be a normal photoresist, such as poly (methyl methacrylate) or PMMA, to a thickness of a few hundred nm. The master stamp is then pressed onto the resist. The pressing should usually be performed under heating, typically to above 50°C. When the temperature is above its glass transform temperature, $T_g$, the PMMA became rubber-like and can fill the pattern on the silicon stamp. When the temperature is reduced to below $T_g$, the PMMA becomes transformed into a high mechanical strength and glass-like material. Pressing will occur at pressures of 50–100 bars. Once the pressing has been performed, demolding removes the stamp (mold). The imprint is usually slightly less than the PMMA thickness, such that that the stamp surface does not have hard contact with the substrate, preventing damage to both. The impression of the stamp pattern will then be subject to etching, typically RIE, which will remove excess resist inside the imprint pattern, leaving exposed areas of the substrate surface after the final stage.

As we can see, the NIL procedure will leave the sample in a similar state to a typical optical photolithography or electron beam lithography process. The resist has been transferred onto the substrate with the desired pattern. The sample can then be processed by etching or deposition and lift-off as with the conventional lithographies.

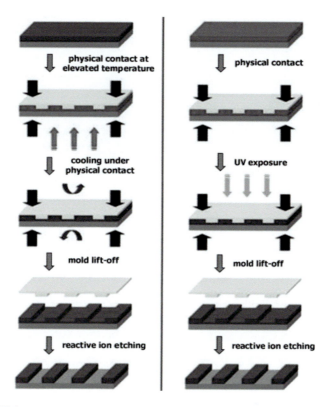

**FIGURE 7.1** Processing steps in nano-imprint lithography. On the left, the process includes a hot-embossing stage, while on the right, UV curing is used at the pressing stage.

The stamp used as the master in NIL must be made of a hard material so that it can withstand repeated printing procedures. It is common for the stamps to be prepared by electron beam lithography in silicon, silicon dioxide, silicon nitride or a metal. The aspect ratio is a little limited and should not exceed 3:1 to facilitate decoupling or demolding. To further aid decoupling, the master patterned surface may be coated with a thin layer of release agent. Examples of NIL formed structures are shown in Figure 7.2.

The NIL method is capable of producing nanostructures with feature sizes of 10s of nm. When imprinting non-uniform patterns, non-uniform filling of the master cavities can occur due to a differentiation of the polymer flow during pressing. This can lead to voids being formed in large feature areas. Improvements can be made by increasing the pressure and temperature at the pressing stage. NIL, as we indicated above, cannot be used for making high

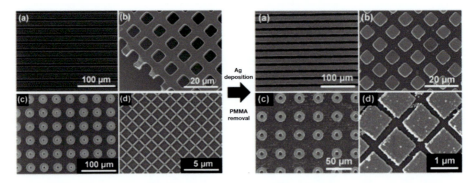

**FIGURE 7.2** Left: SEM images of different PMMA structures fabricated by nano-imprint lithography and used as a passivation layer in the subsequent silver deposition step. Right: The respective silver structures obtained after the removal of the PMMA structures with acetone. Reprinted from G. Mondin et al., (2013). *Mater. Chem. Phys. 137*, 884, ©(2013), with permission from Elsevier.

aspect ratio nanostructures. However, multilayer processing can be used to overcome this issue as is shown in Figure 7.3.

Another NIL technique is to directly coat the imprinting polymer onto the master instead of onto the substrate. In this case, the polymer fills in the structural cavities on the master. The coated master is then pressed onto the sub-

**FIGURE 7.3** High-aspect-ratio pattern imprinted using a double-layered polymer process.

strate to transfer the imprinted polymer. To perform this successfully, the surface binding between the polymer master must be less than that between the polymer and the substrate. To ensure this is the case, the master can be coated to reduce its surface binding energy. Imprinted polymer structures can be transferred to a substrate, which is either flat or has surface topological variations or is curved. This capability allows for further flexibility in the processing of nanostructures. The basic procedure is the same as for printing on a flat surface but has a flexibly master stamp, as illustrated in Figure 7.4 (left). In Figure 7.4 (right), we show the comparison of printing on a flat Si substrate and a curved acrylic substrate.

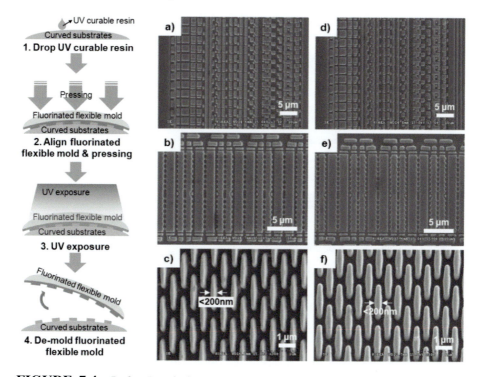

**FIGURE 7.4** Left: Imprinting process using replicated fluorinated polymer-coated flexible PET mold on curved acryl substrates. Right: SEM micrographs of imprinted resist patterns by UV nano-imprint lithography. Using hot-embossed fluorinated polymer-coated PET film, (a, b, c) imprinted resist patterns on a flat Si substrate and (d, e, f) imprinted resist patterns on a curved acryl substrate (Reprinted from Shin et al., 2011 Copyright © 2011, Springer Nature. Open Access.).

## 7.2    Step and Flash Nano-Imprint Lithography

The two principal drawbacks of NIL are the requirements of high tempera-ture and high-pressure working conditions. Also, alignment problems can be included in multistep processing. The step and flash nano-imprint lithogra-phy (SFIL) retains all of the advantages of NIL, while overcoming its main drawbacks. There are two distinctive features of SFIL, one is the master or template, which is made of quartz plate and another is the patterning process, which is not performed by pressing the template into the printing layer, but is done by a UV illumination curing process. This is illustrated in Figure 7.5.

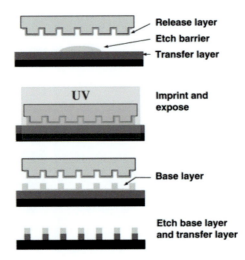

**FIGURE 7.5**    Principal steps in the step and flash imprint lithography method.

The process starts with the spin coating of the resist onto the substrate. The template is then placed on top of the resist coated substrate, but without direct contact. There is a gap of 0.2–0.25 m between the master template and the substrate. A drop of low viscosity photo-curable organo-silicon liquid is introduced into the gap. The liquid is sucked into the gap via capillary forces and rapidly fills the space between the template and substrate. A gentle press of the quartz master squeezes out any excess fluid and brings the master into direct contact with the substrate. The next stage is to illuminate the entire master with UV light. This causes the photo-curable organo-silicon liquid to be solidified. Removal of the quartz template leaves the structure transferred onto the substrate and resist. Since the cured pattern structure is of a silicon-based material, high etch selectivity can be achieved, forming an etch barrier and transfers the pattern to the photoresist after oxygen plasma etching. Finally, the etch barrier is removed by wet chemical etching, thus transferring the final structure onto the substrate.

The key to SFIL is the transparent quartz template, which allows optical alignment and UV illumination. The final pattern resolution, as with NIL, depends on the pattern resolution on the template. This will typically be prepared using electron beam lithography, so it can have very good resolution characteristics. A second key factor in SFIL is the photo-curable organo-silicon liquid. Apart from being photo-curable, it must also be of low viscosity and must be able to whet the quartz template so that it can fill all of the surface relief cavities and still detach from the template after the curing process. An alternative to SFIL is to directly use photoresist instead of the photo-curable organo-silicon liquid. The process will then be similar to that of normal NIL.

## 7.3   Soft Lithography

The soft lithography technique was also derived from the nano-imprinting method. In the context of this method, *soft* refers to the master template material, which is elastic and soft and usually made of a silicone elastomer such as polydimethylsiloxane or PDMS. The fabrication of soft templates is as follows (see also Figure 7.6):

(1) High-resolution patterning, typically via electron beam lithography, of PMMA on to a Si substrate;

(2) PDMS in liquid form is poured onto the PMMA pattern and spread uniformly over the entire patterned surface; (iii) Curing of the PDMS layer cast at a temperature above room temperature for a short period, until it becomes solid and elastic; (iv) The PDMS layer can be easily peeled off from the PMMA; (v) the PDMS sheet with the impression of the PMMA pattern is then immersed in alkanethiol solution, which acts as an ink; (vi) Stamping the alkanethiol coated PDMS template on a substrate, which is coated with a gold film; and (vii) A layer of alkanethiol pattern is transferred to the gold-coated substrate. The alkanethiol layer acts as an etch mask, with an etch rate of around $10^{-6}$ of that of gold. Etching will thus remove the uncoated regions of the gold surface, leaving only those areas protected by the alkanethiol layer. This process is illustrated schematically in Figure 7.7.

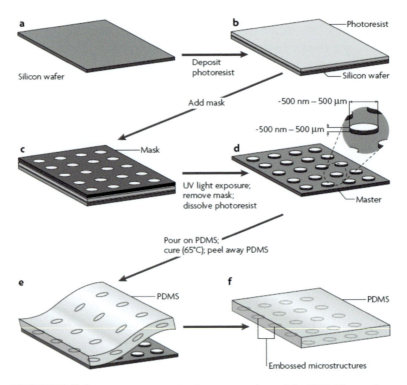

**FIGURE 7.6**    Process steps in the preparation of the PDMS template.

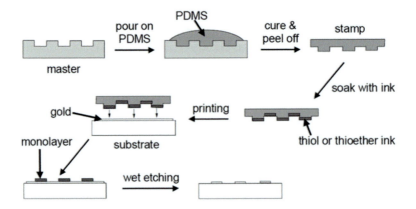

**FIGURE 7.7**    Schematic illustration of the pattern transfer process in soft lithography (Reprinted from Shin et al., 2011 Copyright © 2011, Springer Nature. Open Access.).

Soft lithography has two main characteristics, which set it apart from the other methods:

(1) Since the master is made of an elastomer, the elasticity of the master makes it very good at sticking to uneven surfaces or rough surfaces with particulates. There is no pressure involved in the printing process. PDMS does not swell when immersed in alkanethiol solution and no deformation occurs during printing. All this results in high precision pattern transfer, with a resolution equivalent to EBL in PMMA of $\sim 30$ nm.

(2) The *ink* for soft lithography is for the moment the only alkanethiol. Only this compound can form a 1–2 nm self-assembled monolayer (SAM) on the gold surface. The SAM process takes less than a second to form upon contact with the gold surface. The alkanethiol is very stable in the gold etchant and is the key to the process of soft lithography.

Since PDMS is transparent, multilevel patterning is also possible by alignment through the PDMS template.

## 7.4 Micro-Molding of Plastics

The advent of plastics was an important evolution in our ability to manipulate and process materials, and is considered to be a revolution in 20th-century technology. Indeed, plastics have replaced many materials such as wood and steel traditionally used in construction and other applications. Although silicon is the basic material used in the microelectronics industry and in microelectromechanical systems, it is not the cheapest of materials. If plastics could replace silicon for microdevices, costs could be significantly reduced. In fact, over the last decade or two, significant research activity has been directed to the study of plastic-based electronics such as transistors based on organic materials (OLEDs being a good example), as well as polymer LEDs and organic thin films for displays. In microsystems, macromolecule materials can be used as base materials for microfluid systems, including chemical and biological analysis and lab-on-a-chip devices.

Plastic is a generic name for all types of macromolecule materials. For micro-molding of plastics, there are generally three types of macromolecular materials that can be used:

(1) **Thermoplastic:** This plastic becomes soft when heated to its glass transition temperature, $T_g$. They can be molded into different shapes in their

soft state and maintain it, when they are cooled to below $T_g$. Thermo-plastics are the most widely used of the macromolecule materials in microsystems.

(2) **Elastomer:** This plastic has much weaker molecular bonds and can be easily deformed by external forces, even at room temperature. Once the external force has been removed, they can revert back to the original shape. PDMS and silicone are common examples of elastomers. They are usually shaped by casting in the liquid state and cured into a solid above room temperature.

(3) **Duroplastic:** This type of macromolecular material has strong molecular bonds. They are hard and brittle and cannot be deformed under pressure. They soften very little before reaching their molten temperature. They can be cast into shapes and change little once formed.

The shaping of plastics can be performed by three methods: hot-embossing, injection molding, and casting.

## 7.5   Dip-Pen Nano-Lithography

Dip-pen nano-lithography (DPN) is a nanopatterning method, which uses an AFM like tip as the *pen* and a special *ink* to write directly on a substrate. The AFM probe is first dipped into a container with the special ink and is then moved into position on the surface of the substrate. When tip and substrate are close enough, the ink on the probe tip will touch the surface, and from the capillary forces will be deposited there. This process is schematically illustrated in Figure 7.8(a). The simplest pattern that can be produced by DPN is a dot array. Line patterns are also easily produced, but require a constant supply of ink to the tip, see Figure 7.8 (b). AFM technology is well established and has been adapted for such applications as DPN. The key to the DPN technique is the ink, which must be able to whet the surface of the substrate for a stable SAM of molecules by chemical absorption.

There are a number of advantages that DPN has over other methods. These can be summarized as follows: (i) Simple and low-cost; (ii) Different types of ink can be used, which can also be a biochemical substance; (iii) High-resolution alignment and multilevel patterning can be performed, since the AFM is a high-resolution imaging tool with an accuracy in the range of 5 nm being reported. The high-resolution alignment allows different inks to be written onto the same dot, for example, allowing bottom-up molecular assembly; (iv) High-resolution nano-lithography. Though the ink pattern

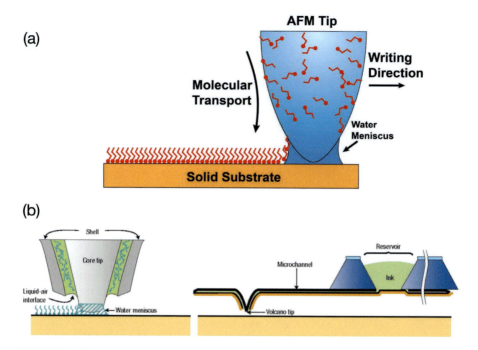

**FIGURE 7.8** (a) Schematic illustration of the working principle of the dip-pen nanolithography technique. (b) Adaptation of the DPN tip with the constant ink supply.

written by DPN is not on the atomic scale, a minimum size 15 nm has been achieved in a dot array; and (v) If an array of AFM tips are used, DPN can perform parallel lithography, which can greatly increase the pattern speed and throughput.

The DPN technique has already been developed into a commercial instrument and could provide low-cost industrial processing for nano-patterning. Some examples of DPN structures are shown in Figure 7.9.

## 7.6  Nano-Sphere and Nano-Stencil Lithography

The function of a mask in lithography is to partially block light transmission, only exposing those regions, that are defined by the desired pattern to be transferred onto the photoresist. Consider the case, where a single layer of closely packed spheres of uniform diameter is lying on the surface of a substrate. The deposition of a material on this substrate surface will only occur in the gaps between the spheres in the form of a regular array of dots. This

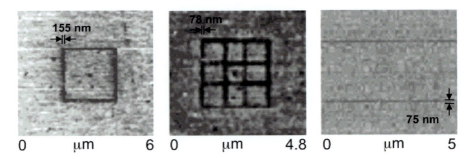

**FIGURE 7.9** Lateral force microscopy (LFM) images of DPN-generated TMS monolayer patterns on semiconductor surfaces. All LFM images were recorded at a scan rate of 4 Hz. In all cases, lighter contrast areas (higher friction) correspond to the hydrophilic (-OH group rich) semiconductor surface and darker areas (lower friction) are due to the deposited silazane via DPN: (a) square TMS pattern (2.4 m edge width) on Si/SiO$_x$ with a writing speed of 0.08 m/s; (b) a TMS grid on a Si/SiO$_x$ substrate (2.4 m edge width) with a writing speed of 0.04 m/s; and (c) a parallel line pattern on oxidized GaAs with a writing speed of 0.02 m/s. Reprinted with permission from Ivanisevic, A. & Mirkin, C. A. (2001). *J. Am. Chem. Soc., 123*, 7887. Copyright (2001) American Chemical Society.

is essentially the principle behind the nanosphere lithography technique. The spheres are made of polystyrene with a diameter of $264 \pm 7$ nm. From the diameter of the spheres, it is possible to calculate the size of the dots formed through the gaps between the spheres as well as the distance between them. For spheres of diameter, $D$, the dots through a single layers nanosphere mask will be $0.233D$, with a separation of $0.577D$. Nanosphere masks can be made of single and double layers of spheres. They can form different sizes of nanodots with variable densities. For the case of a double layer of spheres, the dots formed through the mask will be $0.155D$, with a separation equal to the diameter of the spheres. The patterns generated for single and double layers of nanospheres are shown in Figure 7.10. The smallest dot array, that has been produced by the nanosphere lithography technique, is around 40 nm.

Nanosphere lithography includes coating the nanosphere mask, thin film deposition, and removal of the mask post-deposition. A variety of metal structures and alloys have been formed into regular arrays via this technique, which is both very simple and very cheap. The main limitation of the technique is the size of the nanospheres and the fact, that the nanodot arrays can only be made in two sizes. Some examples of structures produced using this method are illustrated in Figure 7.11.

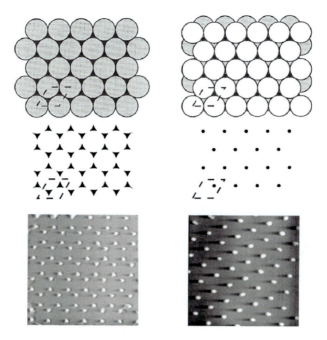

**FIGURE 7.10**  Single and double layer close-packed 265 nm polystyrene spheres on mica.  The deposition of Ag and nanosphere lift-off gives 50 nm particles. Reprinted with permission from J. C. Hulteen, & R. P. van Duyne, *J. Vac. Sci. Technol. A 13*, 1553. ©(1995), American Vacuum Society.

A stencil mask, or shadow mask, for nano-lithography works in the same way as the nanosphere mask.  These are the same type that is also used in electron and ion projection lithography.  A stencil mask is a framed membrane with etch through patterns. Nanostencil lithography uses a stencil mask to partially block material from being deposited on to a substrate. Deposition material forms a pattern on the substrate, which is the same as the etched pattern on the stencil mask.  This shadow mask deposition for nanostencil lithography is illustrated in Figure 7.12.

Nanostencil lithography was first proposed in 1999, originally combining deposition with a scanning probe microscope to deposit sub 100 nm metal nanostructures.  Later, the technique was developed into a large area patterning technique.  As a method for the fabrication of nanostructures, nanostencil lithography offers a number of advantages:

(1)  A thin film pattern can be deposited at any location on the substrate, since the stencil mask is not in direct contact with the substrate. Pattern

**FIGURE 7.11** SEM images of a variety of gold nanostructures based on nanosphere lithography technique (a) nanodisc array, (b) slanted nanocones, (c) nanotriangles, and (d) nanoholes. (Note: 500 nm scale bar in all images.) (Reprinted with permission from Donchev et al., 2014. © Cambridge University Press.).

deposition can also be performed on any type of topographical surface and not just on flat substrates.

(2) Thin-film patterns are formed in one simple step, with no complex photoresist patterning and lift-off process.

(3) Since there is no photoresist in the procedure, the material is directly deposited onto the surface of the substrate. The method can be used in most forms of physical deposition, with some geometrical considerations.

(4) As there is no optical or electron beam lithography involved, there is no proximity effect caused by diffraction phenomena. Also, large and small patterns can be deposited simultaneously.

Nanostencil lithography also has a number of drawbacks and limitations, which need to be taken into consideration. These can be summarized in the following:

(1) This technique can only deposit a single layer. Multilayer deposition will give rise to alignment problems.

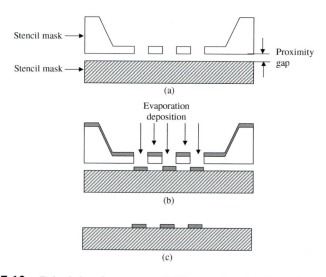

**FIGURE 7.12**  Principle of nanostencil lithography. (a) Stencil mask and substrate configuration, (b) thermal evaporation deposition through stencil mask, (c) pattern transfer to the substrate.

(2) After repeated use of a nanostencil mask, the deposition of material around the edges of the etch pattern will start to clog up and will reduce the feature size. The features will eventually close up altogether. A gradual clogging of the apertures of nanostencils used as miniature shadow masks in metal evaporations can be reduced by coating the stencil with self-assembled monolayers (SAM). An increase in material deposition through the apertures by more than 100% can be achieved with SAM-coated stencils, which increases their lifetime.

(3) Stencil patterns of <1 m will produce rounding of corners. Also, lateral deposition can occur, since the substrate and stencil are not in direct contact.

(4) For a large-sized stencil mask, there can be thermal stability issues. The deposition of metals by thermal evaporation can generate heat from the source. Radiation of heat can cause thermal expansion of the stencil mask, leading to pattern shifting and distortion.

(5) The fabrication of stencil masks involves lithographic techniques and RIE, which limits the minimum feature size possible for the mask. For a large area stencil mask, the minimum mask features can be around 100 nm. For a small-sized stencil mask, FIB sputtering can be used to make very small holes in the membrane. However, this method cannot be applied to large area masks.

Despite its limitations, nanostencil lithography offers a simple and low-cost method of nano-fabrication for simple monolithic features on a substrate. An example of this technique is illustrated in Figure 7.13.

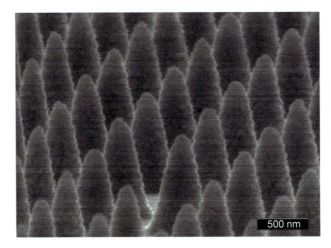

**FIGURE 7.13**   Gold dots (1 m high) obtained by evaporation through a perfluorosilane-coated stencil (magnification × 40k).

## 7.7   Summary

Alternative methods for the fabrication of nanostructures have gained much interest in research over recent decades. These techniques can offer low-cost solutions for preparing good quality reproducible nano-arrays using simple methodologies. Nano-imprinting was one of the first of these techniques and was developed in the 1990s. It requires no complex or expensive optical system and can be used to fabricate structures with feature sizes in the low micron scale. The method is simple and relies on the pressing of a pattern from a stamp master into the resist. This direct transfer can be repeated many times to provide a reproducible method to fabricate multiple samples. Once the pattern is transferred to the resist, processing will occur as with optical lithographic methods to produce a final structure. The NIL method suffers from setbacks such as the requirement of high working pressures and alignment problems. Step and flash NIL overcomes these difficulties by using a transparent template made of quartz and using a UV curing method. This allows alignment and illumination in one step.

So-called soft lithographies are also derived from the NIL technique. In this case, the template is a soft silicone elastomer, typically PDMS (poly-dimethylsiloxane). The advantage of using a flexible master as the template is that the pattern transfer can be made on curved and uneven surfaces and no pressure is involved in the processing. Excellent quality nanostructures can thus be produced on a variety of surfaces.

Dip-pen nanolithography relies on a different approach and uses an AFM like the tip to write directly on the substrate using a specialized ink. Regular arrays of nanodots and line structures can be produced in this manner with feature sizes on the order of 10s of nm.

Nanosphere and nanostencil lithographies are also very simple and cost-effective methods that can be used to produce regular arrays of nanodots. The former uses the close packing of polystyrene spheres to create a periodic array of holes. These holes effectively act as the mask structure for pho-tolithography. Since the spheres can have diameters of the order of a few 100 nm, hole sizes can be well below 100 nm. The nanostencil also acts as a direct shadow mask, allowing direct deposition onto a substrate through the holes in the mask membrane.

## References and Further Reading

Cui, Z., (2008). *Nanofabrication: Principles, Capabilities, and Limits*, Springer Science–Business Media, LLC, New York.

Donchev, E., Pang, J. S., Gammon, P. M., Centeno, A., Xie, F., Petrov, P. K., Breeze, J. D., Ryan, M. P., Jason Riley, D., & McAlford, N., (2014). The rectenna device: From theory to practice (a review). MRS Energy and Sustainability, *A Review Journal, 1*, 1 (Fig. 20).

Ivanisevic, A., & Mirkin, C. A., (2001). *J. Am. Chem. Soc., 123*, 7887.

Kang, S., (2012). *Micro/Nano Replication: Processes and Applications*, John Wiley and Sons, Hoboken, New Jersey.

Lan, H., & Ding, Y. H., (2010). *Lithography*, Michael Wang (Ed.), INTECH, Croatia.

Madou, M., (2002). *Fundamentals of Microfabrication: The Science of Miniaturization*, CRC Press, Boca Raton.

Mondin, G., Schumm, B., Fritsch, J., Hensel, R., Grothe, J., & Kaskel, S., (2013). *Mater. Chem. Phys. 137*, 884.

Shin, J.-H. et al., (2011). Fabrication of flexible UV nanoimprint mold with fluorinated polymer-coated PET film. *Nanoscale Res. Lett. 6*, 458.

Hulteen, J. C., & van Duyne, R., (1995). *J. Vac. Sci. Technol. A, 13*, 1553.

Vigneswaran, N., Samsuri, F., Ranganathan, B., & Padmapriya, (2014). *Proc. Engineering, 97*, 1387.

Xia, Y., & Whitesides, G. M., (1998). *Ann. Rev. Mater. Sci., 28*, 153.

Zhang, H., Elghanian, R., Amro, N. A., Disawal, S., & Eby, R., (2004). *Nano Letters, 4*, 1649.

# Chapter 8

# Nanoparticle and Nanowire Fabrication

The fabrication of nanoparticles and nanowires is of great importance in the preparation of materials in nanotechnology. These are probably the most used objects in applications, ranging from the electronics industry to biomedical sciences. Indeed, there are many techniques that have been employed in the synthesis of materials on the nanoscale. The fabrication strategy will depend on the application and quantity of material required. For small quantities, generally physical methods will be preferred, since they can offer a better control of size and shape distributions. For larger quantities, chemical methods offer a number of routes, which can be low-cost and adapted to industrial processes and quantities.

Nanoparticles are essentially a 0D nanostructure, offering confinement in all spatial directions. Nanowires, on the other hand, are confined in two directions, and extended in the third, forming a 1D nanostructure. In this chapter, we will outline some of the principal characteristics of the most commonly used methods for the fabrication of 0D and 1D nano-objects. Given the nature of this subject, it is beyond the scope of this chapter to provide an exhaustive account of all the methods available for the production of nanoparticles and nanowires.

## 8.1   2D Nanoparticle Assemblies

The manner, in which nanoparticles (NPs) are produced, can give rise to assemblies, which are flat, such as is the case when the particles are deposited onto a substrate surface. One simple method for doing this would be simply to interrupt the growth of a thin film at a stage, where the deposition has created islands. In this case, we could opt for materials in which the deposit grows via the Volmer–Weber growth mode on a particular substrate surface due to poor whetting conditions. Indeed, this method has been used in many studies to obtain discontinuous layers and even multilayers. The size and shape of the islands can be controlled to a certain extent by adjusting the growth conditions (deposition rate, temperature, and nature of the substrate). Some examples are illustrated in Figure 8.1. In Figure 8.2, we show the

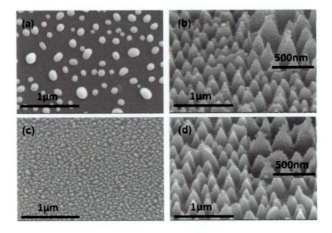

**FIGURE 8.1**    SEM images of (a) Au metal NPs on a flat glass substrate, (b) Au metal NPs on a glass substrate (synthesized by annealing a 10 nm Au film at 600°C for 1 min). (c) Ag metal NPs on a flat glass substrate, (d) Ag metal NPs on a glass substrate (synthesized by annealing of a 8 nm Ag film at 400°C for 1 min) (Reprinted from Tan et al. 2012. Open Access.).

case for the deposition of CoFe nanoparticles in an alumina $(Al_2O_3)$ matrix, which results from the alternating deposition of CoFe (in discontinuous layers, with an equivalent thickness of $t = 13$ Å) and $Al_2O_3$. Such a method allows control of the separation of the particles between the 2D layers. In this case, we see that the islands, which form the nanoparticles, are beginning to coalesce, which makes their appearance more worm-like than particle-like.

Another more specialized method for the deposition of metallic nanoparticles is the sputtering gas aggregation source. The principal component is the aggregation chamber, which contains a magnetron sputtering head and target for the material from which the nanoparticles will be composed. This chamber is known as the growth or aggregation zone. A schematic diagram of the basic set-up is illustrated in Figure 8.3. The length of the growth zone can be adjusted, typically in the range 40–300 mm, by using the linear translation stage attached to the magnetron head. The importance of this length is related to the size of particles formed before they pass through the aperture. The sputtered atoms from the target are carried down the tube, where

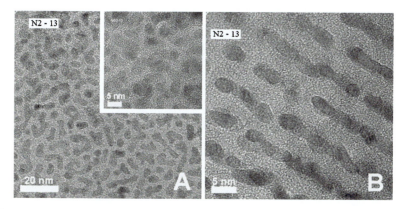

**FIGURE 8.2** HRTEM images of CoFe layers with an equivalent thickness of $t =$ 13 Å, which show the grains inside a layer (a) and a cross-section with the various layers (b). Reprinted from Schmool, D. S., Rocha, R., Sousa, J. B., Santos, J. A. M., Kakazei, G. N., Garitaonandia, J. S., & Lezama, L. (2007). *J. Appl. Phys., 101*, 103907], with the permission of AIP Publishing.

the growth of clusters occurs in the cooled aggregation zone. Cluster growth occurs continuously as the target atoms drift from the target. Conditions of temperature and pressure will also affect the formation of the atom clusters. Once the cluster and particles exit the growth zone, they can be size selected. This can be performed by a number of methods. In Figure 8.3, this is done by the quadrupole size selector, which filters the particles by setting the potential through which the particles pass. Setting larger potentials allows for larger particles to be selected. Other systems have a mechanical chopper, which slits. By setting the frequency of the chopper, it is also possible to filter the sizes of nanoparticles which pass through the system. The clusters or nanoparticles can then be deposited onto a substrate located further down the processing line. At this stage, a protective layer can be deposited onto the particle assembly thus encasing it in a matrix.

The control of the cluster source is achieved through the fluxes of Ar and He gases, the sputter power, the total pressure in the aggregation chamber, the distance between the magnetron head and the aperture at the end of the tube. The principal factors, however, are the argon partial pressure and the length of the growth zone. The other parameters are used to tune the shape of the size distribution and the intensity of the particle beam. When the growth zone is short ($\leq$ 80 mm), the size distribution is dominated by clusters of a few atoms. As the length of the aggregation zone is increased, the second population of larger particles is formed, with around $10^4$ atoms, with sizes

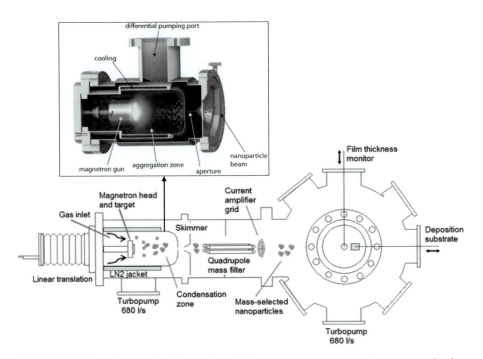

**FIGURE 8.3**     Schematic illustration of the sputter gas aggregation source. In the magnetron/vapor condensation source, sputtered metal atoms enter the aggregation zone, where they undergo collisions with the inert gas and quickly thermalize. Nanocluster ions form and grow, as the mixture moves through the source toward the exit aperture. The ions are filtered by a quadrupole mass analyzer equipped with an ion flux measurement grid and enter the deposition chamber. Reprinted from M. Khojasteh and V. V. Kresin, (2017). *Appl. Nanosci., 7,* 875, under Creative Commons Attribution 4.0 International License.

of around 5–6 nm in diameter. While the distributions of the small clusters are fairly insensitive to the pressure in the chamber, the distance at which larger clusters appear is proportional to the inverse of the Ar partial pressure. Other parameters such as the Ar flux or the presence of He modify the size and shape of the size distribution, but not the point at which they appear. As the target head is withdrawn from the aperture, the very small clusters tend to disappear and the size distribution for the larger clusters moves slowly to smaller sizes. A third regime is then reached, whereby a narrow distribution of almost monodispersity is reached. In Figure 8.4, we show some examples for the size distribution of particles, which have been size-selected using the quadrupole mass filtering potential, $U_{QP}$.

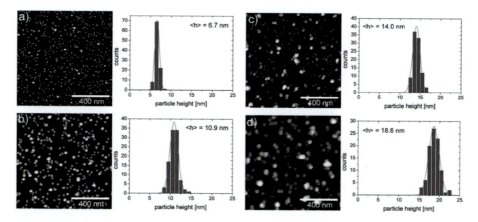

**FIGURE 8.4** AFM images (left) and corresponding height histograms (right) for the clusters deposited after mass filtering at $U_{QP}$ of (a) $\pm100$ V, (b) $\pm400$ V, (c) $\pm950$ V, and (d) $\pm2000$ V. The histograms are fit using Gaussian distributions. Mean particle sizes are shown in the panels. Reprinted from H. Hartmann V. N. Popok, I. Barke, V. von Oeynhausen, and K.-H. Meiwes-Broer, (2012). *Rev. Sci. Instrum., 83,* 073304, with the permission of AIP Publishing.

The size distribution of nanoparticles in the example given in Figure 8.4 was modeled using a Gaussian distribution. However, in many practical situations, the distribution is not symmetric and tends to have a long tail at the high-value end. In this case, the size distribution is more frequently modeled using the log-normal distribution, which has the form:

$$P(D) = \frac{1}{\sqrt{2\pi}\sigma D} e^{-[\ln(D/D_0)]^2/2\sigma^2} \qquad (8.1)$$

Here, $D$ denotes the particle size or diameter, with a mean value, $D_0$ and $\sigma$ is the standard deviation, which signifies the width of the distribution.

The gas aggregation method for the fabrication of nanoparticle assemblies has a number of variants, but has been applied to many nanoparticle systems, most notably in metals and semiconductors. The quality of the particle distribution is very good and the control of size is also a positive point. However, this method is almost entirely used as a research tool and is not easily adapted to large scale production of particles.

## 8.2 3D Nanoparticle Assemblies: Clusters and Colloids

For the production of three-dimensional assemblies, the gas aggregation source can also be used, where the particles are deposited on the substrate at the same time as the deposition of the matrix material. A more common approach to the preparation of larger quantities of nanoparticles and 3D assemblies is that concerning the chemical route. The chemical approach has a number of advantages over other techniques and most notably its cost-effectiveness and relative simplicity, without the need for expensive vacuum systems, etc.

The chemical route to the preparation of nanoparticles of all types of materials and structures has indeed become a major area of research with a broad range of methodologies and applications. In Figure 8.5, we illustrate the range of particle structures that can be produced. This represents a large range of possible structures with specific targeted properties and applications and requires a sophisticated approach for the manufacture of assemblies of like particles, with good control over size and shape distributions for them to be a viable method for use on the industrial scale. That is one of the principal goals of such technologies.

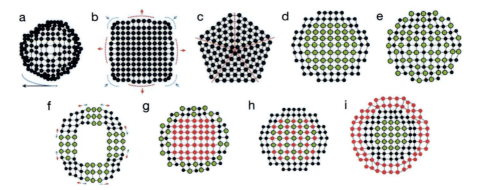

**FIGURE 8.5**   Schematic illustration of the types of nanoparticle structures realizable via chemical approaches. (a) Simple nanoparticle (size control), (b) Cubic structure (shape control), (c) Twinned structure, (d) Core-shell nanoparticle, (e) Alloy nanoparticle, (f) Phase separated heterostructure, (g) Core–alloy shell, (h) Alloy core–shell structure, (i) Core sandwich shell. Republished with permission of The Royal Society of Chemistry, from B. T. Sneed, A. P. Young and C.-K. Tsung., (2015). *Nanoscale (Review), 7,* 12248, permission conveyed through Copyright Clearance Center, Inc.

There are a variety of chemical methods that can be used to make metallic and oxide nanoparticles. In the following, we will describe the general approach without going into too much detail. There are many review papers and chapters that have been dedicated to this topic. There are several types of reducing agents that can be used to produce nanoparticles, for example,

NaBEt$_3$H, LiBEt$_3$H and NaBH$_4$, where Et denotes the ethyl C$_2$H$_5$ radical. For example, if we consider the preparation of Mo nanoparticles, these can be reduced in a toluene solution with NaBEt$_3$H at room temperature, providing a high yield of Mo nanoparticles with dimensions in the range 1–5 nm. The reaction equation is given by:

$$MoCl_3 + 3NaBEt_3H \rightarrow Mo + 3NaCl + 3BEt_3 + \frac{3}{2}H_2 \qquad (8.2)$$

Nanoparticles of Al have been successfully prepared from the decomposition of the compound Me$_3$EtNAlH$_3$ in toluene and heating the solution to 105°C for 2 hours; here Me denotes methyl, CH$_3$. Titanium isopropoxide is added to the solution. The Ti acts as a catalyst for the reaction. The choice of catalyst determines the size of the nanoparticles produced. For instance, 80 nm particles have been made using titanium. A surfactant, such as oleic acid, can be added to the solution to coat the particles and prevent aggregation.

A lot of interest has been generated in the preparation of Au nanoparticles, see for example the review by J. Pérez-Juste et al. (2005), due to their optical properties, which extends from their plasmonic behavior (see Section 3.4 of Volume 2). Indeed, this is one of the most common systems and much work has been done on the fabrication of gold nanospheres dispersed in water using the reduction of HAuCl$_4$ in a solution of boiling sodium citrate. The formation of uniform particles is revealed after about 10 minutes as the solution turns into a deep red wine color. This color change is related to the optical (plasmonic) properties of the nanoparticles as will be discussed in Section 3.4 of Volume 2. The particle diameter can, in fact, be tuned over a broad range of values (10–100 nm) by varying the concentration ratio between the Au salt and type sodium citrate and observing the color variation. It has been noted that for particles with a diameter greater than 30 nm, there is a deviation from the spherical shape, with polydispersion increasing with particle size. The same procedure has been applied to the preparation of Ag nanoparticles, reducing a Ag salt, though control of particle size is more limited. Citrate reduction has also been used in the production of Pt colloids of smaller particle size (2–4 nm), which can be further grown by hydrogen treatment.

Another procedure, which has also gained in popularity for the synthesis of Au nanoparticles is the two-phase reduction method. In this case, the HAuCl$_4$ salt is dissolved in water and subsequently transported into toluene by means of tetraoctylammonium bromide (TOAB), which acts as a phase transfer agent. The toluene solution is then mixed and stirred with an aqueous solution of sodium borohydride (a strong reductant) in the presence of thioalkanes or aminoalkanes, which readily bind to the Au nanoparticles. Depending on the ratio of the Au salt and the capping agent (thiol/amine), the particle size can be tuned between 1 and 10 nm. Several enhancements of the preparation technique have been introduced, with an exploration of the functionalization of the nanoparticles with various coatings.

Several examples exist for the reduction of metal salts by organic solvents. Ethanol has long been used for the preparation of metal nanoparticles such as Pt, Pd, Au, and Rh (used in catalytic applications) in the presence of a protective polymer—usually polyvinylpyrrolidone (PVP). Another important example consists of the refluxing of the solution of the metal precursor in ethylene glycol or larger polls. Ag nanoparticles with high aspect ratios have been grown in the presence of Pt seeds and formed in-situ prior to the addition of the Ag salt. The dimensions of Ag nanowire structures can be controlled by varying the experimental conditions: temperature, seed concentration, the ratio of Ag salts and PVP, etc. N,N-dimethylformamide (DMF) has the ability to reduce $Ag^+$ ions, so that stable Ag nanoparticles can be synthesized using PVP as a stabilizer. In addition, $SiO_2$ and $TiO_2$ coated nanoparticles can be produced by the same method. The size and shape of nanoparticles obtained by this method can depend on several parameters such as the salt and stabilizer concentrations, temperature and reaction times. Specifically, when PVP is used as a protecting agent, spherical nanoparticles form at low concentrations of $AgNO_3$ ($< 1$ mM), while increasing Ag concentration (up to 0.02 M) largely favors the formation of anisotropic nanoparticles. Again, the concentration of PVP and the reaction temperature strongly influence the shape of the final nanoparticles.

Deviation from the spherical shape geometry strongly affects the optical plasmonic response of metallic nanoparticles. For this reason. methods for the synthesis of anisotropic nanoparticles in solution (nanorods, nanowires, nanodiscs, nanoprisms, nanocubes, etc.) are continuously being reported, with a particular emphasis on Au and Ag due to their optical responses.

Excellent aspect ratio control in Au and Ag have been extensively reported, in particular with respect to nanorod formation, in which seeds are used in solution with a mild reducing agent (ascorbic acid), in the presence of cetyltrimethylammonium bromide (CTAB) to promote nanorod formation. The addition of different volumes of the seed solution produces Au nanorods with different aspect ratios.

The controlled synthesis of metallic nanoprisms that have been citrate stabilized from Ag nanospheres, and are truncated, has been achieved by irradiation with a fluorescent lamp in the presence of bis (p-sulfonatophenyl) phenylphosphine. The mechanisms involved in the synthesis of spherical metal nanoparticles are well understood in general, but those leading to the preferential growth in one particular direction are still subject to debate. In any case, such preferential growth is likely related to the surface free energies of particular crystallographic planes. Examples of metal nanoparticle formation with varying shapes and sizes, together with polydispersions and with different optical properties, are illustrated in Figure 8.5. As we can see, there are many shapes that can be formed with good quality dispersion properties also being an important consideration.

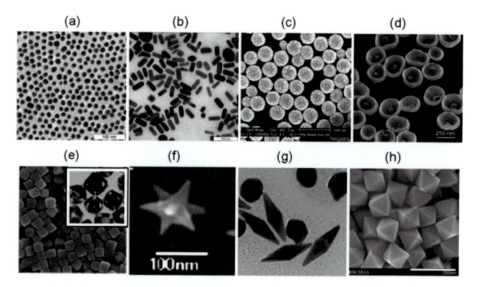

**FIGURE 8.6** TEM images of 15-nm An nanoparticles (a), 15 × 50-nm gold nanorods (b), 160(core)/17(shell)-nm silica/gold nanoshells (c, SEM), 250-nm Au nanobowls with 55-nm Au seed inside (d), silver cubes and gold nanocages (inset) (e), nanostars (f), bipyramids (g), and octahedrals (h). Reprinted from J. Quant., Spectrosc. Radiat. Transfer, N. G. Khlebtsov, & L. A. Dykman, (2010). *"Optical Properties and Biomedical Applications of Plasmonic Nanoparticles,"* 111, 1–35, Copyright (2010), with permission from Elsevier.

Strain-induced at the interface between different components in nanoparticles has received much attention in recent years, since it has a significant impact on the surface electron structure and catalytic properties of nanomaterials. These effects are of great importance in nanostructures, where the surface strain is particularly accentuated, since the proportion of atoms at the surface is a significant number of the total atoms in the structure. This provides important motivation for the synthesis of strain engineered structures with well-defined and strained architectures, as is illustrated in Figure 8.5. Strain in nanoparticles can be induced via a number of approaches. Size and shape are the most obvious in the case of single element NP assemblies. Lattice strain can also be induced through twinning, alloying as well as core-shell epitaxy. This latter case is reminiscent of the interface induced strain in mismatched epitaxial thin films, in which the difference in the lattice constant causes mechanical strain at the boundary of two layers of different materials.

Alloyed nanoparticles of $Pt_3Ni$ can be produced in nano-icosahedra via higher oxygen reduction, leading to strain effects at the edges and corners of the structures, see Figure 8.7. These nanoparticles are fabricated via the co-reduction in oleylamine and oleic acid solutions at high temperatures. Other metallic alloys have also been prepared in a similar manner such as AuPt and PdPt. Carbon monoxide was used as an agent to suppress oxidative etching, to restrict the particle dimensions and to improve monodispersion in the NP assemblies.

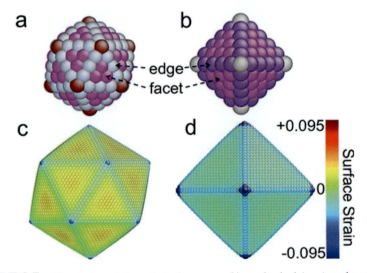

**FIGURE 8.7**   Crystal models and strain maps of icosahedral (a, c) and octahedral (b, d) nanocrystals. Reprinted (adapted) with permission from J. Wu, L. Qi, H. You, A. Gross, J. Li, & H. Yang., (2012). *J. Am. Chem. Soc., 134,* 11880. Copyright (2012) American Chemical Society.

More complex structures, such as core-sandwich-shell nanoparticles, have also been prepared using chemical methods. Such structures can be produced by a solvothermal oleyamine route with several stages to the preparation, coating nanoparticles successively in various solutions, with the appropriate metallic salts. One-step methods have also been shown to be viable, such as that devised by Qiu et al. (2014). In this case, an aqueous solution is used to produce Ag–Co–Fe core-shell-sandwich particles, which relies on ammonia borane as a reducing agent and PVP as a capping agent. Differences in the reduction of potential and magnetic permeability serve as critical parameters in the growth mechanism. A similar procedure was employed in the fabrication of the Pd nanoparticles coated with Ni and Pt, as illustrated in Figure 8.8. These use a catalytic dehydrogenation of ammonia borane.

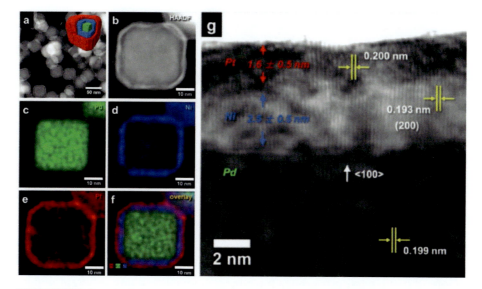

**FIGURE 8.8** (a–f) HAADF/STEM/EDX images of core-sandwich-shell Pd-Ni-Pt nanoparticles and HRTEM image in (g). Reprinted (adapted) with permission from B. T. Sneed, A. P. Young, D. Jalalpoor, M. C. Golden, S. Mao, Y. Jiang, Y. Wang, & C.-K. Tsung, (2014). *ACS Nano, 8,* 7239. Copyright (2014) American Chemical Society.

Extremely good control over the deposition of monolayers of metals on the surface of nanocrystals has also been achieved in low mismatched metallic systems. For example, layers of Pt have been systematically deposited on Pd cubic nanoparticles. The controlled monolayer sequence for this system is illustrated in Figure 8.9.

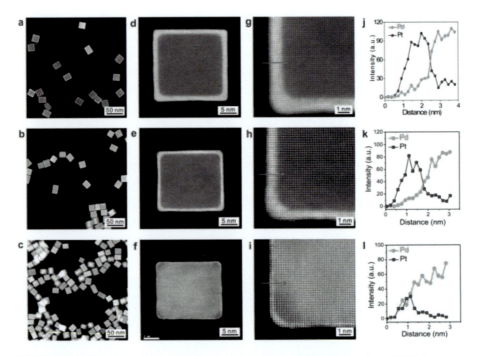

**FIGURE 8.9**   Control of monolayer thickness of Pd-Pt nanocubes. TEM images of 6, 4, and 1 monolayer of Pt on Pd cubes (top to bottom). Reprinted (adapted) with permission from S. Xie S.-Il Choi, N. Lu, L. T. Roling, Jeffrey A. Herron, L. Zhang, J. Park, J. Wang, M. J. Kim, Z. Xie, M. Mavrikakis, & Y. Xia., (2014). *Nano Lett., 14*, 3570, Copyright (2014). American Chemical Society.

As we have seen in the above examples, the chemical routes to the preparation of nanoparticle systems have reached an extraordinary level of sophistication, with excellent control over nanoparticle size and shape. Furthermore, the chemical procedures have been adapted to many different materials, which are often metals and semiconductors, though oxides are also common. Detailed recipes have been produced for the systematic preparation of nanoparticle assemblies with well-defined morphological features including complex core-shell structures designed for specific functionalities. Such is the level of control that in certain materials, nanoparticle assemblies can be prepared commercially to client specifications. This is particularly true of elements such as Au and Ag, which have many applications due to their plasmonic/optical properties. For example, in Figure 8.10, we show a series of images for Ag nanostructures, illustrating the diversity of nanoparticle size and shape that can be produced for the same element.

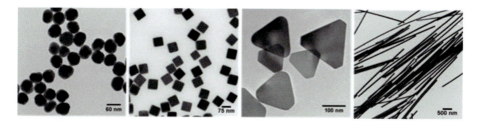

**FIGURE 8.10** Transmission electron micrographs demonstrating the diversity of size and morphology possible by control over the reaction chemistry and kinetics during the solution-phase synthesis of silver nanomaterials: (left to right) uniform 50 nm diameter spheres, 75 nm cubes, 120 nm triangular nanoplates, and silver nanowires.

It is worth noting that the importance of nanoparticles has steadily increased over the years with important applications in biomedical sciences in particular. We will discuss some of these issues with regard to magnetic nanoparticles in Section 4.3.6 of Volume 2. So why are nanoparticles so interesting and useful? The surface properties of nanoparticles can be functionalized in very specific and controlled manners. By this, we mean that the surface chemistry is manipulated such that its binding strength, size, and capping agents are varied to provide a level on control over its behavior in specific environments. For example, the addition of charged species at the surface can introduce electrostatic stability in aqueous solutions. This is the case for Ag particles coated with citrate ions. Alternatively, we may add branched polyethyleneimine (BPEI) to the surface, which creates an amine dense surface with a highly positive charge. Other functionalities can be obtained with other capping agents. Polyethylene glycol (PEG) provides good stability in high salt concentrations, while lipoid acid-coated particles contain a carboxyl group, which can be used for bioconjugation.

Silver ions provide a strong anti-microbial presence due to their interaction with thiol groups for bacterial enzymes and proteins. This can significantly affect cellular respiration and results in cell apoptosis or death. The specific toxicity to bacteria, while being of low toxicity for humans, has been exploited in a range of products, such as wound dressings, packaging materials, and anti-fouling surface coatings. The main mechanism for this property in Ag nanoparticles is due to the large surface area for silver ions. In an aqueous environment, the particles oxidize in the presence of oxygen and protons, which releases $Ag^+$ ions with the gradual dissolution of the Ag. The release rate of the $Ag^+$ ions depends on a number of factors, such as the particle size, shape, the state of aggregation of the particles as well as the environment and the capping agent used. The fastest rate of dissolution of the Ag is for small and elongated particles, since they will have a more curved surface, meaning that the Ag atoms are on average less strongly attached. Increased temperatures also favor the rapid release of the Ag ions, while the presence of chlorine, oxygen, and thiols also favor quick release.

We have already mentioned the optical properties associated with Au and Ag nanoparticles due to plasmonic effects, which makes them good absorbers and scatterers of light. More details of the plasmonic properties of metallic nanoparticles are given in Section 3.4 of Volume 2. The high scattering cross-section can be exploited in bioimaging under darkfield conditions for optical microscopy or hyperspectral imaging. Typically, this is performed by attaching Ag particles to biomolecules, such as antibodies or peptides. Silver nanoparticles can be targeted to specific cells or cellular components via physisorption or covalent bonding. This can be done using ethyl(dimethylaminopropyl) carbodiimide (EDC) to link the free amines to an antibody or carboxyl group on lipoid acid-coated Ag nanoparticles.

Ag is indeed one of the favored materials for biological applications, with *in-vitro* and *in-vivo* applications increasing rapidly. In addition to the tagging, Ag nanoparticles can be used as a thermal source for hyperthermia treatment. By coating the silver nanoparticles with an amorphous silica shell, these can be functionalized for attachment to drug molecules or other high molecular weight molecules for integration within the shell for labeling or drug delivery applications. Since the tendency is for increased use of these nanoparticles in living tissues, it is important that the interaction between the Ag nanoparticle and the biological system be well understood. Clearly, for *in-vivo* applications, we require low toxicity and long circulation times. For this, we need an in-depth knowledge of the interplay between the particle and the target biological system, with particular attention to protein corona, dissolution rate, and bio-distribution. This is also true for the use of magnetic nanoparticles in biomedical applications. We will outline some of the main uses of magnetic systems in Section 4.3.6 of Volume 2.

## 8.3   Ordered Assemblies of Nanoparticles

The ordering of nanoparticles in assemblies has been demonstrated in a number of systems. While this might seem a ridiculously difficult challenge, requiring not only almost perfect monodispersion of nanoparticle sizes, but also the perfect arrangement of these particles in well-defined geometries, there has been some excellent progress made. To obtain an organization between objects, we require some form of interaction between the bodies themselves. In the case of nanoparticles, such an organization can be found via electrostatic and magnetostatic forces.

In a simple example, the magnetic dipolar forces can be employed to coax magnetic nanoparticles to align in long chains. A process of click chemistry under the application of an applied magnetic field has been successfully used in the alignment of iron oxide nanoparticles in single nanoparticle chains, Toulemon et al., (2016). Here, the iron oxide nanoparticles have been assembled onto gold substrates by performing copper-catalyzed alkyne-azide cycloaddition (CuAAC click reaction) under a magnetic field. The classic click reaction is that formed by the copper catalysis of an alkyne and an azide to form a five-membered heteroatom ring: Cu(I)-catalyzed azide-alkyne cycloaddition (CuAAC), (Kolb and Sharpless, 2003). The CuAAC click reaction was performed by immersing the alkyne-terminated self-assembled monolayer (SAM-CC) onto gold substrates in the suspension of NP@N$_3$ in the presence of triethylamine and CuBr(PPh$_3$)$_3$ catalysts. The immobilization of NP@N$_3$ onto SAM-CC proceeds under the reflux of a tetrahydrofuran/triethylamine mixture through the formation of irreversible covalent bonding, e.g., triazole function. It is noted that without applying a magnetic field, nanoparticles assemble in a 2D monolayer, which consists of short chains with random orientation. An applied field gradient of 0.5 T/cm was used to produce the desired chains, as illustrated in Figure 8.11, for the case of 20 nm spherical iron oxide nanoparticles.

2D arrays of nanoparticles have been observed by a number of researchers, where a control over the particle size, for magnetic particles, can be used to determine the interparticle interactions. For example, Krishnan et al., (2006) have used Co nanoparticles with sizes near to the superparamagnetic limit (see Section 4.3.3 of Volume 2), depending on the size and shape of the particles. This means, competing forces (magnetic and van der Waals) will be responsible for the organization of the particles in regular arrays, see Figure 8.12.

The use of photon assisted reduction of gold precursors (HAuCl$_4 \cdot$ 3H$_2$O) deposited in a solid PVA layer has been performed, where the illumination is made via a Mach-Zehnder interferometer to form an interference pattern, (Nadal et al., 2017). The formation of Au nanoparticles was thus shaped into regular stripes, where the Au diffuses to the illuminated regions forming a surface relief grating as well as a nanoparticle grating. AFM images and

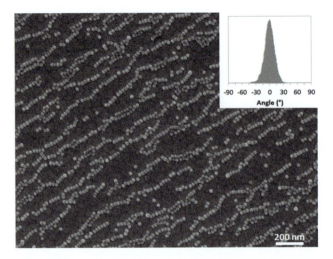

**FIGURE 8.11** SEM images showing nanoparticles assembled by performing the CuAAC click reaction under the magnetic field for 1 h. Inset shows the distribution of orientations of chains. Reprinted from D. Toulemon, M. Rastei, D. S. Schmool, J. S. Garitaonandia, L. Lezama, X. Cattoën, S. Bégin-Colin, & B. P. Pichon, (2016). *Advanced Functional Materials, 26*, 2454, under Creative Commons Attribution License.

profiles are illustrated in Figure 8.13, which demonstrates the formation of both nanoparticles and a surface relief grating, which has the same periodicity as the interference pattern from the interferometer.

A more sophisticated approach has been shown to allow the formation of binary nanoparticle superlattices via electrostatic forces. Superlattices were made by placing a substrate in a colloidal solution of two kinds of nanoparticles. The solvent was evaporated in a low-pressure chamber causing self-assembly of the nanoparticles into ordered structures. Superlattice constituents have included lead selenide (PbSe), palladium, gold, iron oxide, lead sulfide, silver and also triangular nanoplates of lanthanum fluoride. The superlattices had several crystal structures.

The self-assembly process can be directed by setting the charge state of the nanoparticles. This is achieved by adjusting the concentrations of tri-n-octyl phosphine oxide (TOPO), carboxylic acids or dodecylamine to the nanoparticle solution. For example, the use of oleic acid for PbSe nanocrystals converts some negatively charged and neutral nanocrystals into positively charged ones. On the other hand, the addition of TOPO increases the population of negatively charged lead selenide nanocrystals. The addition of oleic acid to gold nanoparticles causes them to become negatively charged. The

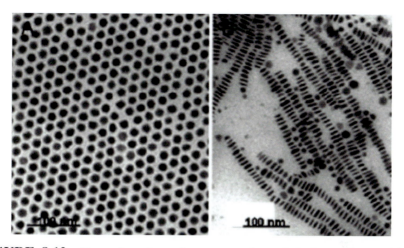

**FIGURE 8.12** Examples of regular patterns formed during self-assembly of magnetic nanoparticles: 2D hexagonal array of 10 nm Co magnetic nanospheres (left) and 1D stacks of 5 × 20 nm anisotropic nanodiscs (right). Reprinted by permission from K. M. Krishnan, A. B. Pakhomov, Y. Bao, P. Blomqvist, Y. Chun, M. Gonzales, K. Griffin, X. Ji, & B. K. Roberts, (2006). *J. Mater. Sci., 41*, 793, ©(2006). Springer Nature.

combination of two or more materials in a superlattice enables a modular approach to material design. Such metamaterials are indeed capable of combining useful properties of the building blocks and generating completely new properties due to the intermixing of components, as with more conventional composite materials.

Complex structures using biological building blocks have provided quite spectacular results. Protein cages, which are electrostatically patchy, have been used to direct the self-assembly of 3D binary superlattices of nanoparticles. The negatively charged cages can encapsulate RNA or superparamagnetic iron oxide nanoparticles, and the nanoparticle superlattices are formed via tunable electrostatic interactions with positively charged gold nanoparticles. Gold nanoparticles and viruses form a $AB_8^{fcc}$ crystal structure, that is not isostructural with any known atomic or molecular crystal structure, (Kostiainen et al., 2012). In Figure 8.14, some binary nanoparticle superlattices are illustrated. Auyeung et al. (2012) have used hollow DNA nanostructures as spacers and employed programmable DNA interactions to assemble highly ordered nanoparticle arrays. Three-dimensional assembly is based on the use of spherical nucleic acid (SNA) nanoparticle conjugates (spherical structures with a layer of densely packed, highly oriented nucleic acid). DNA can serve as a directing ligand, which controls the placement of nanoparti-

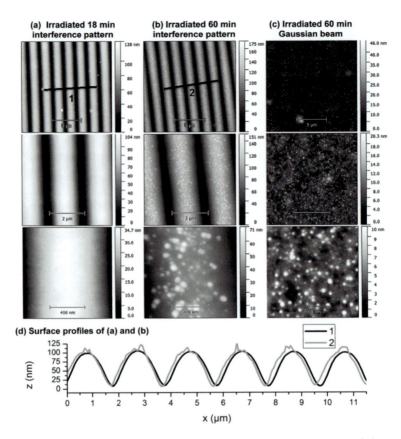

**FIGURE 8.13** Images (a–c) are AFM images of different areas of the sample surface after annealing and irradiation, respectively, for 18 and 60 min with an interference pattern and 60 min with a Gaussian beam. From top to bottom, the scale bars are 5 mm, 2 mm and 400 nm. (d) Surface profile corresponding to images (a) and (b). Republished with permission of The Royal Society of Chemistry, from E. Nadal, N. Barros, H. Glénat, J. Laverdant, D. S. Schmool, & H. Kachkachi, (2017). *J. Mater. Chem. C, 5,* 3553; permission conveyed through Copyright Clearance Center, Inc.

cles into specific positions within the ordered superlattice. Some examples of superstructures formed from Au nanoparticles are illustrated in Figure 8.15. Small-angle X-ray scattering can be used to characterize such arrays, with results that are similar to X-ray diffraction for atomic crystalline structures being observed.

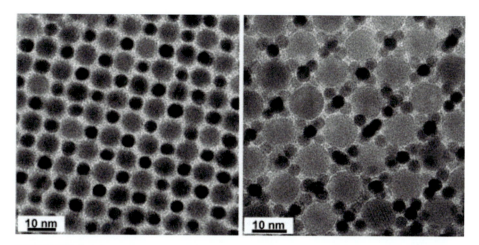

**FIGURE 8.14** Examples of binary nanoparticle superlattices. Left: superlattice assembled from PbSe and Au nanoparticles is isostructural with CuAu intermetallic compound. Right: superlattice assembled from magnetic $Fe_2O_3$ and Au nanoparticles is isostructural with $CaB_6$. (https://www.azonano.com/article.aspx?ArticleID=3217 (accessed on 30 March 2020)).

## 8.4 Template Assisted Nanowire Synthesis

As with the case of nanoparticles, there are many techniques that have been employed to fabricate nanowires. Indeed, there is some overlap in the methods used, as was seen in the previous section. In this and the following section, we will outline some of the most common approaches which have afforded high-quality nanowires with a variety of materials. Again, given the volume of research in this field and the extensive materials studied, we will only give a brief overview of the subject.

An important and highly versatile method of nanowire fabrication is that using template-assisted synthesis. This method is conceptually very simple and intuitive. The template consists of very small cylindrical pores formed in a host material, where these spaces are filled with a chosen material. These will then adopt the morphology of the pores to form the nanowire structures.

We will firstly outline the procedure for the preparation of the template structures and then describe the filling methods. In the template-assisted synthesis of nanowires, the chemical stability and mechanical properties of the templates, as well as the diameter, height, uniformity, and density of the pores, are crucial parameters that all require consideration. The most frequently used template material used in nanowire synthesis is anodic alu-

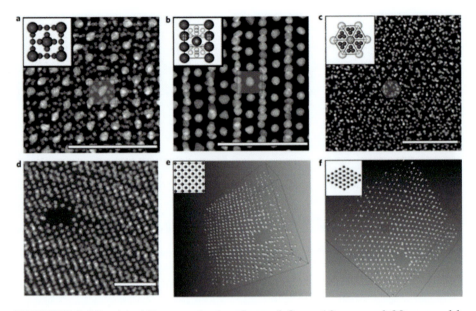

**FIGURE 8.15**    (a) AB$_6$-type lattice formed from 10 nm and 20 nm gold nanoparticles. (b) A bcc lattice formed from 20 nm gold nanoparticles and 10 nm hollow SNA spacers. (c) Lattice X structure formed from 20 nm hollow SNA spacers and 10 nm gold nanoparticles. All scale bars, 200 nm. Lattice projections shown in the insets are outlined in the TEM images. A three-dimensional reconstruction of a thin (∼100 nm) section of the lattice in (b) was obtained from electron tomography. (d) A representative snapshot of TEM images obtained in the tilt series, where the hole was used as a reference point for alignment (scale bar, 200 nm). (e) and (f) Snapshots from the reconstructed lattice along the [100] zone axis (e) and [111] zone axis (f) of a bcc lattice. Insets: perfect bcc lattice along each respective zone axis. A unit cell in each reconstructed lattice is outlined in red for clarity. Reprinted by permission from E. Auyeung, J. I. Cutler, R. J. Macfarlane, M. R. Jones, J. Wu, G. Liu, K. Zhang, K. D. Osberg, & C. A. Mirkin, (2012). *Nat. Nanotechn., 7,* 24, ©(2012), Nature.

minum oxide (AAO) or alumina, Al$_2$O$_3$. Other options include nanochannel glass and mica films.

Porous anodic alumina templates can be produced using a well-defined procedure anodizing Al films in various acids. Under carefully chosen anodization conditions, the resulting oxide film possesses a regular hexagonal array of parallel and nearly perfectly cylindrical channels, as is illustrated in Figure 8.16. The self-organization of the pore structure in an anodic alumina

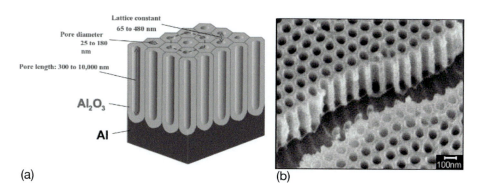

Lattice constant
65 to 480 nm

Pore diameter
25 to 180 nm

Pore length: 300 to 10,000 nm

$Al_2O_3$

Al

(a)

(b)

100nm

**FIGURE 8.16** (a) The preparation of nanoporous membranes of alumina ($Al_2O_3$) by anodization processes, where long-range hexagonal ordering is induced by self-assembling or nano-imprint. Self-aligned nanoporous membranes of titania ($TiO_2$) are also prepared following the same method. The pore diameter, pore length, and lattice parameter can be controlled by the anodization conditions, bath composition and process time. (b) Cross-section view of an $Al_2O_3$ template showing smooth cylindrical nanochannels of about 250 nm in height extending to the bottom of the aluminum substrate, scale bar = 100 nm. Reprinted from E. Gultepe, D. Nagesha, L. Menon, A. Busnaina, & S. Sridhar, (2007). *Appl. Phys. Lett., 90,* 163119, with the permission of AIP Publishing.

template involves two coupled processes: pore formation with uniform diameters and pore ordering. The pores form with uniform diameters because of a delicate balance between the electric field enhanced diffusion, which determines the growth rate of alumina, and dissolution of the alumina into the acidic electrolyte. The pores are believed to self-order because of mechanical stresses at the Al-alumina interface due to volume expansion during anodization. This stress produces a repulsive force between the pores, causing them to form a hexagonal lattice due to energy minimization in this close-packed formation. Depending on the anodization conditions, the pore diameter can be systematically varied from < 10 nm to 200 nm, with a pore density in the range of $10^9 - 10^{11} cm^{-2}$. Many researchers have shown that the pore size distribution and pore ordering can be significantly improved using a two-step anodization process, where the $Al_2O_3$ layer is dissolved after the first anodization in an acidic solution. A second anodization, under the same conditions, of the surface now leads to a more ordered hexagonal structure of pores. A schematic illustration of the steps in the preparation of the AAO templates is illustrated in Figure 8.17.

Another type of porous template used commonly in nanowire fabrication is produced via the chemical etching of polycarbonate membranes. Also,

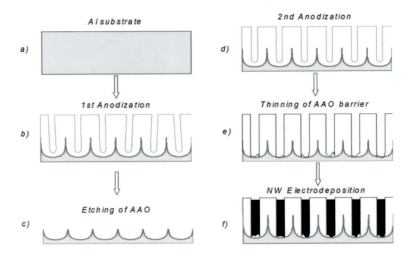

**FIGURE 8.17**    Schematic view of the process flow used for AAO template formation (a) Al foil (b) first anodization step (c) removal of first AAO layer, (d) second anodization step and AAO membranes; before gold sputtering (e), and after gold sputtering (f).

nanochannel glass structures can have densities similar to nonporous alumina, with excellent channel parallelism and channel diameters in the 10 nm region.

The pressure injection method has been successfully employed in the fabrication of highly crystalline nanowires from a low melting point material using porous templates with robust mechanical strengths. In this technique, nanowires are formed by pressure injecting the desired material in liquid form into the evacuated channels of the template. Due to the heating and pressures involved, the templates used must be chemically and physically stable to withstand the high temperatures and pressures. Anodic alumina films and nanochannel glass are two materials that can be used in this technique for growing wires of both metal and semiconductor materials.

The pressure required to overcome the surface tension for the liquid material to fill the pores of diameter, $d_w$, is determined by the Washburn equation:

$$d_w = -\frac{4\gamma\cos\theta}{P} \tag{8.3}$$

where $\gamma$ is the surface tension of the liquid, $\theta$ is the contact angle between the liquid and the template and $P$ is the pressure required. To reduce the necessary pressure and maximize the filling factor, surfactants can be used, which decrease the surface tension and contact angle, though care is required to avoid contamination problems. Nanowires produced by the pressure injection method can possess high crystallinity with preferential orientation along the wire axis.

Electrochemical deposition (ECD) has also been frequently used as a means of nanowire fabrication in conjunction with template structures. Traditionally, ECD has been used as a method to grow thin films on metallic surfaces. Since electrochemical growth is usually controllable in the direction normal to the substrate surface, this method can be readily extended to 0D or 1D nanostructures, if the deposition is confined within the pores of an appropriate template. In ECD, a thin conducting metal film is first coated on one side of the porous membrane to act as the cathode for electroplating. The length of the deposited wires can be controlled by varying the duration of the electroplating process. This method has been applied to the synthesis of a wide variety of metal and semiconducting nanowires as well as multi-layer wire structures. An example of Fe/Cu nanowires grown in the AAO templates is shown in Figure 8.18.

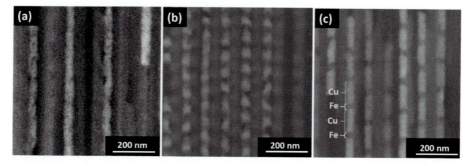

**FIGURE 8.18** Cross-sectional Scanning Electron Microscopy (SEM) images of Fe/Cu nanowires (NWs) grown in Anodic Aluminum Oxide (AAO) membranes. The average diameter and length of the Fe segments were kept constant to $(45 \pm 5)$ nm and $(30 \pm 7)$ nm, respectively, while the Cu segment length, $L_{Cu}$, was varied: (a) $(15 \pm 5)$ nm; (b) $(60 \pm 6)$ nm; (c) $(120 \pm 10)$ nm. Reprinted from S. Moraes, D. Navas, F. Béron, M. P. Proença, K. R. Pirota, C. T. Sousa, & J. P. Araújo, (2018). *Nanomaterials, 8,* 490, under Creative Commons Attribution License.

In addition to the preparation of wires inside the AAO templates, Sousa et al., (2009), have also shown that it is possible to prepare nanotubes by coating the walls inside the nanopores of the AAO. This process is illustrated in Figure 8.19.

In the ECD procedure, the chosen template must be chemically stable in the electrolyte during the electrolysis process. Cracks and defects must be

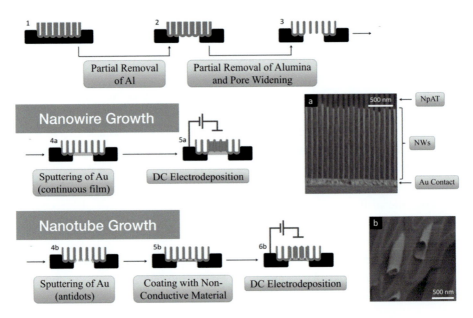

**FIGURE 8.19** Schematic illustration of the template preparation for nanowire (NW) and nanotube (NT) growth by DC electrodeposition. SEM cross-sectional images of Ni (a) NWs with $D_{NW} \sim 50$ nm and (b) NTs with $D_{NT} \sim 300$ nm, fabricated by the DC electrodeposition method. Reprinted from C. T. Sousa, D. C. Leitao, M. P. Proença, J. Ventura, A. M. Pereira, & J. P. Araújo, (2014). *Appl. Phys. Rev., 1,* 031102, with the permission of AIP Publishing.

avoided as much as possible, since they will be detrimental to the nanowire growth. To use anodic $Al_2O_3$ films in dc ECD, the insulating barrier layer that separates the pores from the bottom Al substrate has to be removed and a metal film is then evaporated onto the back of the porous template membrane. Surfactant use has also been reported in this procedure. It is possible to use ac ECD in AAO templates without the removal of the barrier layer by utilizing the rectifying properties of the oxide barrier. In ac ECD, although the applied voltage is sinusoidal and symmetric, the current is greater during the cathodic half-cycle, making deposition dominant over the stripping, which occurs in the anodic half-cycle. Since no rectification occurs at defect sites, deposition and stripping will be equal and no resultant material will be deposited in these sites.

In contrast to wires grown using the pressure injection method, nanowires prepared by ECD are typically polycrystalline with no preferential crystalline orientation. Some exceptions have been found, such as the case of CdS nanowires. Also, Pb crystalline nanowires have been formed by pulsed ECD under conditions of over-potential. One advantage of the ECD method is the possibility of fabricating multilayered structures within the nanowires. By varying the cathodic potentials in the electrolyte that contains two different types of ions, different metal layers can be controllably deposited, as seen in Figure 8.18.

Anodization has been applied to other systems such as titanium and has been successfully employed to fabricate high quality $TiO_2$ nanotubes, which can be made in very regular hexagonal arrays with very uniform wall thicknesses. An example is illustrated in Figure 8.20. The various parameters shown in the figure are key geometrical factors, which can be controlled to a certain extent via the nanotube preparation conditions.

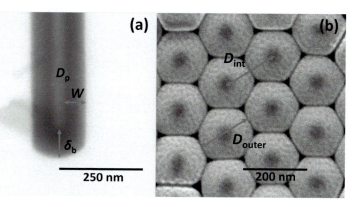

**FIGURE 8.20** (a) STEM and (b) SEM images of $TiO_2$ nanotube bottoms, with the respective geometrical parameters barrier layer thickness, $\delta_b$, pore diameter, $D_p$, and outer diameter, $D_{outer}$, interpore distance, $D_{int}$, and wall thickness, $W$. (Apolinário, 2015) with permission.

Vapor deposition is another alternative for nanowire growth. Techniques such as physical vapor deposition (PVD), chemical vapor deposition (CVD) and metal-organic chemical vapor deposition (MOCVD) can all be applied as the deposition method. As with ECD, vapor deposition is usually capable of preparing nanowires of a diameter inferior to 20 nm, since it does not rely on high pressure and the surface tension to insert the material into the pores. For PVD, the material is usually heated to produce a vapor, which is then introduced through the pores of the template and cooled to solidify. Single crystal wires of Bi have been produced in this manner with diameters of down to 7 nm.

Compound materials, that result from two reacting gases have also been prepared by CVD, such as single crystals of GaN nanowires. Nanowires of GaAs and InAs have also been prepared in nanochannel glass arrays. Here, the nanochannels are filled with a liquid precursor, such as trimethyl-gallium ($Me_3Ga$) or triethyl-indium ($Et_3In$), via the capillary effect and the nanowires are formed within the template by reactions between the liquid precursor and another gas reactant, such as arsine gas ($AsH_3$).

It has also bee demonstrated that carbon nanotubes (CNT) can be fabricated in the pores of AAO templates using CVD to form highly ordered 2D CNT arrays. A small amount of catalyst metal, such as Co, is first electroplated into the bottom of the pores. The templates are then placed in a furnace and heated to $\sim 700 - 800°C$ with a flowing gas mixture of $N_2$ and acetylene ($C_2H_2$) or ethylene ($C_2H_4$). The hydrocarbon molecules are pyrolyzed to form carbon nanotubes in the pores of the template with the aid of the metal catalyst. Well-aligned nanotube arrays have attracted much interest due to their potential in various applications. We will discuss carbon nanotubes in more detail in the following chapter, see Section 9.1.

## 8.5    The VLS Method of Nanowire Growth

The vapor–liquid–solid (VLS) method of nanowire growths, which promotes a strongly anisotropic crystal growth, has achieved much success in the synthesis of semiconductor nanowires. This mechanism of nanowire growth was initially used for the preparation of single-crystal silicon whiskers of 100 nm to hundreds of microns in diameter, and was first proposed by Wagner and Ellis, (1964). The growth mechanism involves the absorption of source material from the gas phase onto a liquid droplet, which acts as the catalyst for growth. In the original work of Wagner and Ellis, they used molten Au on a silicon substrate. The source material can be delivered via a number of techniques, typically a CVD related method such as MOCVD. The catalyst can be deposited via evaporation and by heating will form droplets on the substrate surface. The supersaturated liquid alloy is formed, when the source molecules arrive at the droplet to form the alloy droplet. For Si nanowire growth, the source is typically a silane ($SiH_4$) molecule for CVD processes. The pathways for the arrival of source material for the formation of the nanowire are illustrated in Figure 8.21(a).

In VLS growth, the liquid droplet consists not only of the catalytic material, but also of a certain amount of the semiconductor material to be deposited in the nanowire. The mixed material has a much lower melting temperature than each of the two constituents, for example, gold and silicon. The lowest possible melting temperature for a specific composition is called the

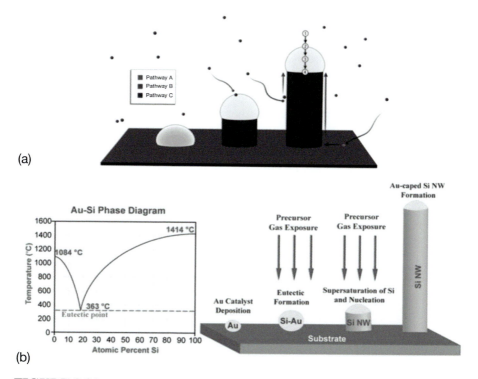

**FIGURE 8.21** (a) Precursor materials (blue spheres) are transported to the growing nanowire by three different pathways. In Pathway A, precursor materials adsorb directly to the surface of the catalyst. In Pathway B, precursor materials adsorb to the nanowire sidewalls, where adatoms diffuse to the catalyst surface. In Pathway C, precursor materials adsorb to the substrate, where adatoms diffuse along the surface to the nanowire sidewalls and catalyst. Transport processes are driven by surface concentration gradients. The numbered mechanism shows the vapor-liquid-solid growth method. Step 1 of this process is represented by Pathway A, B, and C. After being transported to the catalyst, precursor molecules adsorb, step 2, and diffuse through the liquid, alloyed catalyst droplet, step 3. Upon reaching the surface, atomic precursors crystallize and are incorporated at the NW tip, step 4. (b) The Au-Si eutectic phase diagram showing the eutectic point and temperature, $T_E$, at the 18.6 % Si concentration. Also illustrated is the schematic for the growth of Si nanowire for the VLS method. (a) Reprinted from M. J. Crane, & P. J. Pauzauskie, (2015). *J. Mater. Sci. Technol., 31*, 523 (2015), ©(2015) with permission from Elsevier and (b) Reprinted from R. Ghosh, & P. K. Giri, (2015). *Sci. Adv. Today, 2*, 25230, ©(2016) Lognor.

eutectic temperature, $T_E$. For the gold–silicon system, the eutectic temperature is about 360°C, as is shown in the Au–Si eutectic phase diagram, Figure 8.21(b). Also illustrated in this figure is the schematic representation of the VLS process for the Au-Si system. Upon supersaturation of the liquid alloy, nucleation, and growth at the interface between the droplet and the substrate generates a solid precipitate of the source material. This seed serves as a preferential site for further deposition of the material at the substrate–droplet interface, thus promoting the anisotropic growth for elongated nanowires and whiskers. This will also suppress further nucleation events on the droplet. The diameter of the wire or whisker will be dictated by the diameter of the liquid droplet. The nanowires obtained by this method are of high purity, except for the end of the wire, which contains the solidified catalyst as an alloy particle. Figure 8.22 shows SEM and HRTEM images of Si nanowires grown using Au and Cu catalysts. These micron-sized wires exhibit good uniformity on size and a high-quality crystalline structure. We can also note the droplets at the top of the wires.

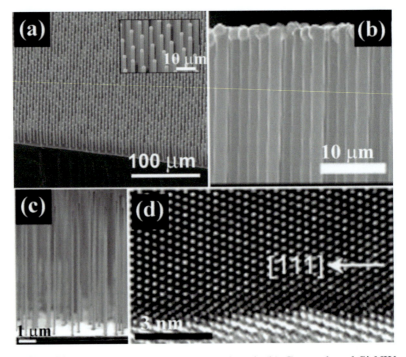

**FIGURE 8.22**     SEM image of (a) Au-catalyzed, (b) Cu-catalyzed Si NW array having nearly 100% fidelity over a large ($> 1 cm^2$) area. (c) Cross-sectional image of Si NW grown from 50 nm Au colloids and resulting Si NW diameters. (d) HRTEM image of a single crystalline Si NW. Reprinted from R. Ghosh, & P. K. Giri, (2015). *Sci. Adv. Today, 2*, 25230, ©(2016) Lognor.

The reduction of the average wire diameter to the nanometer scale requires the generation of nano-sized catalyst droplets. Due to the balance between the liquid-vapor surface free energy and the free energy of condensation, however, the size of a liquid droplet, in equilibrium with its vapor, is usually limited to the micrometer range. This obstacle has been overcome using several innovations: (i) Advances in the synthesis of metal nanoclusters have made the availability of monodisperse nanoparticles commercially viable. These can be dispersed on a solid surface in high dilution so that, when the temperature is raised above the melting point, the liquid clusters do not aggregate; (ii) Metal islands of nm size can self-form, when a strained thin layer is grown or heat-treated on a non-epitaxial substrate; (iii) Laser-assisted catalytic VLS growth is a method used to generate nanowires under non-equilibrium conditions. PLD of a target containing both catalyst and source materials produced the nucleation of nanoclusters of the catalyst. Single crystal nanowires will grow as long as the catalyst remains liquid; (iv) By optimizing the material properties of the catalyst–nanowire system, conditions can be found for which nanocrystals nucleate in a liquid catalyst pool supersaturated with the nanowire material. The growing atoms migrate to the surface due to the large surface tension and continue growing as nanowires perpendicular to the liquid surface. In this case, the supersaturated nanodroplets are sustained at the other end of the nanowire due to the low solubility of the nanowire material in the liquid.

A wide variety of elemental, binary and compound semiconductor nanowires have been made using the VLS technique, where good control of wire diameter and diameter size distribution has been accomplished. Some modifications of the VLS set-up can allow for the growth of modulated structures, a variation of doping and composition in the nanowires. In Figure 8.23, we show a selection of nanowire structures produced using the VLS method.

## 8.6   Other Methods of Nanowire Synthesis

There are several other methods, which have been employed in the synthesis of nanowire structures. We will consider some of these, which have not been previously discussed.

A solution-phase synthesis of nanowires with controllable diameters is possible without the use of templates, catalysts or surfactants. Here, use is made of the anisotropy of the crystalline structures of trigonal Se and Te, which can be viewed as rows of 1D helical atomic chains. This approach is

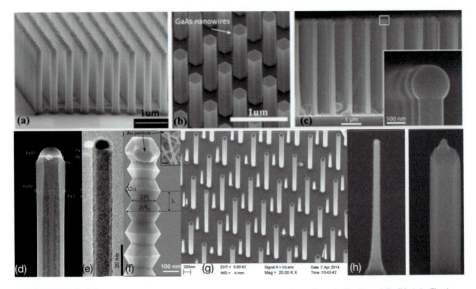

**FIGURE 8.23**    A selection of VLS grown nanowires (a–c) GaAs, (d) Si (e) GaAs, (f) GaN, (g) Si, (h) A single InAsP nanowire core grown with a 20 nm gold catalyst (left) and a clad nanowire (right). (a–c) Reprinted from Y. Zhang, J. Wu, M. Aagesen, & H. Liu, (2015). *J. Phys. D: Appl. Phys. 48*, 463001, under Creative Commons Attribution License; (f) Z. Ma, D. McDowell, E. Panaitescu, A. V. Davydov, M. Upmanyu, & L. Menon, (2013). *J. Mater. Chem. C, 1*, 7294, Published by The Royal Society of Chemistry.; (h) Reprinted (adapted) with permission from D. Dalacu, K. Mnaymneh, J. Lapointe, X. Wu, P. J. Poole, G. Bulgarini, V. Zwiller, & M. E. Reimer, (2012). *Nano Lett., 12*, 5919, Copyright (2012) American Chemical Society.

based on the mass transfer of atoms, during an aging step, from a high-energy solid phase to a seed that grows preferentially along one crystallographic direction. The lateral dimension of the seed, which dictates the diameter of the resulting nanowire, can be controlled by the temperature of the nucleation step.

More often, the use of surfactants is necessary for the promotion of anisotropic growth (1D) of the nanocrystals. Quantum dots, for example, have been fabricated using this method. Surfactants are required to stabilize the interfaces of the nanoparticle and retard the aggregation and oxidation processes. Growth conditions can be optimized to manipulate the growth of these nanocrystals, which usually generate nanorods with aspect ratios in the region of 10:1.

Stress-induced crystalline Bi nanowires have been grown from sputtered films of Bi and CrN. The nanowires are thought to grow from defects and cleavage fractures of the film and can be up to several mm in length with diameters ranging from 30–200 nm.

Selective electrodeposition (ED) along step edges in highly-oriented pyrolytic graphite (HOPG) has been successfully used to grow nanowires of Mo, Cu, Ni, Au, and $MoO_2$. The mechanism of growth is illustrated in the following Figure 8.24 along with some images of Cu wire growth at various growth times.

The growth of $MoO_2$ nanowires has been particularly studied, with excellent results. In this case, the solution for the electrolyte is $MoO_4^{2-}$, where the reduction follows the reaction:

$$MoO_4^{2-} + 2H_2O + 2e^- \rightarrow MoO_2 + 4OH^- \tag{8.4}$$

Site selectivity is achieved by applying a low overpotential to the electrochemical cell, in which the HOPG serves as the cathode. This minimizes the nucleation events on less favorable sites, i.e., on the plateaus. A pseudo-steady state growth can be achieved and leads to a growth rate, in which the wire radius can be expressed as (Penner et al., 2003):

$$r(t) = \sqrt{\frac{2I_{dep}t_{dep}V_m}{\pi n F L}} \tag{8.5}$$

Here, $I_{dep}$ is the deposition current, $t_{dep}$ the deposition time, $V_m$ is the molar volume of the deposited material (19.8 cm$^3$ mol$^{-1}$ for $MoO_2$), $n$ is the number of electrons transferred for the deposition of each $MoO_2$ unit, and $L$ is the total length of nanowires present within a 1.0 cm$^2$ area of the graphite surface. The square root dependence of the nanowire radius with deposition time is confirmed by experiment, as illustrated in Figure 8.25.

While the $MoO_2$ nanowires cannot be released from the substrate, they can be reduced, see Figure 8.24 (left), to metallic Mo nanowires, which can then be lifted from the substrate as free-standing nanowires. In this method, the substrate defines only the position and orientation of the nanowire, but not its diameter. In this context, other surface morphologies, such as self-assembled grooves in etched crystal planes, have been used to synthesize nanowire arrays via gas shadow deposition. The cross-section of artificially prepared superlattice structures has also been used for site-selective ED of parallel and closely spaced nanowires.

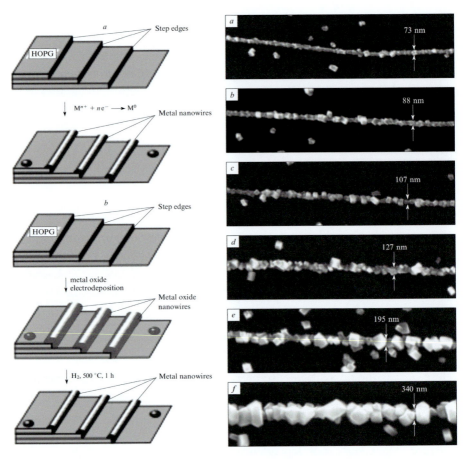

**FIGURE 8.24**   Left:  Illustration of two ways of deposition (a, b) of metal nanowires on HOPG, based on the electrochemical decoration of steps. Right: SEM images of copper nanowires deposited on HOPG at the nucleation potential of $-0.8$ V vs. normal calomel electrode and the growth potential of $-5$ mV. The time of nanowire growth (in seconds) is (a) 120, (b) 180, (c) 300, (d) 600, (e) 900, (f) 2700. Reprinted from E. C. Walter, M. P. Zach, F. Favier, B. J. Murray, K. Inazu, J. C. Hemminger, & R. M. Penner, (2003). *Chem. Phys. Chem., 4*, 131, under Creative Commons Attribution License.

## 8.7   The Hierarchical Arrangement of Nanowires and Nanowire Superstructures

Ordering nanowires into useful structures is another challenge that requires addressing in order to harness the full potential of nanowires for applications. We have already discussed the cases, where nanowires with varying composi-

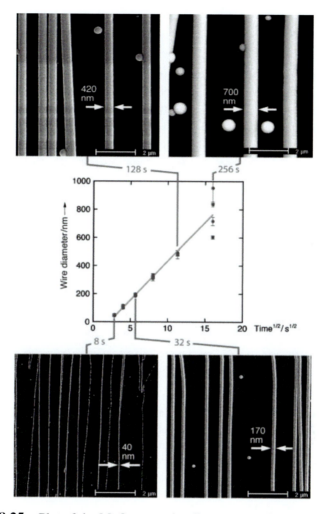

**FIGURE 8.25** Plot of the $MoO_2$ nanowire diameter as a function of the square root of the deposition time. The growth conditions in this experiment were $E_{dep} = -0.90\ V_{SCE}$, 0.16 mM $Na_2MoO_4$, 1.0 M NaCl, 1.0 M $NH_4Cl$, pH 8.5. The linearity of this plot is consistent with the predictions of Eq. (8.5). Also shown are SEM images (scale bars 2 m) showing HOPG surfaces after the deposition of $MoO_2$ for various durations as indicated. Reprinted from E. C. Walter, M. P. Zach, F. Favier, B. J. Murray, K. Inazu, J. C. Hemminger, & R. M. Penner, (2003). *Chem. Phys. Chem., 4*, 131, under Creative Commons Attribution License.

tion, or superlattice structures along with their axes, have been performed by controlling the gas-phase chemistry as a function of time during the nanowire growth process in VLS and other methods. Control of composition along the axial dimension was also demonstrated for the template grown nanowires. Alternatively, the composition can be varied in the radial direction by first growing a nanowire by the VLS method and then switching the synthesis conditions to grow different material on the surface of the nanowire by CVD. Indeed, this strategy has been adopted for the synthesis of Si/Ge and Ge/Si coaxial (core-shell) nanowires. It has been shown that thermal annealing can produce an outer shell, which can be epitaxial with the inner core.

A different category of non-trivial nanowire structures is that of a non-linear structure, which results from multiple growth steps. In such a process, a tetrahedral quantum dot core can be synthesized. The growth conditions can then be altered to induce the 1D growth of a nanowire from each of the tetrahedral facets. This can lead to hierarchical structures with hyperbranching. Examples of some hierarchical nanowire structures are illustrated in Figure 8.26.

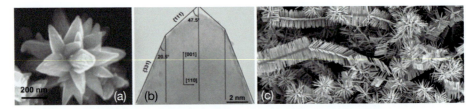

**FIGURE 8.26**   SEM images of TiO$_2$ synthesized by hydrothermal treatment of a titanium-glycolate complex in the presence of different amounts of picolinic acid 4 mmol, showing a pyramidal branch of a flower-like particle; (b) HR-TEM image of the top apex of a pyramidal branch from which branching can occur. (c) Core nanowires of single-crystal In$_2$O$_3$ with 6, 4, and 2 facets, and the secondary nanorods formed of single-crystal hexagonal ZnO, which grow either perpendicular on or slanted to all the facets of the core In$_2$O$_3$ nanowires. (a, b) Reprinted from Q. Duc Truong, H. Kato, M. Kobayashi, M. Kakihana, (2015). *J. Cryst. Growth, 418,* 86, ©(2015), with permission from Elsevier; (c) Reprinted (adapted) with permission from J. Y. Lao, J. G. Wen, & Z. F. Ren, (2002). *Nano Letters, 2,* 1287. ©(2002) American Chemical Society.

Control of the position of a nanowire in the growth process is important for device preparation, especially for long structures. Post-growth methods to align and position nanowires include microfluidic channels, Langmuir–Blodgett assemblies and electric field assisted assembly. The first controls nanowire orientation by the injection of a nanowire solution into microfluidic channels. The second involves the alignment of the nanowires at the liquid–gas or liquid–liquid interface by the application of compressive forces on the interface. The third technique is based on di-electrophoretic forces that pull polarizable nanowires towards regions of high electric field strength. The nanowires align between two isolated electrodes, that are capacitively coupled to a pair of buried electrodes biased with an ac voltage. Once a nanowire shorts the electrodes, the electric field is eliminated and none more will deposit. Alternatively, alignment and positioning of nanowires can be specified and controlled during their growth by applying the appropriate synthesis conditions. Lithographies can be used as a catalyst to position and define the nanowire array. The orientation of the nanowires can be achieved by selecting a substrate with a lattice structure matching that of the nanowire material to facilitate epitaxial growth. Such conditions will result in an array of nanowire posts at predetermined positions, all vertically aligned with the crystal growth orientation. Similar structures can be obtained using template-grown ECD nanowires, particularly this with anodic alumina templates, with their parallel and ordered channels. The control over the location of the nucleation of nanowires in the ECD process is determined by the pore positions and the back electrode geometry. The pore positions can be precisely controlled by imprint lithography, for example. By growing the template on a patterned conductive substrate that serves as the back electrode, different materials can be deposited into the pores at different regions of the template.

## 8.8  Summary

Nanoparticle and nanowire structures are of enormous interest for both fundamental research and for a wide range of technological applications, notably in biomedical use and optoelectronic and electronics applications. There are a large number of techniques that have been developed to prepare size-controlled structures in metallic, semiconducting, oxides and other systems. Controlled environmental conditions are a basic requirement in all methods of preparation. For the physical preparation of nanoparticles, this will usually mean in a high or ultra-high vacuum system. Temperature control can also be of vital importance in the formation of the nanostructure. We have already discussed lithographic means of preparations in a previous chapter. Thin-film techniques can be employed for the formation of nanoparticle structures since island formation in the early stages of growth can be used as a simple method in which arresting the growth at this point will typically result in a two-dimensional array of disorganized nanostructures. This doesn't generally provide much control over the size and shape distributions of the particle assembly. Dedicated growth systems such as the gas aggregation method allow a much better control of these parameters, with reduced distribution widths.

Chemical methods have proven to be spectacular in their results in the diversity of control of particle size distributions and also on the particle shapes that can be produced by chemically controlling the solution from which the particles form and precipitate. The formation of nanoparticles from solution allows further the preparation of core-shell type structures with good shape and size monodispersion. It is possible to prepare alloyed and oxide nanoparticle systems and even multilayer onion-like structures. In addition to these advantages, the chemical routes to nanoparticle production are relatively cheap and can produce large quantities of high-quality assemblies of nanoparticles.

The interaction between nanoparticles can allow the nanoparticles to form chain-like structures, as is found for the case of magnetic nanoparticles, which form long chains due to the magnetic dipolar interaction between the particles. Ordered structures of nanoparticles have also been formed using a number of strategies, including dipolar interactions and light interference to selectively form particles in the regions of high-intensity illumination. However, the use of biological building blocks, such as RNA and DNA, have been employed to dramatic effect, producing binary superlattice arrays of nanoparticles with long-range order.

Nanowires can also be prepared using a number of approaches. The use of templates is a popular method for the formation of nanowire structures. To these ends, nanoporous alumina templates provide a cheap and very effective method. $Al_2O_3$ can be readily formed into hollow pore structures by anodizing Al in acid solution. The control of potential coupled with a two-step procedure has been shown as an efficient means of producing high-quality arrays of hexagonally organized nanopores. Pore widths of from 10s to 100s of nm can be formed into large area substrates. The filling of the pores can be performed by electrochemical deposition. Such a method is capable of producing nanowires of a variety of metals. Alternating the anode between two different metals can be used to fabricate multilayered wires as well as core-shell structures if the cathode structures are correctly prepared. The anodization methods has also been applied to materials other than Al.

The vapor-liquid-solid (VLS) technique is a specialized method for preparing nanowires of a variety of materials. It can be used to produce well-controlled structures with a range of aspect ratios in a number of materials. The growth mechanism involves the adsorption of source materials from the gas phase on to a liquid droplet, acting as a catalyst for the growth process. This droplet is often formed by liquid Au deposited onto a Si substrate. The source materials are typically provided by CVD or MOCVD. Regular arrays of Au droplets can be formed using lithographic techniques, such as the nanosphere method. These can be heated to the eutectic point, about 360°C for Au-Si. The control of environmental conditions during growth as well as composition and materials can allow a number of deformations of the wire surface to produce a broad range of morphological structures, which rely on the surface energies of the different crystalline planes to preferentially form particular surfaces.

## References and Further Reading

Apolinário, A. (2015). PhD Thesis, University of Porto, Portugal.

Auyeung, E., Cutler, J. I., Macfarlane, R. J., Jones, M. R., Wu, J., Liu, G., Zhang, K., Osberg, K. D., & Mirkin, C. A., (2012). *Nat. Nanotechn., 7,* 24.

Crane, M. J., & Pauzauskie, P. J., (2015). *J. Mater. Sci. Technol. 31,* 523.

Cui, Z., (2008). *Nanofabrication: Principles, Capabilities, and Limits,* Springer Science–Business Media, LLC, New York.

Dalacu, D., Mnaymneh, K., Lapointe, J., Wu, X., Poole, P. J., Bulgarini, G., Zwiller, V., & Reimer, M. E., (2012). *Nano Lett., 12,* 5919.

Duc Truong, Q., Kato, H., Kobayashi, M., Kakihana, M., (2015). *J. Cryst. Growth, 418,* 86.

Ghosh, R., & Giri, P. K., (2016). *Sci. Adv. Today, 2,* 25230.

Gultepe, E., Nagesha, D., Menon, L., Busnaina, A., & Sridhar, S., (2007). *Appl. Phys. Lett., 90,* 163119.

Hartmann Popok, H. V. N., Barke, I., von Oeynhausen, V., & Meiwes-Broer, K.-H., (2012). *Rev. Sci. Instrum., 83,* 073304.

Khlebtsov, N. G., & Dykman, L. A., (2010). *J. Quant., Spectrosc. Radiat. Transfer, 111,* 1–35.

Khojasteh, M., & Kresin, V. V., (2017). *Appl. Nanosci., 7,* 875.

Kostiainen, M. A., Hiekkataipale, P., Laiho, A., Lemieux, V., Seitsonen, J., Ruokolainen, J., & Ceci, P., (2013). *Nat. Nanotechn., 8,* 52.

Kolb, H. C., & Sharpless, B. K., (2003). *Drug Discov. Today, 8,* 1128.

Krishnan, K. M., Pakhomov, A. B., Bao, Y., Blomqvist, P., Chun, Y., Gonzales, M., Griffin, K., Ji, X., & Roberts, B. K., (2006). *J. Mater. Sci., 41,* 793.

Lao, J. Y., Wen, J. G., & Ren, Z. F., (2002). *Nano Letters, 2,* 1287.

Ma, Z., McDowell, D., Panaitescu, E., Davydov, A. V., Upmanyu, M., & Menon, L., (2013). *J. Mater. Chem. C, 1*, 7294.

Madou, M., (2002). *Fundamentals of Microfabrication: The Science of Miniaturization*, CRC Press, Boca Raton.

Moraes, S., Navas, D., Béron, F., Proença, M. P., Pirota, K. R., Sousa, C. T., & Araújo, J. P. (2018). *Nanomaterials, 8*, 490.

Nadal, E., Barros, N., Glénat, H., Laverdant, J., Schmool, D. S., & Kachkachi, H., (2017). *J. Mater. Chem. C, 5*, 3553.

Pérez-Juste, J., Pastoriza-Santos, I., Liz-Marzan, L. M., & Mulvaney, P., (2005). *Coordination Chem. Rev., 249*, 1870.

Petrii, O. A., (2015). *Russ. Chem. Rev., 84*, 159.

Qiu, F., Liu, G., Li, L., Wang, Y., Xu, C., An, C., Chen, C., Xu, Y., Huang, Y., Wang, Y., Jiao, L., & Yuan, H., (2014). *Chem. Eur. J., 20*, 505.

Schmool, D. S., Rocha, R., Sousa, J. B., Santos, J. A. M., Kakazei, G. N., Garitaonandia, J. S., & Lezama, L., (2007). *J. Appl. Phys., 101*, 103907.

Sneed, B. T., Young, A. P., Jalalpoor, D., Golden, M. C., Mao, S., Jiang, Y., Wang, Y., & C.-Tsung, K., (2014). *ACS Nano, 8*, 7239.

Sneed, B. T., Young, A. P., & C.-K. Tsung., (2015). *Nanoscale (Review), 7*, 12248.

Sousa, C. T., Leitao, D. C., Proença, M. P., Ventura, J., Pereira, A. M., & Araújo, J. P., (2014). *Appl. Phys. Rev., 1*, 031102.

Tan, C. L., Jang, S. J. & Lee, Y. T., (2012). Localized surface plasmon resonance with broadband ultralow reflectivity from metal nanoparticles on glass and silicon subwavelength structures, *Opt. Express 20*, 17448–17455.

Toulemon, D., Rastei, M., Schmool, D. S., Garitaonandia, J. S., Lezama, L., Cattoën, X., Bégin-Colin, S., Pichon, B. P., (2016). *Advanced Functional Materials, 26*, 2454.

Wagner, R. S., Ellis, W. C., (1964). *Appl. Phys. Lett., 4*, 89.

Walter, E. C., Zach, M. P., Favier, F., Murray, B. J., Inazu, K., Hemminger, J. C., & Penner, R. M., (2003). *Chem. Phys. Chem., 4*, 131.

Wu, J., Qi, L., You, H., Gross, A., Li, J., & Yang., H. (2012). *J. Am. Chem. Soc., 134*, 11880.

Xie, S., Il Choi, S., Lu, N., Roling, L. T., Jeffrey Herron, A., Zhang, L., Park, J., Wang, J., Kim, M. J., Xie, Z., Mavrikakis, M., & Xia., Y. (2014). *Nano Lett., 14*, 3570.

Zhang, Y., Wu, J., Aagesen, M., & Liu, H., (2015). *J. Phys. D: Appl. Phys. 48*, 463001.

# Chapter 9

# Other Fabrication Techniques and Technologies

## 9.1 Fullerenes

Fullerenes and carbon nanotubes (CNT) are relatively recent discoveries of stable structures of carbon that can occur under certain preparation conditions. Fullerenes were discovered in 1985, winning the 1996 Nobel prize in Chemistry for the discovery by H. W. Kroto, R. F. Curl, and R. E. Smalley. These have been extensively studied since the 1990s as have the later discovered and related nanotube structures. Indeed. these discoveries shed new light on the nature of carbon as an element, and has led to wide-ranging research and development in the fundamental properties of these structures, their unique physical properties, and applications of many carbon-based nanostructures. We note that the folded structures, which form fullerene and CNT structures, are intimately related to graphene, which was also the subject of the Nobel prize in Physics in 2010 for A. Geim and K. Novoselov.

In its natural state, carbon exists as a solid with two possible crystalline structures, which are very familiar: graphite and diamond. Graphite is a crumbly and brittle black-colored mineral, which has been used for many centuries for writing (Indian ink and pencil lead). In contrast, the diamond is transparent and extremely hard. It was only at the end of the 18th century, that diamond was identified as a carbon crystalline phase by Lavoisier and Tennant. In the normal diamond structure, each atom is connected to four nearest neighbors in a regular tetrahedron by forming hybrid bonds of the $sp^3$ type. Tetrahedral symmetry is an indication of a dense and isotropic medium. The distance between neighboring atoms is 0.136 nm. Graphite is a layered or lamellar structure, made up of parallel planes, each of which has a regular array of hexagons, forming a honeycomb pattern. The carbon atoms are arranged at the vertices of the hexagons. Each atom is connected to three neighbors via hybrid bonds of the $sp^2$ type, making an angle of $120°$. The bonds in the plane are very strong, with an interatomic separation of 0.142 nm. However, each atom is only weakly bonded to atoms in neighboring planes, where the intercolumnar separation is 0.34 nm. This structure has a density, which is only one-third of that of a diamond. Graphite is, therefore, a highly anisotropic, quasi-2D solid, since the weakly connected planes slide very easily relative to one another. In Figure 9.1, we show a schematic representation of the crystalline structures of diamond and graphite.

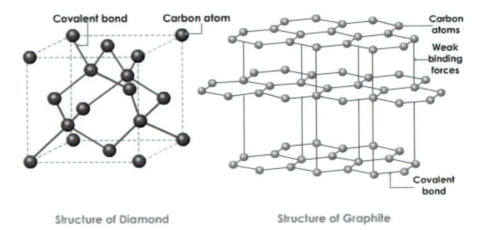

**FIGURE 9.1**   Crystalline structures of diamond and graphite.

Observations of interstellar space by radio astronomers revealed the existence of chains of carbon atoms in certain stars; the red giants. While seeking to produce similar conditions as those in such stars, Kroto, Curl, and Smalley went on to revolutionize our understanding of carbon. When examining the carbon clusters produced in a very hot plasma, which were obtained by vaporizing graphite using a high-powered laser, they observed specific peaks in the mass spectra, showing stable configurations of carbon atoms, notable for $C_{60}$ and $C_{70}$, which had larger than expected intensities. It was found, that the size distribution of carbon clusters could be drastically affected by increasing the degree of chemical boiling in the inlet nozzle to the vacuum chamber. Clusters with 60 and 70 carbon atoms could be produced. In Figure 9.2, we show a mass spectrum of carbon clusters. These structures were eventually identified as the cage-like structures which are shown in the inset of Figure 9.2 and in Figure 9.3. The $C_{60}$ is a football-like structure, while the $C_{70}$ fullerene has an additional 10 carbon atoms added in the central region, forming a slightly elongated ball. It will be noted, that the peak at 60

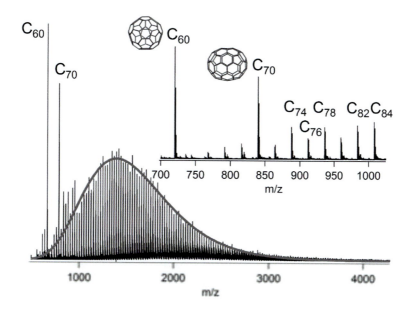

**FIGURE 9.2** LDI-TOF mass spectrum of fullerene soot after toluene extraction (500 shots 11–14 mJ/cm$^3$), Fullerene soot after Soxhlet extraction with toluene to remove most of the C$_{60}$/C$_{70}$. Magic number fullerenes are known to be relatively stable due to symmetry (C60, C70, C74, C76, C82, and C84), see inset. Reprinted from J. W. Martin, G. J. McIntosh, R. Arul, R. N. Oosterbeek, M. Kraft, & T. Söhnel, (2017). *Carbon, 125*, 132, ©(2017), with permission from Elsevier.

carbon atoms is the largest, indicating its relative stability with respect to the other configurations. The structure was dubbed as a fullerene after the Architect Buckminster–Fuller, who made similar such architectural structures. The fullerene also gained the name *bucky ball* for a similar reason. In Figure 9.3, we show some of the main allotropes of carbon.

There is an important similarity in many of the allotropes. The planar graphite sheets and graphene are made up of purely hexagonal rings. However, we note that by converting some of these into pentagons, the planar structure of the graphene sheet will curl up making a three-dimensional structure. With the pentagons placed in the correct places, the C$_{60}$ and C$_{70}$, etc. molecules will be formed. The C$_{60}$ ball has 12 pentagons and 20 hexagons. We can generalize these observations by stating that the fullerenes are molecules containing $2(10+n)$ carbon atoms, which form 12 pentagons and $n$ hexagons. A hexagon forms a planar lattice, as we saw earlier. Geometrically the pentagons, due to the reduced number of C atoms, will transform the lattice into an open cone with an apex angle of 112°. Introducing further

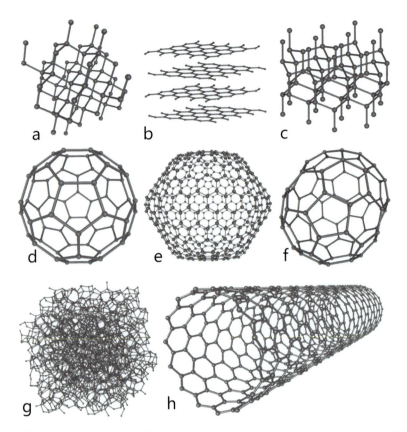

**FIGURE 9.3**    Allotropes of carbon: (a) diamond; (b) graphite; (c) lonsdaleite; (d) $C_{60}$ (Buckminster-fullerene); (e) $C_{540}$; (f) $C_{70}$; (g) amorphous carbon; (h) single-walled carbon nanotube.

pentagons, the structure will close in upon itself, forming the shell or cage-like structure. Euler's theorem tells us, that 12 pentagons suffice, to close the shell, forming the closed polyhedron of the $C_{60}$ molecule. The smallest structures, that can be formed in this way, will be the $C_{20}$, which has just 12 pentagons and no hexagons, i.e., $n = 0$. The $C_{60}$ molecule is the smallest stable Buckminsterfullerene. The shape is a truncated icosahedron with a regular structure, that corresponds to an Archimedean solid, that can be inscribed within a sphere. The carbon atoms are all equivalent in terms of their symmetry, which is not the case with other fullerenes. It should be noted that there are two types of carbon–carbon bond in this molecule; bonds that join two hexagons, which are the 6–6 bonds, while those at the join of a hexagon and a pentagon are 5–6 bonds. These can be distinguished as follows: for

**FIGURE 9.4** $C_{60}$ Buckminster-fullerene, showing the two types of bond: 6–6, indicated as double lines and 5–6, shown as single lines.

each hexagon, the sides are made up of alternating 6–6 and 5–6 bonds, while the pentagons only have 5–6 bonds along their edges, see Figure 9.4. The 6–6 bonds are shorter than the 5–6 bonds and hence have the character of a double bond, while the 5–6 bonds have the character of a single bond. The localization of the $\pi$ electrons arises due to the pyramidization of the $sp^2$ carbon atoms, which in turn results from the fact that the spherical structure prevents full orbital overlap. $C_{60}$ molecules are therefore not aromatic.

The second stable form of the fullerene is the $C_{70}$ molecule. Its structure also respects an isolated pentagon rule and has a more oval profile. At its poles, $C_{70}$ has a structure, which is similar to the $C_{60}$ molecule, but the equatorial belt reveals the difference between the two, being made up of chains of hexagons. A number of other stable fullerenes exist with magic numbers as shown in Figure 9.2.

## 9.2 Carbon Nanotubes

The carbon nanotube (CNT) was discovered in 1991 by S. Iijima, while studying the by-products of $C_{60}$ synthesis. The production of CNTs can be performed using an electric arc technique, which was used in the 1960s for making carbon filaments and whiskers. In all likeliness, carbon nanotubes had been produced, but not identified, at the time.

Carbon nanotubes typically have diameters in the range of 1–10 nm, which at the lower end is of the same order of magnitude as the DNA double helix molecule ($\sim 1$ nm). It will be noticed, that the basic crystal structure of the CNT is closely related to that of graphite. The atomic structure of the nanotube is in fact obtained by rolling up a sheet of graphene to form a cylinder. The cylinder is open at the ends and can be closed by introducing the sp$^2$ bonds at the ends to form the pentagonal rings. The C$_{70}$ molecule can be seen as the shortest possible nanotube structure. In fact, the topology of the end structures depends on the distribution of pentagons and the normal hexagonal rings. Six pentagons are required for each end closure. For larger nanotube dimensions, different distributions of the pentagons will be required. A regular distribution will define a hemispherical end, while in the general case, one obtains a conical tip shape.

To give a complete description of the nanotube structure we must examine the manner in which the graphene sheet is rolled up. In Figure 9.5, we illustrate the different nanotube structures and their orientational relationship to the hexagon structure of the graphene sheet from which they originate. This can be envisaged as choosing two hexagons of the graphene lattice, which are then made to superpose. The type of CNT, which results will depend entirely and uniquely on the choice of the two hexagons. This will fix the diameter of the nanotube and also the *chiral angle*, $\theta$, which will determine the helicity of the CNT, since it will specify the direction of rolling. We choose the direction of the reference from which the angle is measured as illustrated in Figure 9.5. The chiral angle can vary between $0°$ and $30°$, given the symmetry of the hexagonal lattice. This then allows a complete classification of all possible configurations into three categories, which are designated as: armchair, zig-zag and chiral. The zig-zag and armchair have angles of $0°$ and $30°$, respectively. Since in these cases the hexagons remain always in the same orientation, such nanotubes are said to be achiral. When the chiral angle differs from these two limiting values, this will not be the case. We can state formally, that the structure of the CNT is derived from the graphene sheet. The chiral vector, also referred to as the *roll-up* vector, defined as: $\mathbf{C} = n\mathbf{a}_1 + m\mathbf{a}_2$, characterizes the CNT, see Figure 9.5. The figure illustrates the three types of nanotubes: armchair $(n,n)$; zig-zag $(n,0)$ and chiral $(n,m)$. The diameter, $d_{NT}$, of the nanotube, the chiral angle, $\theta$, and the vector, $\mathbf{T}_a$, which defines the unit cell, are given by the following relations:

$$d_{NT} = \frac{1}{\pi}|\mathbf{C}| = \frac{\sqrt{3}}{\pi}a_0\sqrt{m^2 + mn + n^2} \tag{9.1}$$

$$\theta = \arctan\left(\frac{\sqrt{3}m}{m + 2n}\right) \tag{9.2}$$

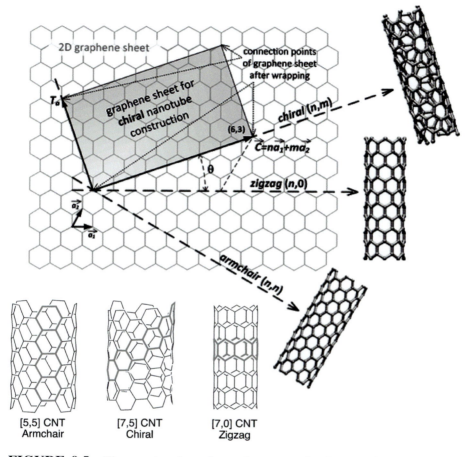

**FIGURE 9.5** The construction of a carbon nanotube from a single graphene sheet. The chiral vector denotes different types of nanotubes. Three different CNT structures can be distinguished: chiral, zig-zag and armchair. Some specific examples of the three forms of CNT are illustrated in the lower part of the figure. Reproduced with permission of Royal Society of Chemistry, in the format Book via Copyright Clearance Center: J. Prasek, J. Drbohlavova, J. Chomoucka, J. Hubalek, O. Jasek, V. Adam, & R. Kizek, (2011). *J. Mater. Chem., 21*, 15872.

and

$$\mathbf{T}_a = \frac{1}{d_r}(2m+n)\mathbf{a}_1 - (2n+m)m\mathbf{a}_2 \qquad (9.3)$$

where $d_r = \text{GCD}(m,n)$, if $(m-n)$ is not a multiple of $3 \times \text{GCD}(m,n)$; or $d_r = 3\text{GCD}(m,n)$, if $(m-n)$ is a multiple of $3 \times \text{GCD}(m,n)$. Here, GCD denotes the *greatest common divisor*. The full set of ingredients defining a nanotube can be stated as follows: the nanotube has a structure derived from graphite, in which a simple curvature has been introduced along with several topological defects, and to which a one-dimensional aspect and molecular size have been attributed.

Depending on the conditions, under which the CNTs are synthesized, they can self-organize during synthesis according to two possible self-assembly modes. In the first, see Figures 9.6(a) and (b), the CNTs are nested one inside another like Russian dolls. This form of structure is called a multi-walled nanotube (MWNT). The number of walls and their diameter can vary widely. In the second mode, see Figure 9.6(c) and (d), the nanotubes are single-walled nanotubes (SWNT), with very similar diameters in such a way, that they assemble to form bundles or ropes. These can be twisted. In each bundle, the tubes pack together compactly to form a periodic arrangement with triangular symmetry, as can be seen in Figure 9.6(c). There can be as many as a few dozen SWNTs in one bundle, with diameters ranging from 0.6–1.5 nm, depending on the specific conditions used in the synthesis procedure.

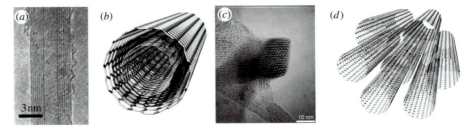

**FIGURE 9.6** Two basic morphologies of CNTs with the corresponding schematic. (a) Transmission electron microscope (TEM) image of a multi-walled carbon nanotube (MWNT); (b) Schematic of a multi-walled nanotube. (c) TEM image of a bundle of single-walled CNTs (d) A schematic of a bundle (rope) of single-walled nanotubes (SWNT). (a) Reprinted by permission from Nature: S. Iijima, (1991). *Nature 354*, 56, ©(1991); (b, c) and (d) Reproduced with permission of ROYAL SOCIETY in the format Republish in a book via Copyright Clearance Center; A. Kis & A. Zettl, (2008). *Phil. Trans. R. Soc A, 366*, 1591.

In both types of self-assembly, the distance between the neighboring tubes is roughly equal to the distance between two sheets in graphite. This suggests, that such an assembly of nanotubes does not modify the type of chemical bonds, since these remain the same as those in graphite.

## 9.3 Synthesis of Carbon Nanotubes

While it is possible, that CNTs exist in the natural state, at the present time, they have only been observed as a result of synthetic processes. Since the original discovery of CNTs by Iijima in 1991, various devices have been tested for the synthesis procedure, with the two-fold aim of producing new objects and of finding methods for the large scale production of nanotubes in a precise and controlled manner.

### 9.3.1 High Temperature Synthesis

This method involves high process temperatures as a matter of principle, since it consists in vaporizing graphite carbon, which sublimes at 3200°C, and then condensing it in a vessel with a strong temperature gradient in an inert gas such as helium or argon with a partial pressure of around 600 mbar. There are several different variants of the method, which apply these basic principles and which are generally distinguished in their manner of vaporizing the graphite source material.

In the Krätschmer–Huffmann process, which was that originally used by Iijima, an electric arc (30–40 V, 10 A) is set up between two electrodes of graphite. One electrode, the anodes, is consumed to form a plasma at a temperature of around 6000°C. The plasma then condenses on the cathode, in a deposit which contains nanotubes. This process is very simple and low-cost, and by changing the operating conditions, it is possible to obtain different types of nanotubes with different dimensions. For cathodes of pure graphite, the CNTs are multi-walled and form directly in the vapor phase in the hottest regions of the arc (at least 3000°C). In order to obtain SWNTs, a metal catalyst must be used; a few percents with transition metals such as Ni, Co, Pd or Pt or rare-earths such as Y or La. In Figure 9.7(a), we show the schematic of the arc discharge method.

A second vaporization process can be used. This consists of bombarding a target of graphite with a high-energy laser, i.e., laser ablation, see Figure 9.7(b). The conditions of synthesis and the nature of the nanotubes produced depend on whether the laser is pulsed, such as with a Nd:YAG laser, or continuous wave, such as with a 1–5 kW $CO_2$ laser. In general, for pulsed lasers, only SWNTs are produced, while for CW mode lasers, higher temperatures are attained and MWNTs can also be synthesized.

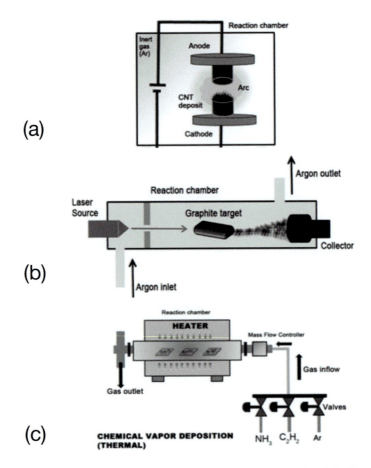

**FIGURE 9.7**   Schematic diagrams for the principle methods of CNT synthesis. (a) Arc discharge; (b) Laser ablation; (c) Chemical vapor deposition (CVD).

### *9.3.2   Moderate Temperature Synthesis*

Another approach to the formation of nanotubes uses more moderate operating temperatures. Such methods stem from catalytic or pyrolytic techniques, which are traditionally used in the synthesis of carbon fibers. Such a CVD process decomposes a carbon-bearing gas at the surface of particles of a metal catalyst in an oven heated to a temperature in the range $500 - 1100°C$, depending on the gas source used. A schematic illustration is shown in Figure 9.7(c). Typical gases, that are employed in these procedures, are carbon monoxide (CO) or hydrocarbons such as acetylene and methane, while the catalysts commonly used are Ni, Fe, and Co. The size of these particles is critical and for the formation of nanotubes, the particles must be of nanometric size. Depending on the operating conditions used such as oven temperature, gas pressure, flow rate, particle size, etc., the synthesis can produce either MWNTs or SWNTs. Generally, despite the diversity of the parameters, SWNTs are produced at higher temperatures than the MWNTs and are favored by using smaller catalyst particle sizes. The MWNTs obtained by these methods often exhibit the inferior quality of graphitization than those grown using higher temperature methods. However, they have more uniform geometric characteristics of shape, length, and diameter. This can be a distinct advantage for applications. Additionally, CNT growth can be oriented and localized by synthesizing the nanotubes on dots of catalyst arranged with some definite geometry on a substrate, which can be of interest for certain applications. Moreover, medium temperature processes can be scaled up to the production levels of carbon fibers, which is more difficult to achieve using the high-temperature techniques.

### 9.3.3   Growth Mechanisms for CNTs

The two self-assembly modes for CNT synthesis, see Figure 9.6, are mutually exclusive and occur under very different growth conditions. In the high-temperature approach, bundles of SWNTs can only be obtained by mixing a few percents of a metal catalyst to the graphite powder. In the moderate temperature scheme, the type of CNT assembly appears to be controlled by the temperature and size of the metal catalyst particles.

In a schematic way, we can say, that an open SWNT will close spontaneously by establishing bonds between atoms on the rim of the nanotube. This evolution is rapid and does not allow new carbon atoms to be incorporated into the CNT lattice. This instability makes it impossible for the nanotube to grow in the vapor phase. However, such growth is possible for MWNTs, because the dynamic bonds form between the rims of adjacent nested nanotubes, stabilizing the open tubes and allowing for the incorporation of new carbon atoms from the vapor.

While this does not give a mechanism for the formation and growth of SWNTs, it does provide some explanation as to why a catalyst is required. Detailed electron microscopy has been used to study the morphology of SWNT ropes from different synthesis methods and has allowed a more in-depth assessment to be made as to the role of the catalyst. CNTs are formed from small catalyst particles, where two distinct situations have been observed. In certain cases, the diameter of the nanotube is conditioned by that of the catalyst particle, which is generally encapsulated at the end of the tube. The CNT growth is characterized as being parallel to the particle surface. In the second case, there is no correlation between the particle size and the CNT diameter and several tubes may grow from the same catalyst particle. Here, the CNT growth occurs perpendicular to the particle surface. The former case is illustrated in Figure 9.8. In addition to the formation at the base of

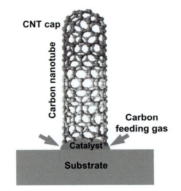

**FIGURE 9.8**   Schematic illustration of the formation of a SWNT, where the diameter is limited by the size of the catalyst particle.

the CNT, there have also been reports of tip-growth as well as a combination of both tip- and base-growth (Zhao et al., 2012). These mechanisms are illustrated in Figure 9.9.

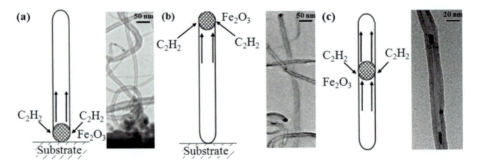

**FIGURE 9.9**   CNT synthesis mechanisms: (a) base-growth model, (b) tip-growth model, and (c) combination of base- and tip-growth model. Reprinted from W. Zhao, B. Basnet, & I. J. Kim, (2012). *J. Adv. Ceramics, 1,* 179, under Open Access License.

The growth mechanism is similar to that of carbon nanowhiskers and the SWNT must be considered as the smallest possible filament. The catalyst favors the chemisorption of carbon-bearing gases, which then segregate the carbon in the solid phase, allowing the formation of graphitic structures at lower temperatures. This explains the correlation between the CNT diameter and that of the particle. This mechanism does not apply to the case, where the growth is perpendicular to the particle surface, mainly observed using high-temperature techniques. Observations by electron microscopy suggest a formation mechanism of the type found in VLS growth. Here, the carbon concentration gradient provides continuous growth of the CNT. The nanotube formation follows a specific kinetic pathway, whereby carbon is segregated at the surface of small liquid particles of the catalyst (10–20 nm) that are saturated with carbon. The first step is the condensation of the carbon and metal vapors. Given the difference in vaporization temperatures of the carbon and metal catalysts such as Ni or Co, it is the carbon that condenses first to form low-density clusters, followed by the metal, which forms small liquid-metal particles with sizes that can be controlled by local cooling conditions. In the liquid state, these particles can dissolve large quantities of carbon. During cooling, the solubility threshold falls off to practically zero at the solidification point ($\sim 1200°C$). The supersaturated carbon segregates out by diffusing to the surface of the particle and thus forms graphitic structures. Depending on the local segregation conditions, it can either form continuous layers of graphite which encapsulate the particle in a shell, or it can form islands, which become the seeds of nanotubes. This nucleation mechanism implies that the formation of the CNTs is correlated with the solidification temperature of the metal, and seems to be borne out experimentally. This involves the CNT formation growth at one end of the tube so that the length of the nanotube should be directly related to the stability of the local conditions, i.e., the temperature and rate of carbon arrival.

The structural and chemical analysis of MWNTs can be performed by electron microscopy, some examples are illustrated in Figure 9.10. The walls appear to be equidistant concentric layers. Also shown in the figure is an example of the end of a multi-walled nanotube as well as a helical MWNT. The tube may have a cross-section that is either circular or polygonal with extended flat regions. Tubes of the circular cross-section may contain from 1 to 9 or more different helix angles depending upon the number of graphitic sheets in the tube. The helix angle changes after every three to five graphitic sheets. For most tubes, the helix angles increase by regular increments, probably due to the residual memory effect of the successive sheet groups on the underlying helical structure. An analysis of the way, in which the hexagonal

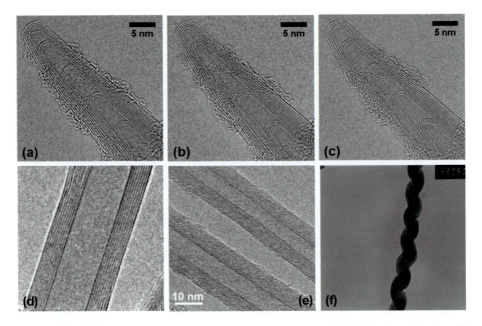

**FIGURE 9.10**    TEM images of MWCNTs. (a–c) MWNT showing the tip of the nanotube; (d) and (e) MWNT in cross-section; (f) Helical multi-walled carbon nanotube.

configuration of carbon atoms in successive sheets may fit together, suggests that neighboring sheets may maintain the same helix angle by changing the pitch of the helix, and that the helix angles within symmetrical tubes are limited to a well-defined set of possible values in agreement with experimental observations.

The structure of a SWNT bundle or rope is illustrated in Figure 9.11. Unlike the MWNTs, which are generally quite straight, the SWNT ropes are flexible and bend easily to form a tangled mass. This allows them to be viewed in parallel projection on end-on. In this latter case, we can see the close-packed formation of the bundle in a hexagonal formation.

The STM technique has also been used to observe carbon nanotubes. Some examples are shown in Figure 9.12. In the first image, Figure 9.12(a), we see a STM micrograph of SWNT coiled into a helical formation. This is illustrated in an atomic representation in Figure 9.12(b). In the other two images, Figure 9.12(c) and (d) we show a single SWNT in high resolution, showing the chiral structure of the SWNT.

In the upper image shown in Figure 9.13, we show a SWNT, which has been filled with $C_{60}$ molecules. Capillary forces can be exploited to fill the

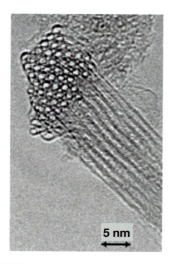

5 nm

**FIGURE 9.11**    TEM images a single-walled carbon nanotube bundle.

carbon nanotubes with molecules, such as shown for $C_{60}$, or other particles, such as $Fe_3O$ nanoparticles, as is shown in the lower section of Figure 9.13.

## 9.4   Self-Assembly

There are many processes, which are defined as *self-assembly*. In fact, we have already discussed quite a few of these in the previous chapter and in the sections above, with the ordering of nanoparticles, the production of non-porous alumina and with the formation of CNTs, all of which rely on self-assembly to produce ordered structures.

The term self-assembly is used in a very broad manner to encompass a plethora of processes. It is generally used to mean any spontaneous process resulting in the organization of individual components without the intervention of the experimenter. The main goals of fabrication via self-assembly are two-fold: firstly, it is to be able to build structures on the nanoscale, using building blocks such as molecules and nanoparticles, and secondly, it is to fabricate such structures at a low cost so that the processes can be used in mass fabrication for commercial purposes. This is indeed a real challenge: to assemble to the various components of molecules and particles into the desired order to build a microscopic device that functions as designed. We must recognize that the current state of nanofabrication and the technologies that have been developed are still at a primitive level. As we noted earlier, the assemblies that have been constructed rely on basic forces related to the electric

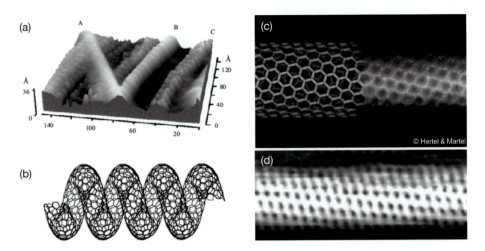

**FIGURE 9.12**     (a) 3D view of a constant-current STM image showing an assembly of several carbon nanotubes on HOPG. The assumption that the objects B and C are single-wall coiled nanotubes provides a straightforward interpretation of their regular, periodic structures along their axis. (b) Atomic model of a SWNT made from segments of (6,6) and (10,0) nanotubes (0.8 nm diameter). A unit cell of the helix is composed of 440 atoms and contains 5 octagons on the inner side of the spires and 10 pentagons on the outer side. The external diameter of the structure is 2.3 nm and the pitch is 1.17 nm. (c) and (d) Scanning tunneling microscopy images of single-walled carbon nanotubes. (a) and (b) Reprinted from L. P. Biró, S. D. Lazarescu, P. A. Thiry, A. Fonseca, J. B. Nagy, A. A. Lucas, (2000). *Ph. Lambin, Europhys. Lett., 50*, 494. ©(2000) EDP Sciences.

and magnetic fields generated by the constituent molecules and nanoparticles. In many cases, the self-assembly has to be guided or takes place on a template, requiring aid from conventional top-down fabrication processes. Combining top-down and bottom-up strategies, where top-down meets bottom-up, is an important step in this direction. We are still some way of reaching our goal and only by building on past progress and developing innovative new approaches will we realize this central goal of modern nanotechnology. In the following, we will add to the description of self-assembly principles and practices that have already been outlined in the previous sections.

Self-assembly as a strategy for nanofabrication has made some important progress over the past decades and processes deemed self-assembly can take place at the molecular and meso-/macroscopic levels. At the lower end of the scale, the forces involved can be covalent to non-covalent bonding, electrostatic and hydrophobic or interfacial hydrogen bonding. On the larger scale, gravitational and capillary forces may come into play as will external

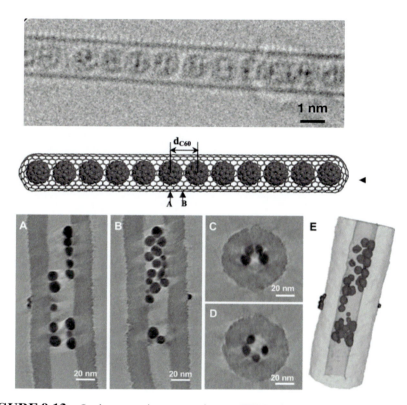

**FIGURE 9.13**    In the upper image, we have a TEM micrograph of a single-walled nanotube filled with $C_{60}$ molecules. The middle image shows the schematic diagram for this situation. In the lower images, we show TEM tomography micrographs of the $Fe_3O_4$@CNT showing the selective filling of the CNT channels by homogeneous $Fe_3O_4$ NPs. (A and B) section views along the XY axis, (C and D) section views along the XZ axis.

electromagnetic forces. In order for self-assembly to be realizable, the component building blocks must be mobile, and this frequently means the use of a liquid environment or on a smooth surface, such that the basic components for assembly can move and interact more or less freely. The formation of the self-assembled components in some form of organization is commonly a periodic or ordered configuration, which has been influenced by the environment, such as via the application of a force or the input of energy into the system. The final state is usually a result of the arrival at an equilibrium state for the system as a whole.

We can classify the form of self-assembly into three categories: chemical, physical and colloidal. In the case of chemical self-assembly, molecular ordering is attained via the ordering of molecules in the formation of non-covalent bonds. A typical example of such a process would be self-assembled monolayers (SAM). A classic example of this process is given by the ordering of alkanethiolate molecules on a gold surface. The ordering process is schematically illustrated in Figure 9.14. The alkanethiolate SAMs assemble into their lowest energy configuration on Au, with the general formula: $T(CH_2)_n$, where H denotes the head, with a certain substrate specificity, T indicates the surface property and $n$ is the chain length. A good example of

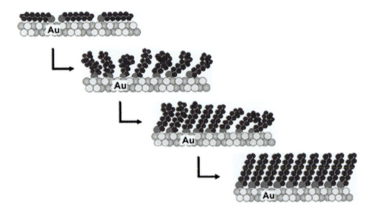

**FIGURE 9.14**     Alkanethiolate self-assembled monolayers.

this process is shown in Figure 9.15, which shows the formation of $CuPcF_{16}$ and CuPc molecules on HOPG. The SAM preparation is relatively simple and the SAMs form quickly in solution. These are well ordered and robust, representing a good thermodynamic stability and control the film thickness to within about 1 nm. In the physical self-assembly process, we can consider the preparation of a thin epitaxial film, as prepared say by MBE. The process involves the ordering of atoms in specific positions to form a perfectly order crystalline layer. Colloidal self-assembly refers to the organization of particles into ordered arrays, as were illustrated in Section 8.3, see Figure 8.14.

The complexity of any particular self-assembly process depends on the forces involved and anyone process may depend on a combination of forces. For example, even in the simple case of the organization of nanospheres in a liquid suspension, not only are there capillary forces in action that can predicate the aggregation of the particles, but electrostatic and van der Waals forces can interfere with the normal Brownian motion of the particles. Surface tensions in liquids are also known to have important effects on the dynamics of self-assembly processes in liquids. The surface tension in photore-

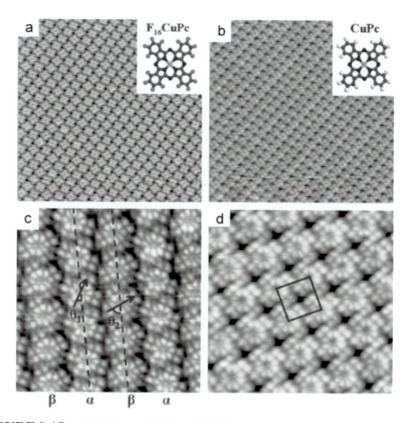

**FIGURE 9.15** CuPcF$_{16}$ and CuPc on HOPG. STM images of (a) CuPcF$_{16}$ monolayer and (b) CuPc monolayer (30 × 30 nm$^2$); panels (c) and (d) are the corresponding 8 × 8 nm$^2$ close-ups. Reprinted (adapted) with permission from Y. L. Huang, H. Li, J. Ma, H. Huang, W. Chen, & A. T. S. Wee, (2010). *Langmuir, 26,* 3329. ©(2010) American Chemical Society.

sists can be utilized in the positioning of lithographed structures such as in polysilicon MEMS components. Such a process works via the heating of the photoresist into its liquid state, this causes the photoresist to contract into a spherical shape due to surface tension, which forces the liquid to minimize its surface free energy and hence its surface area. In this process, the contraction of the photoresist can pull small objects.

In the previous discussion, we only considered those processes, which are governed by intermolecular or interparticle forces, which are responsible for connecting the basic building blocks together. Such structures tend to be lattice or chain-like ordered assemblies. Another strategy to the self-ordering process is that, which governs larger building blocks, such as this found in nature. This approach aims to harness biological processes such as the use of DNA molecules, which can act as scaffolds for the construction of more complex nanosystems. Carbon nanotubes are also a good example of larger building blocks that can be envisaged as forming larger nanostructures, as we saw from the previous section.

The structure of DNA (deoxyribonucleic acid) was deduced in 1953 by James Watson and Francis Crick and was based on X-ray diffraction studies performed notably by R. Franklin and R. Gosling. Important contributions by M. Wilkins who led the team including Watson and Crick at Cambridge. In 1962, after Franklin's death, Watson, Crick, and Wilkins jointly received the Nobel Prize in Physiology or Medicine. The structure of DNA is based on a double chain of nucleotides, which coil around each other to form the famous double helix structure. Each nucleotide is composed of one of four nitrogen-containing nucleobases (cytosine [C], guanine [G], adenine [A], or thymine [T]), a sugar called deoxyribose, and a phosphate group. The nucleotides are linked via covalent bonds to the bases, which link at the center of the helical assembly in a specific manner: only A with T and G with C, which forms the base-pairing rules. Links between these base pairs are formed by hydrogen bonds. The structure, illustrated in Figure 9.16, is an elegant assembly and leads to the deduction of the replication mechanism of DNA, elaborated by Crick. The weaker interpair coupling allows the chain to uncoil and split down the center. The base pair rule then ensures that two identical strands of DNA can naturally form. In its natural form DNA is supercoiled, meaning that the double helix is further wound into a coil. Other forms of DNA are the A, B, and Z forms, with the B form being the most commonly found in living cells.

The DNA chain is robust and flexible and can be stretched from its coiled state. The hydrogen bonds that link the base pairs are also responsible for the bonding of other biological molecules. This can be used to provoke the linking of other organic molecules and even inorganic molecules to form functional nanostructures. In this way, the DNA chain can be harnessed to form templates or scaffolds.

The base-pairing rule can be exploited to trick the selected strands to join together. Strands of DNA can be artificially sequenced with a specific order of bases and can form a single-stranded tile (SST), which can then fold into a block-like structure. The SST can form the basis for a building block for larger structures. The blocks can join to other blocks, but only if they have the correct sequence to complement its own sequence, allowing them to

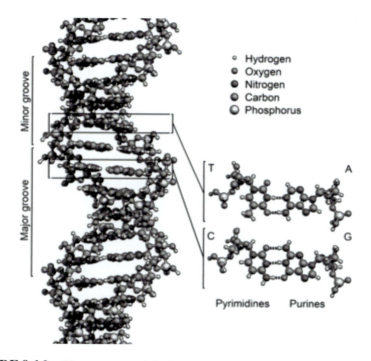

**FIGURE 9.16** The structure of the DNA double helix. The atoms in the structure are color-coded by element and the detailed structures of two base pairs are shown in the bottom right.

bond. By preparing the correct set of DNA strands, it is possible to provoke the self-assembly of specific structures. These can consist of any number of blocks so long as the correct blocks are brought together. Therefore, each block will have a specific sequence and will occupy a specific position within the assembly of all the DNA SSTs. By withholding certain SST blocks that are not required, specific assemblies can be fabricated in a reliable manner, since only those SSTs that are required are available. This method has been successfully applied to make a range of DNA structures, as shown in Figure 9.17. Recent work by Wagenbauer et al., (2017) have taken the DNA origami method to new levels with the construction of 3D objects with dimensions of up to 450 nm. Some examples are shown in Figure 9.18 along with the models of the structures, which are made from strands of DNA.

The metallization of DNA nanowires is possible using the molecular recognition feature of DNA. Here, Ag ions are found to be able to bind to DNA by Ag-Na ion exchange. This is then reduced to form a nanometer-sized metallic silver aggregate, which is bound to the DNA skeleton. These

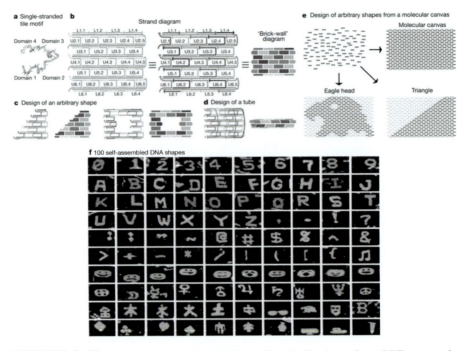

**FIGURE 9.17**   (a) The canonical SST motif. (b) Design of an SST rectangle structure. Left and middle: two different views of the same secondary structure diagram. Each standard (full) tile has 42 bases (labeled U), and each top and bottom boundary (half) tile has 21 bases (labeled L). Right: a simplified brick-wall diagram. Standard tiles are depicted as thick rectangles, boundary tiles are depicted as thin rectangles and the unstructured single-stranded portions of the boundary tiles are depicted as rounded corners. Each strand has a unique sequence. Colors distinguish domains in the left panel and distinguish strands in the middle and right panels. (c) Selecting an appropriate subset of SST species from the common pool in b makes it possible to design the desired target shape, for example, a triangle (left) or a rectangular ring (right). (d) Design of a tube with prescribed width and length. (e) Arbitrary shapes can be designed by selecting an appropriate set of monomers from a pre-synthesized pool that corresponds to a molecular canvas (top right). To make a shape, the SST strands corresponding to its constituent pixels (dark blue) will be included in the strand mixture and the remainder (light blue) will be excluded. (f) AFM images of 100 distinct shapes, including the 26 capital letters of the Latin alphabet, 10 Arabic numerals, 23 punctuation marks, and other standard keyboard symbols, 10 emoticons, 9 astrological symbols, 6 Chinese characters, and various miscellaneous symbols. Each image is 150 nm × 150 nm in size. Reprinted by permission from B. Wei, M. Dai, & P. Yin, (2012). *Nature, 485,* 623, ©(2012). Nature.

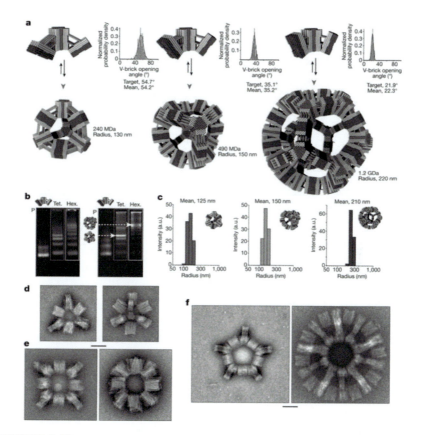

**FIGURE 9.18** (a) Cylinder models of reactive vertices designed for self-limiting self-assembly into a tetrahedron (left), a hexahedron (middle) and a dodecahedron (right). Histograms (red) of the vertex angles measured in single-particle micrographs for the three V-brick variants used in the respective vertex variant are also shown; the solid lines are Gaussian fits. The target and average measured angles are listed below each histogram. (b) Laser-scanned fluorescent images of 0.4% agarose gels, on which the reactive vertex and the assembly products for the tetrahedron (Tet.) and the hexahedron (Hex.) were electrophoresed, after incubation for 12 h (left) and after 14 d (right). P, gel pocket. Red and blue boxed regions of interest were auto-leveled separately. White arrows indicate the fully assembled tetrahedron and hexahedron bands. (c) Dynamic light-scattering intensity (a.u., arbitrary units) histograms as a function of particle radius, for the tetrahedron (red, left), the hexahedron (light blue, middle) and the dodecahedron (purple, right) (see insets). The numbers above each panel represent the mean measured particle radius. d-f, Typical negative-stained TEM micrographs of intermediate assembly products (left), reflecting the designed symmetry of assembly products (three-fold, d; four-fold, e; five-fold, f), and fully assembled closed polyhedra (right). (d) tetrahedron; e, hexahedron; f, dodecahedron. The projection directions of the micrographs of the closed polyhedra are indicated by gray arrowheads in a. Scale bars, 50 nm. Reprinted by permission from K. F. Wagenbauer, C. Sigl, & H. Dietz, (2017). *Nature, 552*, 78. ©(2017). Nature.

are then developed in an acidic solution of hydroquinone containing Ag ions under low light conditions. The silver acts as a catalyst, speeding up the process. The final metalized DNA nanowire can measure about 100 nm in diameter and can be several microns in length.

The DNA structures can be used as a scaffold for other materials such as CNTs. The key process is to immerse the DNA-template into 1-pyrenemethylamine hydrochloride (PMA) solution. With the PMA treatment, single-walled CNTs in the CNT suspension are able to anchor along the DNA strand.

Much interest has been generated by the use of CNTs due to their unique and varied properties. In terms of mechanic strength, they are stronger than high-carbon steel and have electrical conductivities far superior to those of silver and copper. Their thermal conductivities are also superior to those of Cu. The difficulty with CNTs is that unlike the DNA strands, they have no molecular recognition properties making them more difficult to manipulate. To connect a CNT to an electrode, it is more common to let the electrode find the CNT and not the other way round. This means dispersing the CNT solution over the surface of a substrate, which has been pre-patterned with electrodes in the hope that one of the CNTs will land fortuitously on an electrode. This is a random hit-or-miss process with a relatively low probability of success ($< 10\%$). An alternative, though costly method, is to use a FIB to deposit contact electrodes onto the CNT.

A new class of structures can be formed using block copolymers. These differ from other assembly techniques, since the structuring occurs within the bulk or thin film of copolymer materials. Block copolymers (BCP) are chemically dissimilar polymer chains, which are covalently linked to form a single molecule. By establishing specific conditions, the different blocks will segregate into different spatial domains much like oil and water. This is referred to as a microphase. The spatial separation is limited by the block connectivity to other blocks. The competition between these two effects allows the block copolymers to self-assemble into periodic structures. The formation of a structure will depend on the relative volume ratio between blocks and chain architecture as well as the persistence lengths of the different blocks. Some simple architectures that are common in BCPs are linear, branched (grafted and star) and cyclic molecular structures, which are illustrated in Figure 9.19. The versatility of composition and structural formation makes BCP technologies an important class of materials for the design and construction of man-made nanoscale structures. More complex architectures are possible by joining the more BCPs using *living anionic polymerization* and coupling chemistry (Feng et al., 2017). This technique, while being experimentally demanding, is probably the best polymerization technique available for the preparation of well-defined BCP structures based on vinyl monomers, such as styrenes, dienes, (meth)acrylates, vinyl pyridines, acrylonitriles, as well as cyclic monomers such as lactones, oxiranes, and siloxanes. Indeed, it is widely used in the BCP industry to produce BCPs on a very large scale.

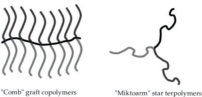

| Linear block terpolymers | "Comb" graft copolymers | "Miktoarm" star terpolymers | Cyclic block terpolymers |

**FIGURE 9.19** Representative architectures of linear block terpolymers, comb graft polymers, miktoarm star terpolymers, and cyclic block terpolymers. Reprinted from H. Feng, X.i Lu, W. Wang, N.-G. Kang, & J. W. Mays, (2017). *Polymers, 9*, 494, under Open Access License.

Self-assembly with BCPs is driven by the unfavorable mixing enthalpy and low mixing entropy. The composition, $f$, number of repeat units, $N$ and the Flory-Huggins interaction parameter, $\chi$, are the principal parameters that determine the morphology of the structures formed by BCPs. These structures include spheres (S), cylinders (C), gyroids (G) and lamellae (L). External forces, which can be applied mechanically via an electric field, can also influence the formation of structures. These morphologies are illustrated in Figure 9.20 for linear diblock BCPs and evolve as a function of $f$ and $\chi N$. This figure also shows the theoretical and experimental phase diagrams for the diblock structures.

There are many preparation protocols that have been applied, from slow cooling to rapid quenching and a broad range of polymers that can be employed in the fabrication of ordered BCP structures. Heating is one key element that has been thoroughly explored. The thermal annealing of a BCP to above its glass transition temperature facilitates the mobility of block polymers allowing them to interact. It has also been found that uniform heating is not necessarily the best approach. With the use of temperature gradients via zonal heating, the BCPs can more easily develop ordered states. Mechanical flow can also create conditions for the directional alignment of microdomains. For BCPs with differing dielectric constants, an electric field

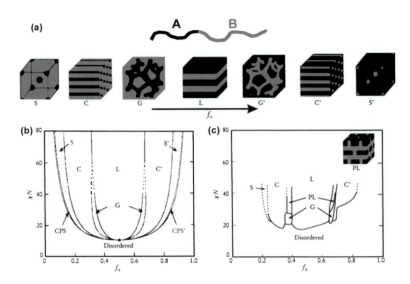

**FIGURE 9.20** (a) Equilibrium morphologies of AB diblock copolymers in bulk: S and S' = body-centered- cubic spheres, C and C' = hexagonally packed cylinders, G and G' = bicontinuous gyroids, and L = lamellae. (b) Theoretical phase diagram of AB diblock predicted by the self-consistent mean-field theory, depending on volume fraction ($f_A$) of the blocks and the segregation parameter, N; CPS and CPS' = closely packed spheres. (c) Experimental phase diagram of polystyrene-b-polyisoprene copolymers, in which $f_A$ represents the volume fraction of polyisoprene, PL = perforated lamellae. Reprinted from Y. Mai & A. Eisenberg, (2012). *Chem. Soc. Rev., 41*, 5969. Reproduced with permission of the Royal Society of Chemistry in the format Book via Copyright Clearance Center.

can also orient the microdomains in BCPs. Template self-assembly has also been employed to generate topographic structures. With the range of polymers available and the variation in methodologies for preparation, block copolymers represent an extremely important class of materials for the development of architectures in nanofabrication.

Applications of BCP structures are numerous. One of the most important being their use as a thermoplastic elastomer (TPE). These are cross-linked glassy domains with a continuous rubbery domain, which offer the elasticity of rubber, but are also suitable for injection molding and melt extrusion. TPEs can be found in adhesives, coatings, food packaging among others. Advanced nanoscale systems formed by self-assembled BCPS have shown enormous progress in applications for drug delivery. Such BCPs offer effective control of morphology and surface chemistry. The most studied being micelles and vesicles. In the former, hydrophobic drugs can be entrapped in the core and transported at concentrations that exceed their intrinsic water solubility. Also, hydrophilic blocks can form hydrogen bonds with aqueous surroundings, forming a tight shell around the micellar core. For drug delivery applications, it is important that the delivery agents are non-toxic or at least of low toxicity. BCPs such as biocompatible PEO, PLA, and PCL can all be used to form vesicles to perform the controlled release of active dyes and anti-cancer drugs over periods of up to two weeks. Furthermore, BCP vesicles modified with a stimuli-responsive functionality are being studied for use as a smart drug delivery system. While progress has been promising, there is still a need to improve performance to attain better control of site recognition and to accurately control the rate of drug release. BCPs have also found potential use in soft lithography, where the ordered features have dimensions as low as a few nanometers. This could be used for the next generation of patterning, which could be applied to the fabrication of bit patterned media for hard disk drive applications, FinFET fabrication and contact holes for microelectronics. One promising BCP, poly(styrene-b-methyl methacrylate) (PS-b-PMMA), is of particular interest, since it can form lamellar and cylindrical morphologies, which are perpendicularly oriented with respect to the substrate. By controlling the value of the $\chi$ interaction, it is possible to design feature sizes to under 10 nm.

## 9.5 Summary

Carbon allotropes have become important materials in nanosystems due to their particular set of unique properties. In particular carbon nanotubes (CNT) and graphene structures, which form a hexagonal array of atoms in rolled up or single sheet form, respectively. These forms of carbon exhibit extraordinary mechanical and electronic properties, having a very large value of Young's constant and also excellent electrical mobility, making them candidate materials for a number of nanotechnological applications. The ball-like structures formed in the fullerenes are also of interest, but applications have been more limited. In the latter structures, the formation of pentagons, instead of hexagons, allow the structure to curve in the ball shape and allows for the closed structures at the ends of carbon nanotubes. Given the properties and structures that can be formed by CNTs, they could be important materials for NEMS structures, as will be discussed in the following chapter.

The rolled-up structures of the CNTs have three distinct types, depending on the direction of rolling with respect to the principal axes of the graphene base. These are referred to as armchair, chiral and zigzag structures, which are related to the cross-section edge of the cut-off tube. In addition to single-walled tubes, multi-walled structures also exist and form a more rigid rod-like object. There are a number of methods for the fabrication of CNTs. These include high-temperature synthesis via the vaporization of graphite at 3200°C, CVD, arc discharge, laser ablation, and VLS type growth. Depending on specific growth conditions single-walled or multi-walled nanotubes can be produced. Generally speaking, high-temperatures favor the single-walled structures.

Self-assembly covers a broad range of processes, such as the organization of nanoparticles, the production of nanoporous alumina, as well as more complex organizations among large molecular structures and DNA chains. In general, this involves a spontaneous reorganization of the individual building blocks of the structure under the action of interactions between the constituents, such as electrostatic forces. While progress has been significant, we are still a long way from being able to fully control these processes. The combination of top-down and bottom-up methodologies can be of use in these processes. DNA, in particular, has been extensively studied as a means or preparing larger structures and scaffolds, with CNTs also attracting much attention. Block copolymers are another material of particular interest and can be manufactured to specific designs for specific applications. While this subject is still in its infancy, we can expect much progress and developments in the coming years and decades.

## 9.6    Problems

(1) Consider a single-walled carbon nanotube formed of 20 hexagonal units around its circumference and 50 along its length. If the interatomic distance is 1.38Å and we consider this to be unchanged under the action of a force applied along the length of the CNT. The only changes to the structure are this due to change in the angles $\alpha$ and $\beta$, see figure below. Determine the force necessary to increase the angle $\alpha$ by 5°, with a corresponding reduction in $\beta$ of 10°. Use Young's modulus of 0.62 TPa for the SWNT.

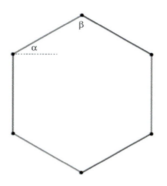

(2) The orientation and circumference of a nanotube can be characterized by the so-called chiral vector $\mathbf{C} = n\mathbf{a}_1 + m\mathbf{a}_2 \equiv (n, m)$ which joins two atoms that are identified upon rolling up the graphene sheet into a tube (see Figure 9.5). Depending on the direction of $\mathbf{C}$, we obtain different geometries: armchair, zigzag and chiral nanotubes. The names derive from the pattern observed along the section of the tube. The armchair nanotubes correspond to a chiral vector for which $n = m$. The zigzag nanotubes correspond to $m = 0$. For the chiral nanotubes it is enough to consider $0 \le |m| \le n$. The unit cell of a nanotube is spanned by the vectors $\{\mathbf{C}, \mathbf{T}\}$, where the translation vector $\mathbf{T} = t_1\mathbf{a}_1 + t_2\mathbf{a}_2$ is the vector perpendicular to the chiral vector joining two equivalent lattice sites.

For each of the following chiral vectors $\mathbf{C} = (5, 5)$, $(7, 5)$, and $(10, 0)$, determine the diameter $d_h$, the chiral angle, $\theta$ and the translation vector $\mathbf{T}$ of the resulting nanotube.

## References and Further Reading

Barish, R. D., Schulman, R., Rothemund, P. W. K., & Winfree, E., (2009). *PNAS, 106,* 6054.

Biró, L. P., Lazarescu, S. D., Thiry, P. A., Fonseca, A., Nagy, J. B., & Lucas, A. A., (2000). *Ph. Lambin, Europhys. Lett., 50,* 494.

Cui, Z., (2008). *Nanofabrication: Principles, Capabilities, and Limits,* Springer Science–Business Media, LLC, New York.

Feng, H., Lu, X. I., Wang, W., Kang, N.-G., & Mays, J. W., (2017). *Polymers, 9,* 494.

Gottfried, J. M., (2015). *Surf. Sci. Rep., 70,* 259.

Huang, Y. L., Li, H., Ma, J., Huang, H., Chen, W., Wee, A. T. S., (2010). *Langmuir, 26,* 3329.

Iijima, S., (1991). *Nature, 354,* 56.

Kis, A., & Zettl, A., (2008). *Phil. Trans. R. Soc A, 366*, 1591.

Madou, M., (2002). *Fundamentals of Microfabrication: The Science of Miniaturization*, CRC Press, Boca Raton.

Mai, Y., & Eisenberg, A., (2012). *Chem. Soc. Rev., 41*, 5969.

Martin, J. W., McIntosh, G. J., Arul, R., Oosterbeek, R. N., Kraft, M., & Söhnel, T. (2017). *Carbon, 125*, 132.

Nuzzo, R. G., & Allara, D. L., (1983). *J. Am. Chem. Soc. 105*, 4481.

Park, S., Lee, D. H., Xu, J., Kim, B., Hong, S. W., Jeong, U., Xu, T., & Russell, T. P., (2009). *Science, 323*, 1030.

Prasek, J., Drbohlavova, J., Chomoucka, J., Hubalek, J., Jasek, O., Adam, V., & Kizek, R., (2011). *J. Mater. Chem., 21*, 15872.

Thess, A., Lee, R., Nikolaev, P., Dai, H., Petit, P., Robert, J., Xu, C., Lee, Y. H., Kim, S. G., Rinzler, A. G., Colbert, D. T., Scuseria, G. E., Tom‡nek, D., Fischer, J. E., Smalley, R. E., (1996). *Science, 273*, 483.

Wagenbauer, K. F., Sigl, C., & Dietz, H., (2017). *Nature, 552*, 78.

Wei, B., Dai, M., & Yin, P., (2012). *Nature, 485*, 623.

Yang, N., Youn, S. K., Frouzakis, C. E., & Park, H. G., (2018). *Carbon, 130*, 607.

Zhao, W., Basnet, B., & Kim, I. J., (2012). *J. Adv. Ceramics, 1*, 179.

# Index